Interdisciplinary Biotechnological Advances

Series Editors

Jayanta Kumar Patra, Research Institute of Integrative Life Sciences, Dongguk University, Ilsandong, Goyang, Korea (Republic of)

Gitishree Das, Research Institute of Integrative Life Sciences, Dongguk University, Goyang, Korea (Republic of)

This series is an authoritative resource that explains recent advances and emerging directions in biotechnology, reflecting the forefront of research clearly and reliably, without excessive hype. Each volume is written by authors with excellent reputations and acknowledged expertise in the topic under discussion. The volumes span the entire field from an interdisciplinary perspective, covering everything from biotechnology principles and methods to applications in areas including genetic engineering, transgenic plants and animals, environmental problems, genomics, proteomics, diagnosis of disease, gene therapy, and biomedicine. The significance of these applications for the achievement of UN Sustainable Development Goals is highlighted. The series will be highly relevant for Master's and PhD students in Biotechnology, Nanochemistry, Biochemical Engineering, and Microbiology, medical students, academic and industrial researchers, agricultural scientists, farmers, clinicians, industry personnel, and entrepreneurs.

Luis María Vaschetto
Editor

Omics Technologies in Plant Science

Unlocking the Potential of Crops

Springer

Editor
Luis María Vaschetto
Ronin Institute (RIIS 2.0)
Sacramento, CA, USA

ISSN 2730-7069 ISSN 2730-7077 (electronic)
Interdisciplinary Biotechnological Advances
ISBN 978-981-92-0532-5 ISBN 978-981-92-0533-2 (eBook)
https://doi.org/10.1007/978-981-92-0533-2

This Springer imprint is published by the registered company Springer Nature Singapore Pte Ltd.
The registered company address is: 152 Beach Road, #21-01/04 Gateway East, Singapore 189721, Singapore

Preface

The worldwide agricultural industry is currently facing a critical challenge, which involves the dual issue of feeding a growing population and developing more resilient crops capable of thriving in a changing climate. Consequently, in order to reach food security, it is imperative to holistically explore omics layers in breeding programs. *Omics Technologies in Plant Science: Unlocking the Potential of Crops* is a comprehensive guide to this aim. The book explains in detail critical insights into major omics, namely, genomics, transcriptomics, epigenomics, metabolomics, proteomics, and phenomics.

The book follows a natural flow of biological information, starting at the foundation with plant genomics, exploring how next-generation sequencing and genome-wide association studies allow us to pinpoint the genetic variation responsible for superior traits. Next, it explores gene regulatory layers, specifically examining the roles of transcriptomics, noncoding RNAs, and epigenetic modifications that shape molecular response mechanisms against stressful conditions. Subsequently, it analyzes in depth plant metabolomics and proteomics. A specific focus is made on posttranslational modifications, which act as dynamic regulators of protein function. Finally, it discusses data integrated into phenomic prediction, which is fundamental for breeders looking to select more resilient and high-yield crops.

I am sincerely thankful to Dr. Adrián González Ortega-Villaizán, Prof. Vincenzo Cunsolo, Dr. Wenrao Li, Dr. Seyedeh Najmeh Banihashemian, Dr. Karel Doležal, Dr. Dr. Jameel R. Al-Obaidi, Prof. Manasi Dash, and Dr. Pratyoosh Shukla, whose critical revisions have greatly enhanced the quality and scientific rigor of this book.

Sacramento, CA, USA — Luis María Vaschetto

Contents

Contributors

M. N. Ashwath Department of Environmental Science and Agroforestry, College of Agriculture, University of Agricultural Science, Gangavathi, Karnataka, India

Muralidhar S. Aski Division of Genetics, Indian Agricultural Research Institute, New Delhi, India

Ankita Department of Forestry, Post Graduate College of Agriculture, Dr. Rajendra Prasad Central Agricultural University, Samastipur, Bihar, India

Bishal Binaya Bhatta Department of Crop Physiology, Institute of Agricultural Sciences (IAS), Siksha 'O' Anusandhan, Bhubaneswar, Odisha, India

Mallikarjun Biradar ICAR-National Bureau of Plant Genetic Resources, New Delhi, India

Bulbul Boblina Department of Plant Pathology, Faculty of Agriculture, Sri Sri University, Cuttack, Odisha, India

Christin Bündig Leibniz Universität Hannover, Institute of Plant Genetics, Section Reproduction and Development, Hannover, Germany

Nibedita Chakraborty Department of Botany, UGC Centre for Advanced Studies, The University of Burdwan, Golapbag Campus, Burdwan, West Bengal, India

Roberta Pena da Paschoa Laboratório de Biotecnologia, Centro de Biociências e Biotecnologia (CBB), Universidade Estadual do Norte Fluminense Darcy Ribeiro (UENF), Campos dos Goytacazes, Rio de Janeiro, Brazil

Setor de Genômica e Proteômica, Unidade de Biologia Integrativa, Universidade Estadual do Norte Fluminense Darcy Ribeiro (UENF), Campos dos Goytacazes, Rio de Janeiro, Brazil

Carlos Eduardo Assis da Silva Laboratório de Biotecnologia, Centro de Biociências e Biotecnologia (CBB), Universidade Estadual do Norte Fluminense Darcy Ribeiro (UENF), Campos dos Goytacazes, Rio de Janeiro, Brazil

Setor de Genômica e Proteômica, Unidade de Biologia Integrativa, Universidade Estadual do Norte Fluminense Darcy Ribeiro (UENF), Campos dos Goytacazes, Rio de Janeiro, Brazil

Laura Eliza de Oliveira Alves Laboratório de Biotecnologia, Centro de Biociências e Biotecnologia (CBB), Universidade Estadual do Norte Fluminense Darcy Ribeiro (UENF), Campos dos Goytacazes, Rio de Janeiro, Brazil

Setor de Genômica e Proteômica, Unidade de Biologia Integrativa, Universidade Estadual do Norte Fluminense Darcy Ribeiro (UENF), Campos dos Goytacazes, Rio de Janeiro, Brazil

Harsh K. Dikshit Division of Genetics, Indian Agricultural Research Institute, New Delhi, India

G. Harish Kumar Department of Genetics and Plant Breeding, School of Agriculture, SR University, Warangal, Telangana, India

Mahipal Singh Kesawat Department of Genetics and Plant Breeding, Faculty of Agriculture, Sri Sri University, Cuttack, Odisha, India

K. N. Kiran ICAR-Indian Institute of Horticultural Research, Bengaluru, Karnataka, India

Manju Kohli Division of Genetics, Indian Agricultural Research Institute, New Delhi, India

Manchikatla Arun Kumar Department of Genetics and Plant Breeding, School of Agriculture, SR University, Warangal, Telangana, India

Pream Kumar Division of Genetics, Indian Agricultural Research Institute, New Delhi, India

Garima Mishra Division of Genetics and Tree Improvement, ICFRE-Forest Research Institute, Dehradun, Uttarakhand, India

Biodiversity and Climate Change Division, ICFRE, Dehradun, Uttarakhand, India

Gyan P. Mishra Division of Seed Science and Technology, Indian Agricultural Research Institute, New Delhi, India

Udit Nandan Mishra Department of Crop Physiology and Biochemistry, Faculty of Agriculture, Sri Sri University, Cuttack, Odisha, India

Ankita Mohanty Department of Crop Physiology and Biochemistry, Faculty of Agriculture, Sri Sri University, Cuttack, Odisha, India

Sourav Ranjan Mohapatra Department of Forest Biology and Tree Improvement, Odisha University of Agriculture and Technology, Bhubaneswar, Odisha, India

Arnab Mukherjee Department of Botany, UGC Centre for Advanced Studies, The University of Burdwan, Golapbag Campus, Burdwan, West Bengal, India

Pavan K. Naik Division of Genetics, Indian Agricultural Research Institute, New Delhi, India

Shreyashi Nandi Department of Botany, UGC Centre for Advanced Studies, The University of Burdwan, Golapbag Campus, Burdwan, West Bengal, India

Debashis Oraon Department of Botany, UGC Centre for Advanced Studies, The University of Burdwan, Golapbag Campus, Burdwan, West Bengal, India

Daniel Dastan Rezabala Pacheco Laboratório de Biotecnologia, Centro de Biociências e Biotecnologia (CBB), Universidade Estadual do Norte Fluminense Darcy Ribeiro (UENF), Campos dos Goytacazes, Rio de Janeiro, Brazil

Setor de Genômica e Proteômica, Unidade de Biologia Integrativa, Universidade Estadual do Norte Fluminense Darcy Ribeiro (UENF), Campos dos Goytacazes, Rio de Janeiro, Brazil

Bidwan Rath Division of Genetics, Indian Agricultural Research Institute, New Delhi, India

Shashidhar B. Reddappa Krishi Vigyan Kendra, ICAR-Vivekananda Parvatiya Krishi Anusandhan Sansthan, Almora, Uttarakhand, India

Sujit Roy Department of Botany, UGC Centre for Advanced Studies, The University of Burdwan, Golapbag Campus, Burdwan, West Bengal, India

Debanjana Saha Department of Genetics and Plant Breeding, Faculty of Agriculture, Sri Sri University, Cuttack, Odisha, India

Suman Kumar Saha Department of Botany, UGC Centre for Advanced Studies, The University of Burdwan, Golapbag Campus, Burdwan, West Bengal, India

Claudete Santa-Catarina Laboratório de Biologia Celular e Tecidual, CBB-UENF, Campos dos Goytacazes, Rio de Janeiro, Brazil

Krystle Angelique Aguda Santiago School of Science, Monash University Malaysia, Bandar Sunway, Selangor, Malaysia

Rohit Sarkar Division of Genetics, Indian Agricultural Research Institute, New Delhi, India

Nandakumar Shekharappa ICAR-Central Arid Zone Research Institute, Jodhpur, Rajasthan, India

Vanildo Silveira Laboratório de Biotecnologia, Centro de Biociências e Biotecnologia (CBB), Universidade Estadual do Norte Fluminense Darcy Ribeiro (UENF), Campos dos Goytacazes, Rio de Janeiro, Brazil

Setor de Genômica e Proteômica, Unidade de Biologia Integrativa, Universidade Estadual do Norte Fluminense Darcy Ribeiro (UENF), Campos dos Goytacazes, Rio de Janeiro, Brazil

Tiziana M. Sirangelo Division Biotechnologies, ENEA—Italian National Agency for New Technologies, Energy and Sustainable Economic Development, Casaccia Research Center, Rome, Italy

Natasha Damiana Spadafora Department of Environmental and Prevention Sciences, University of Ferrara, Ferrara, Italy

Tushal Srivastava Division of Genetics and Tree Improvement, ICFRE-Forest Research Institute, Dehradun, Uttarakhand, India

Sukirti Division of Genetics and Tree Improvement, ICFRE-Forest Research Institute, Dehradun, Uttarakhand, India

Adeline Su Yien Ting School of Science, Monash University Malaysia, Bandar Sunway, Selangor, Malaysia

Shabir H. Wani National Seed Project Laboratory, Sher-e-Kashmir University of Agricultural Sciences and Technology of Kashmir, Srinagar, Jammu and Kashmir, India

Narkhede Gopal Wasudeo Department of Genetics and Plant Breeding, School of Agriculture, SR University, Warangal, Telangana, India

Lucas Rodrigues Xavier Laboratório de Biotecnologia, Centro de Biociências e Biotecnologia (CBB), Universidade Estadual do Norte Fluminense Darcy Ribeiro (UENF), Campos dos Goytacazes, Rio de Janeiro, Brazil

Setor de Genômica e Proteômica, Unidade de Biologia Integrativa, Universidade Estadual do Norte Fluminense Darcy Ribeiro (UENF), Campos dos Goytacazes, Rio de Janeiro, Brazil

Next-Generation Sequencing for Plant Genomics

Garima Mishra, Tushal Srivastava, Sukirti, Sourav Ranjan Mohapatra, M. N. Ashwath, and Ankita

Abstract Advances in next-generation sequencing (NGS) have transformed plant genomics by enabling high-resolution analysis of genome structure, function, and evolution. This chapter provides a thoughtful review of the development of first-, second-, and third-generation sequencing platforms, the principles on which they are based, technological advancements, and potential developments in plant science. The chapter outlines how successive sequencing generations have progressively improved genome assembly quality, gene expression profiling, variant discovery, and the detection of epigenetic modifications. Key applications, including whole-genome sequencing, RNA–Seq, SNP discovery, epigenome analysis, and integrative multi-omics approaches, are the dominant themes, showing how genomics can fold with proteomics, metabolomics, and phenomics. Moreover, case studies and examples demonstrating the potential of NGS in both model and non-model

G. Mishra
Division of Genetics and Tree Improvement, ICFRE-Forest Research Institute, Dehradun, Uttarakhand, India

Biodiversity and Climate Change Division, ICFRE, Dehradun, Uttarakhand, India

T. Srivastava · Sukirti
Division of Genetics and Tree Improvement, ICFRE-Forest Research Institute, Dehradun, Uttarakhand, India

S. R. Mohapatra (✉)
Department of Forest Biology and Tree Improvement, Odisha University of Agriculture and Technology, Bhubaneswar, Odisha, India
e-mail: srmohapatra.forestry@ouat.ac.in

M. N. Ashwath
Department of Environmental Science and Agroforestry, College of Agriculture, University of Agricultural Science, Gangavathi, Karnataka, India

Ankita
Department of Forestry, Post Graduate College of Agriculture, Dr. Rajendra Prasad Central Agricultural University, Samastipur, Bihar, India

L. M. Vaschetto (ed.), *Omics Technologies in Plant Science*, Interdisciplinary Biotechnological Advances, https://doi.org/10.1007/978-981-92-0533-2_1

plant species are included to illustrate its utility in addressing complex biological questions, advancing crop improvement, and supporting conservation efforts. While describing challenges of polyploidy, repetitive DNA, data management, and bioinformatics bottlenecks, future advances in the field, and artificial intelligence datasets are also included.

Keywords Plant genomics · Next-generation sequencing · Transcriptomics · Epigenomics · Crop improvement

Abbreviations

NGS	Next-generation sequencing
WGS	Whole-genome sequencing
RNA–Seq	RNA sequencing
SNP	Single nucleotide polymorphism
SSR	Simple sequence repeat
QTL	Quantitative trait locus
GWAS	Genome-wide association study
EST	Expressed sequence tag
cDNA	Complementary DNA
PCR	Polymerase Chain reaction
DNA	Deoxyribonucleic acid
Bp	Base pair
Mb	Megabase
Kb	Kilobase
MAS	Marker-assisted selection
FGS	First-generation sequencing
SGS	Second-generation sequencing
TGS	Third-generation sequencing
PacBio	Pacific biosciences
ONT	Oxford Nanopore technologies
mRNA	Messenger RNA
DEGs	Differentially expressed genes
NCBI	National Center for biotechnology information
TF	Transcription factor
C	Cytosine
T	Thymine
G	Guanine
A	Adenine
TPM	Transcripts per million

1 Introduction

Plant genomics involves the thorough analysis of plant genetic material, including genome mapping, sequencing, and gene annotation. It also encompasses the study of the biochemical functions of genes and their products, ultimately revealing how these molecular components affect plant traits, development, and adaptation (Altmann 2005). Exploring plant genomes has revolutionized our understanding of plant biology, enabling insights into growth mechanisms, adaptation, stress response, and evolutionary diversification (Song et al. 2023). Although Sanger sequencing was pivotal in the foundational stages of molecular biology, its drawbacks became more evident as the demands of contemporary genomics increased. This technique is inherently time-consuming, labor-intensive, and costly, since it sequences one DNA fragment at a time with comparatively short read lengths of 800 to 1000 base pairs. It is therefore not suitable for high-throughput analysis or for deciphering complex, large genomes. In addition, its low sensitivity also makes it less capable of detecting infrequent genetic variations, particularly in the case of mixed or heterogeneous specimens. The unique complexity of plant genomes, often being large, polyploid, and containing high amounts of repetitive sequences, worsens these issues in plant biology, highlighting the need for efficient and scalable technology (Song et al. 2023). These limitations ultimately gave rise to next-generation sequencing (NGS), which has gone on to take over the landscape of genomics research.

Next-generation sequencing is a set of high-throughput technologies that allow rapid sequencing of entire genomes or transcriptomes to an unprecedented scale and resolution. Compared to the first-generation methods, NGS enables concurrent parallel sequencing of hundreds of thousands of DNA fragments at less cost and speed. Accelerated innovation since its inception in the mid-2000s has transformed NGS from whole-genome shotgun sequencing to the application of multi-omics approaches. In plant biology, its application spans from model plants to ecologically and economically significant non-model organisms, revolutionizing genetic study (Michael and VanBuren 2020). Forest trees tend to present many challenges to genomics by having large genome sizes, high levels of heterozygosity, and frequent polyploidy. NGS, and especially PacBio and ONT long-read technologies, have greatly enhanced the potential for high-fidelity de novo genome assemblies and thereby created a platform for sustainable forestry, genetic improvement, and climate-resilient breeding programs (Plomion et al. 2018).

NGS has revealed hitherto unimaginable opportunities for research and utilization of plant genomes. WGS allows for the detection of structural variants, delivers information on evolutionary history, and supports comprehensive genome assembly and annotation. For organisms such as *Quercus robur*, *Populus trichocarpa*, and *Eucalyptus grandis*, high-quality genome assemblies have been constructed, significantly enabling functional genomics studies and trait mapping. (Plomion et al. 2018; Myburg et al. 2014; Tuskan et al. 2006). One should consult RNA sequencing (RNA–Seq) data that reveal stress responses, developmental processes, and patterns in gene expression. Transcriptome mapping of *Camellia sinensis*, *Dalbergia sissoo*, and *Shorea robusta* are prime examples that provided insights into the mechanisms through which plants respond to biotic and abiotic stresses (Li et al. 2020).

The typical pipeline for next-generation sequencing in plant genomics is a series of steps coordinated together, beginning with sample preparation and progressing through library construction, sequencing, and data analysis, to eventually reach useful applications in trait improvement, disease diagnosis, and biodiversity conservation (Fig. 1). Restriction site-associated DNA sequencing (RAD–seq) and genotyping-by-sequencing (GBS) have played key roles in population genomics and diversification studies. They have helped in the discovery of genetic diversity among forest trees economically and environmentally valuable. Also, through the facilitation of genome skimming and plastid genome sequencing, NGS has sped up phylogenomics. With increasing evidence supporting the application of NGS in the field, our knowledge about the ancient phylogenetic relationships and biogeographic histories of *Ficus benghalensis*, *Abies pindrow*, and *Pinus taeda* has become better due to this study (Liu et al. 2020a, b). In conservation genomics, NGS offers an opportunity to assess genetic diversity, composition, and adaptive potential in threatened and endemic species. Such genomic information is crucial to guide effective conservation strategies like genetic rescue, habitat reconstruction, and planning for ex situ conservation (Allendorf et al. 2010; Supple and Shapiro 2018).

Outside the nuclear genome, NGS technologies have also advanced our understanding of epigenomics and metagenomics in plants. In epigenomic research, methods such as ChIP–seq and methylation sequencing (Methyl–seq) have begun to unravel the epigenetic modifications that regulate plant development, gene expression, and stress responses, although such studies are still in their infancy for

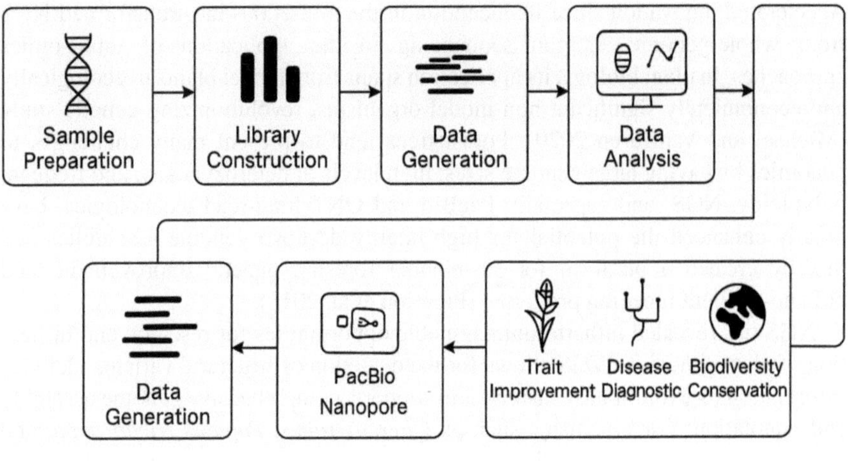

Fig. 1 Workflow of next-generation sequencing (NGS) in plant genomics

non-model plants (Springer and Schmitz 2017). In parallel, metagenomic sequencing approaches are providing a holistic view of plant-associated microbiomes, particularly in the rhizosphere and phyllosphere, offering valuable insights into plant–microbe interactions that play critical roles in forest ecosystem processes (Bulgarelli et al. 2013). Despite these revolutionary developments, issues remain. The absence of good-quality reference genomes for most non-model organisms, the problem of handling and processing large-scale datasets, and problems in assembling very repetitive plant genomes are still major challenges. With pangenomics, multi-omics platforms, and artificial intelligence poised to overcome these constraints, the future is bright. Precision breeding, synthetic biology, and landscape genomics will be made possible by these advancements, which will drive the next stage of plant genomics to meet the demands of forestry, agriculture, and environmental restoration (Michael and VanBuren 2020).

Comparing the dominant sequencing platforms is crucial to comprehending the technological backbone of NGS in plant genomics. The salient features of the widely used sequencing technologies, such as read length, precision, output, and expense, along with their optimal applications, are listed in Table 1. The comparison helps researchers choose the most appropriate platform for their experimental objectives and resource constraints.

NGS has generally transformed the field of plant sciences by providing a more comprehensive and in-depth understanding of the functional complexity of plant genomes. It has a cross-disciplinary impact on breeding plans, ecology, biotechnology, and evolutionary biology. Specifically, within the realms of forest tree research and conservation work, NGS technology's relentless evolution continues to shed light on the genetic underpinnings of plant diversity, adaptation, and resilience.

Table 1 Comparison of DNA sequencing technologies

S. No.	Parameter	Sanger sequencing	Illumina (NGS)	PacBio (HiFi)	Oxford Nanopore
1.	Read length	~800–1000 bp	100–300 bp	10–25 kb (HiFi reads)	Up to 2 Mb
2.	Accuracy	>99.99%	>99.9%	>99.8%	90–98%
3.	Turnaround time	Slow	Fast	Moderate	Real time
4.	Cost per Mb	High	Low	Medium	Low
5.	Scalability	Low	Very high	Moderate	High
6.	Preferred use case	Small gene fragments	WGS, RNA–Seq, GBS	De novo assembly	Structural variation, field use

2 Generations of Sequencing Technologies

Sequencing technologies have seen remarkable development over the last few decades. The sequencing technique has been characterized into three main generations. Sanger's first generation, which emerged in the 1970s, defined chain-termination methods that, although precise, were constrained by low yields and scalability issues (Sanger et al. 1977). The introduction of second-generation or next-generation sequencing (NGS) technologies in the early 2000s revolutionized genomics by allowing fast, massively parallel sequencing, thus lowering the costs and time of extensive genome analyses by tens of thousands (Mardis 2008). More recently, third-generation sequencing systems from Pacific Biosciences and Oxford Nanopore Technologies have made it possible to perform single-molecule level long-read sequencing, allowing improved precision in the demarcation of challenging genomic regions and facilitating epigenetic modification analysis directly without amplification (Van Dijk et al. 2018). Generally, these developments have pushed the boundaries and depth of genomic explorations, particularly in the direction of non-model plants like forest tree species. Below is an overview summarizing the most notable features and accomplishments gained through all three generations of sequencing technologies that focus on their contribution to plant genomic research.

2.1 *First-Generation Sequencing (FGS) Technologies in Plant Genomics*

The advancement of DNA sequencing has transformed our understanding of the structure and functions of the genomes of all living organisms. Use of sequencing technologies in plant genomics has enabled the characterization of complex genomes, the discovery of genes for agronomic traits, and the speeding up of crop improvement programs. The genomic revolution was made possible by first-generation sequencing technologies, i.e., Maxam–Gilbert and Sanger sequencing (Verma et al. 2016; Eren et al. 2022). Developed in the 1970s, these were the first techniques for subsequent high-throughput sequencing methods (Mardis 2008). Although considered obsolete for large-scale sequencing, first-generation sequencing remains useful in certain areas of plant genome study, especially in the verification and accurate sequencing of individual DNA fragments (Barba et al. 2014).

Prior to the initiation of DNA sequencing technology, knowledge regarding DNA was more or less confined to its conventional double-helical structure, as explained by Watson and Crick in 1953. But the order of nucleotide bases (adenine, thymine, guanine, and cytosine) within a DNA sequence was not known (Levinson and Gutman 1987). This stopped the advancement of gene function and regulation research. The initial sequencing breakthrough was the Maxam–Gilbert method (Maxam and Gilbert 1977), which appeared soon after the Sanger method (Sanger 1981) and ultimately became routine for sequencing protocols (Eren et al. 2022).

These first-generation technologies were introduced at a time when scientists were just starting to become familiar with molecular biology. The Sanger method, developed in 1977, was the most widely used sequencing method and remained so for close to three decades. It played a key role in the early stages of the *Arabidopsis thaliana* genome project, the first model organism whose genomic sequence was completely sequenced in 2000, as well as the chloroplast genome and expressed sequence tags (ESTs) sequencing (Weber et al. 2007).

2.1.1 Maxam–Gilbert Sequencing (Chemical Cleavage Method)

Allan Maxam and Walter Gilbert in 1977 developed the Maxam–Gilbert sequencing method, or chemical cleavage method, which was one of the first methods for sequencing DNA. DNA is altered chemically, and then it is broken up into its constituent nucleotide bases—adenine (A), guanine (G), cytosine (C), and thymine (T). First, one end of the DNA molecule is radioactively labeled, often using phosphorus-32 (32P) (Maxam and Gilbert 1977). Next, the labeled DNA is divided into four reaction tubes, each of which is exposed to a different set of reagent combinations that cleave the DNA backbone at particular bases (Maxam and Gilbert 1980). For instance, guanine residues are methylated by dimethyl sulfate, while cytosine and thymine are cleaved by hydrazine (Kochetkov 2012). Each of these processes yields a group of DNA fragments with varied lengths depending on where the base was cleaved.

Polyacrylamide gel electrophoresis, a method by which DNA fragments are sized and separated into groups based on size, is utilized to separate the resulting fragments. The gel is subjected to X-ray film after electrophoresis for autoradiography, and the band patterns that result are visually inspected manually to determine the DNA sequence (Maxam and Gilbert 1980). Each band on the gel corresponds to a particular nucleotide position in the sequence, and the precise base configuration can be reconstructed from comparing the banding patterns in the four different reactions.

When Maxam–Gilbert sequencing was initially invented, it had some advantages. As automated sequencing technology had not yet been developed, it was extremely useful in the early years of molecular biology. It enabled scientists to sequence double-stranded DNA directly without a requirement for prior cloning. Its numerous flaws soon made it obsolete in favor of more efficient methods (Heather and Chain 2016). The method involved the handling of radioactive isotopes, hydrazine, piperidine, and other toxic and hazardous chemicals, and it was time-consuming. The sequencing reaction also involved a lot of human intervention and precision, and therefore, human error was a likelihood. The method also produced too short read lengths for genomic research on a large scale of 100 to 200 base pairs.

In plant genomics, the Maxam–Gilbert sequencing had limited but pioneering applications. It was first utilized to sequence short DNA fragments from plant species, especially in initial studies that aimed to grasp gene structure and regulatory components in plant nuclear and organellar genomes. Nonetheless, owing to its labor-intensive, toxic, and low-throughput characteristics, the technique was soon phased out and substituted with the more resilient and user-friendly Sanger sequencing technique. Maxam–Gilbert sequencing is no longer involved in contemporary plant genomics; it remains of historical value as one of the earliest viable techniques to open the sequence of DNA and set the stage for subsequent sequencing developments (Eren et al. 2022).

2.1.2 Sanger Sequencing (Chain-Termination Method)

Sanger sequencing, or the chain-termination technique, which was first developed by Frederick Sanger in 1977, is regarded as the foundation of molecular genetics and genomics for almost decades. It revolutionized DNA sequencing with a systematic procedure by using enzymatic DNA synthesis and selective termination (Pareek et al. 2011).

The sequencing technique begins by binding a short primer to a single-stranded DNA template. DNA synthesis is performed using DNA polymerase, four common deoxynucleotide triphosphates (dNTPs), and a minimal amount of dideoxynucleotide triphosphates (ddNTPs), in which each ddNTP is labeled with a unique fluorescent dye or radioactive tag (Lee et al. 1992). Although these ddNTPs have the same structure as dNTPs, they lack the 3'-OH group necessary for the addition of the next nucleotide; however, chain termination occurs at the point of inclusion (Chen 2014). When DNA polymerase extends the strand, ddNTPs are incorporated at random, resulting in a collection of fragments with various lengths that end at the same base (Wang 2021). Capillary electrophoresis successfully isolates fragments depending on their size, facilitating synthesis. Once the fragments are passed through the laser detectors, the nucleotide sequence is displayed as a chromatogram, which extracts the mean fluorescence signal associated with every ddNTP. Sanger sequencing is the gold standard for DNA sequencing when exact base recognition is critical. It generally provides read lengths of 800–1000 base pairs, having an exceptional precision of more than 99.99% (Gupta and Verma 2019).

Sanger sequencing was critical to the initial discovery and annotation of plant genes and genomes. Sanger sequencing was the primary technology used in the 2000 completion of the *Arabidopsis thaliana* genome project, the first higher plant genome to be sequenced completely (Meinke et al. 1998). This achievement earned *A. thaliana* is currently exploited as a model system in plant biology and genomics research. Besides whole-genome sequencing, the Sanger method has been instrumental in the generation of expressed sequence tag (EST) libraries for numerous crops, including rice (Yamamoto and Sasaki 1997), maize (Davis et al. 1999), soybean (Shoemaker et al. 2002), and wheat (Peng and Lapitan 2005). Genes associated with significant agronomic traits such as seed development, flowering time, disease resistance, and drought tolerance were identified with these ESTs.

This technique has also found long-term applications in molecular breeding and plant biotechnology. It remains the method of choice for confirming gene constructs in genetic transformation, such as transgene integrity, orientation, and presence.

This technique is often employed for DNA barcoding of plant species through the use of matK and rbcL, aiding in species identification and biodiversity studies (Zischewski et al. 2017). Furthermore, Sanger sequencing is also commonly used to validate gene alterations and confirm qPCR amplicons in desired regions due to the increased use of CRISPR/Cas9 genome editing (Sentmanat et al. 2018). Despite the arrival of next-generation sequencing (NGS), which gives a much higher throughput, Sanger sequencing continues to be a valuable tool in plant genomics, particularly for marker-assisted selection (MAS) and transgenic verification, because of its precision, relatively inexpensive cost for small-scale sequencing, and ease of data analysis.

Given the intricate composition of plant tissues, which are typically highly enriched with polysaccharides, polyphenols, and secondary metabolites, the process for DNA sequencing entails the use of aggressive DNA extraction methods, followed by target gene PCR amplification of genes such as rbcL, matK, or ITS. After purification, the DNA goes through a second cyclization reaction that has fluorescently labeled dNTPs (Shendure et al. 2008). These steps produce fragments of DNA of a range of sizes, which are subsequently sorted in a computer-controlled sequence using capillary electrophoresis. The sequence of nucleotides is next established by interpreting the fluorescent signals with expert software such as Chromas, Geneious, or MEGA. This method ensures the accuracy and durability of plant DNA sequencing, irrespective of complex or deteriorated material (Fig. 2).

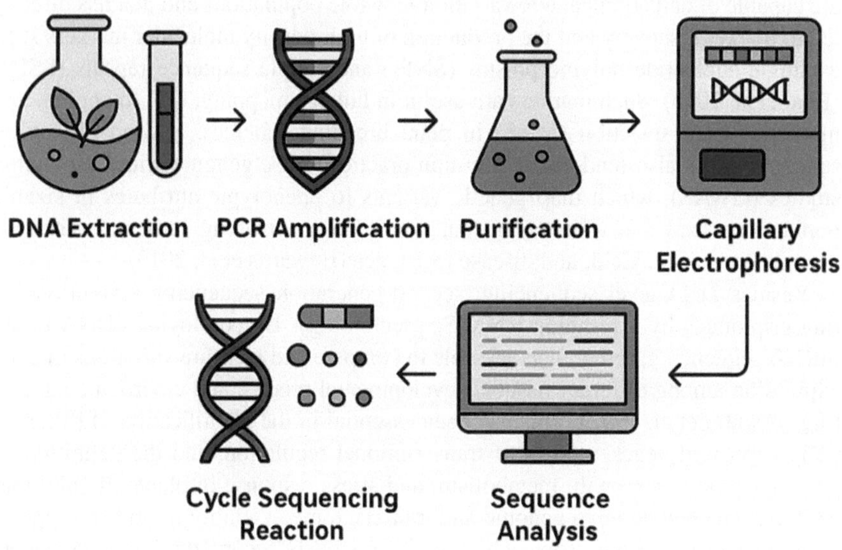

Fig. 2 Workflow and methodology of Sanger sequencing

2.2 Second-Generation Sequencing (SGS) Technologies in Plant Genomics

Second-generation sequencing platforms, also known as next-generation sequencing (NGS), represent a significant improvement over first-generation methodology. Inventions of the mid-2000s, NGS platforms revolutionized genomics by making massively parallel sequencing, i.e., simultaneous sequencing of millions to billions of DNA fragments, possible (Goodwin et al. 2016). This revolution reduced DNA decoding time, labor, and expense significantly, thereby making genome-scale information available in a broad spectrum of organisms, including plants. Kumar et al. (2024) suggest that NGS platforms have, in fact, accelerated genomics research and facilitated discoveries that were not possible or very costly before.

In plant genomics, NGS technologies have played a major role. In contrast to the largely small and compact genomes of most of the model animals, however, the plant genomes are more extensive, highly repetitive, and structurally complex, posing tremendous obstacles to genome assembly and annotation (Claros et al. 2012). Second-generation sequencing shattered such obstacles by producing vast quantities of data that could be used to produce in-depth reference genomes for numerous crop plants. Maize, wheat, rice, soybean, cotton, and numerous other economically important crops, for instance, have had their genomes sequenced on NGS platforms to provide the root materials for crop improvement (Berkman et al. 2012).

One of the major advantages of SGS technologies in plant genomics is that they are capable of detecting genetic variation in whole populations and that has directly led to marker discovery and the production of high-density molecular markers such as single-nucleotide polymorphisms (SNPs) and simple sequence repeats (SSRs) (Bhat et al. 2016). Such markers are useful in linkage mapping, QTL mapping, and marker-assisted selection (MAS) in plant breeding schemes. Second-generation sequencing has also rendered it common practice to use genome-wide association studies (GWAS), which map genetic variants to phenotypic attributes in sizable populations, and that offers good data on genes controlling intricate traits like drought resistance, yield, and disease resistance (Edwards et al. 2013).

Besides DNA-level sequencing, second-generation sequencing revolutionized transcriptomics by leveraging RNA–Seq technology. By employing cDNA-based mRNA sequencing, researchers are able to compare and measure variations in gene expression among different tissues, developmental phases, and environmental settings (Okubo et al. 1992). This has been essential in the identification of differentially expressed genes, studies of transcriptional regulation, and the definition of gene networks for growth, metabolism, and stress response in plants. In addition, NGS has enabled de novo genome and transcriptome assembly in orphan crops or wild relatives of famous crops, which were previously accorded minimal consideration in research (Ye and Fan 2021). This has unveiled the genetic basis of crop improvement and agrobiodiversity conservation. In plant evolutionary biology and plant systematics, NGS has made it possible to sequence organellar genomes

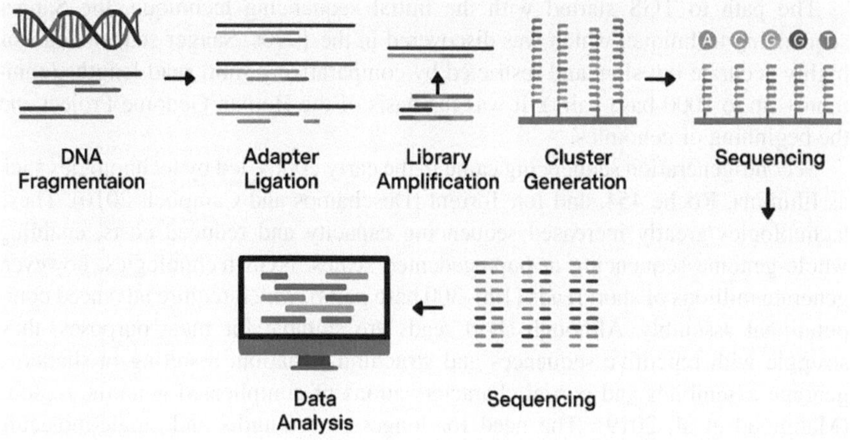

Fig. 3 Workflow of second-generation DNA sequencing

(chloroplasts and mitochondria) and nuclear markers at high throughput, which is useful in phylogenetic research as well as the identification of species (Wicke and Schneeweiss 2015).

Contrary to Sanger sequencing, in which single fragments of DNA are sequenced one after the other, SGS platforms sequence millions of fragments simultaneously, for instance, in genome-wide studies (Fig. 3). The major SGS platforms include Illumina, Roche 454, Ion Torrent, and SOLiD, each using a different strategy, i.e., synthesis, pyrosequencing, semiconductor detection, and ligation-based methodologies, respectively. These technologies are now the mainstay of plant genomics, allowing scientists to sequence whole genomes, study transcriptomes, identify genetic variation, and explore epigenetic modifications in a variety of crop and model plant crops (Deschamps and Campbell 2010).

2.3 *Third-Generation Sequencing (TGS) Technologies*

Third-generation sequencing has a huge advantage in DNA sequencing technology, with the potential to read long DNA stretches from single molecules in real time. TGS technologies directly sequence native DNA or RNA molecules, eliminating amplification bias and native nucleotide changes compared to first- and second-generation platforms that are largely made up of amplification steps and yield relatively short reads (Schadt et al. 2010; Hu et al. 2021). Particularly useful in plant genomics applications, the single-molecule resolution significantly improves the capability to resolve complicated genomic regions, define structural variations, and conserve epigenetic information.

The path to TGS started with the initial sequencing technique, the Sanger sequencing technique, which was discovered in the 1970s. Sanger sequencing was highly accurate but slow and restricted by comparatively short read lengths (commonly up to 1000 base pairs). It was the basis of the Human Genome Project and the beginning of genomics.

Second-generation sequencing came in the early 2000s, led by technologies such as Illumina, Roche 454, and Ion Torrent (Deschamps and Campbell 2010). These technologies greatly increased sequencing capacity and reduced costs, enabling whole-genome sequencing at unprecedented scales. NGS technologies, however, generate millions of short reads (100–300 base pairs), which require advanced computational assembly. Although short reads are suitable for most purposes, they struggle with repetitive sequences and structural variation, resulting in shattered genome assemblies and partial characterizations of complicated genomic regions (Mahmoud et al. 2019). The need for longer read lengths and single-molecule sequencing capability resulted in the emergence of third-generation sequencing (TGS) technologies. TGS was created and released in the early 2010s. However, the first attempt at single-molecule sequencing dates back to 2009.

Among the benefits of TGS is that it can generate ultra-long reads of up to over two million base pairs in some cases. This is compared to second-generation sequencing (NGS), where the typical reads are 100–300 base pairs. The long-read capability of TGS is invaluable in the assembly of complex genomes with high repetition, polyploidy, and structural variation—characteristics of the majority of plant species (Li et al. 2017). Long reads are simpler to assemble by extending repeating sequences and ambiguous regions, which have a tendency to shatter contiguity in short-read assemblies. This translates into more contiguous, accurate, and complete genome assemblies that are invaluable in the detection of gene clusters, chromosomal rearrangements, and regulatory elements (Wee et al. 2019).

An important advantage of TGS is that it is capable of sequencing the DNA in real time without any need for PCR amplification. Apart from making library preparation easier and time-saving, it also eliminates the PCR-induced errors and bias, providing a clear representation of the native DNA or RNA (Wee et al. 2019). TGS platforms also possess the ability to directly visualize epigenetic marks such as DNA methylation and base modification without the need for further chemical modification (Searle et al. 2023). This is particularly relevant to plants, in which epigenetic regulation has a significant role to play in regulating gene expression in accordance with environmental signals and developmental processes (Fig. 4).

While it has many benefits, TGS initially suffered from high error rates, especially compared to the very high accuracy of the Sanger and Illumina sequencing technologies (Hassan et al. 2023). As sequencing chemistries, hardware, and bioinformatics algorithms continue to evolve, however, the quality and reliability of TGS data have vastly improved. For example, consensus-based sequencing technologies, in which the identical DNA molecule is sequenced multiple times, have significantly boosted accuracy rates to typically well over 99% in high-fidelity modes (Salk et al. 2018).

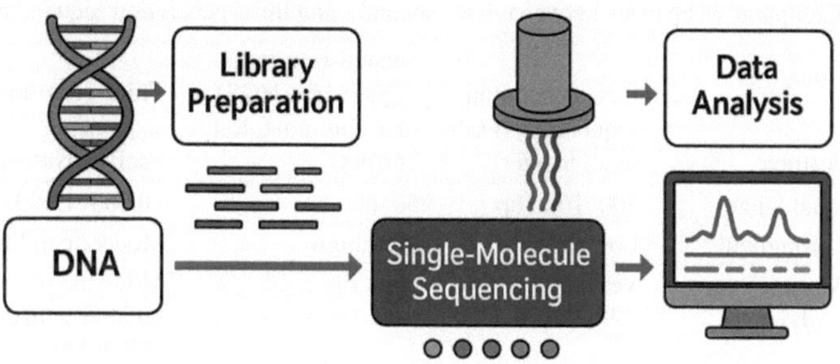

Fig. 4 Workflow of third-generation DNA sequencing

Two of the leading TGS platforms are Pacific Biosciences (PacBio) and Oxford Nanopore Technologies (ONT). PacBio Single Molecule Real-Time (SMRT) sequencing utilizes a patented chip called a zero-mode waveguide (ZMW), which observes DNA polymerase function while it incorporates fluorescently labeled nucleotides in real time (Iizuka et al. 2022). The latest PacBio technology generates HiFi (High Fidelity) reads, which utilize the advantages of long read length and high base accuracy, and are therefore appropriate for genome assembly, isoform detection, structural variant calling, and epigenetic profiling (Lang et al. 2020).

In contrast, ONT presents a very different solution: DNA or RNA molecules are forced through a biological nanopore inserted in a membrane in ONT sequencing, and the changes in the ionic current are measured to read out the nucleotide sequence (Ying et al. 2022). ONT platforms such as MinION, GridION, and PromethION are extremely portable, scalable, and versatile and provide researchers with the ability to sequence in field labs or far-away labs (Krehenwinkel et al. 2019). The convenience and mobility of ONT platforms have made them extremely accessible for real-time diagnosis, monitoring of biodiversity, and rapid genotyping in agricultural contexts.

To summarize the discussion on the history of sequencing technologies, a comparative table of the first-, second-, and third-generation sequencing platforms is presented in Table 2 (Sanger et al. 1977; Mardis 2008; Deschamps and Campbell 2010; Goodwin et al. 2016; Plomion et al. 2018; Van Dijk et al. 2018; Michael and VanBuren 2020). The table accentuates the important features of read length, throughput, accuracy, cost, and applicability across different applications in plant genomics. By outlining the strengths and limitations of each generation, this table serves as a useful reference for understanding how sequencing technologies have progressed and how they are strategically applied based on specific research needs in plant science.

Table 2 Comparative analysis between first-, second-, and third-generation sequencing

S. No.	Feature	First-generation sequencing (FGS) (e.g., Sanger)	Second-generation sequencing (NGS) (e.g., Illumina, Ion Torrent)	Third-generation sequencing (TGS) (e.g., PacBio, Nanopore)
1.	Read length	500–1000 bp	50–600 bp	10,000–1,000,000+ bp
2.	Throughput	Low	Very high	Moderate to high
3.	Accuracy (raw reads)	Very high (>99.99%)	High (99.5–99.9%)	Moderate (85–95% raw, >99.9% with consensus/HiFi reads)
4.	Turnaround time	Days	Hours to days	Real time or hours
5.	Cost per base	High	Low	Moderate (decreasing rapidly)
6.	PCR amplification requirement	Required	Required (bridge/emulsion PCR)	Not required (less bias)
7.	Structural variant detection	Poor	Moderate	Excellent
8.	De novo assembly	Not feasible for large genomes	Challenging for repetitive/large genomes	Ideal for large, complex, repetitive genomes
9.	Polyploid genome handling	Poor	Moderate	High-resolution haplotyping for polyploid crops
10.	Transcriptome (full-length RNA)	Only cDNA via PCR	Fragmented, bias-prone	Direct full-length transcript sequencing (Iso–Seq, ONT RNA–Seq)
11.	Epigenetics/base modifications	Not supported	Requires separate assays (e.g., bisulfite)	Native detection of 5mC, 6 mA, etc.
12.	Portability	No	No	Yes (e.g., ONT MinION)
13.	Sample preparation time	Long	Moderate	Short (no amplification/labeling needed)
14.	Application in plant genomics	Small gene regions, validation	WGS, RNA–Seq, GWAS, marker discovery	High-quality genomes, isoform sequencing, epigenome profiling
15.	Suitability for field use	No	No	Yes (ONT is field-deployable)
16.	Homopolymer error issues	No	Yes (especially ion torrent, 454)	Less severe with consensus reads

3 Applications of NGS in Plant Genomics

Prior said that high-throughput analysis of DNA and RNA was impossible, but it is possible now with next-generation sequencing (NGS). NGS altered genomic research in plants and allows for deep examination. Even non-model plant species can be easier to study due to the flexibility offered by NGS. In plant genomics,

Table 3 Common plant NGS applications and expected outputs

S. No.	Application	NGS method	Target molecule	Typical output	Bioinformatics tools
1.	Gene expression profiling	RNA–Seq	mRNA	Expression levels (TPM/FPKM), DEGs	STAR, HISAT2, DESeq2, edgeR
2.	Genome assembly	WGS (short/long)	Genomic DNA	Contigs, scaffolds, genome draft	SPAdes, Canu, Flye, MaSuRCA
3.	SNP discovery	WGS, GBS	Genomic DNA	Variant calls (VCF files)	BWA, GATK, SAMtools
4.	Methylation analysis	Bisulfite sequencing	DNA (methylated)	CpG methylation sites	Bismark, BS-seeker, MethGo
5.	Protein–DNA interactions	ChIP–Seq	DNA	Peak regions for TF binding	MACS2, HOMER
6.	Phylogenomics	Targeted/WGS	DNA	Tree topologies, clade divergence	IQ-TREE, RAxML, MEGA

RNA–Seq focuses on gene expression profiling, de novo assembly, and SNP discovery. All of these techniques have to be paired with proper molecular targets and analytical methods. For the purpose of study planning, NGS methods and relevant bioinformatics tools, along with the types of biological molecules and expected results, can be found in Table 3 (Wang et al. 2009; Kakumanu et al. 2012; Edwards et al. 2013; Zhang et al. 2014; Ye and Fan 2021).

The following sections elaborate on the major uses of NGS in plant genomics, which gainfully advance conservation, crop productivity, and functional genomics, and offer fresh perspectives on plant biology. The scope of these advanced technologies must be defined against the foundational work laid by NGS in plant genomics. NGS's multiple applications range from providing opportunities for resolving genomes to complex gene expression and epigenetic modifications, encompassing various levels of genomic scrutiny. These core applications illustrate how NGS technologies are driving advancements in genetics, breeding, and conservation of plants.

3.1 Whole-Genome Sequencing (WGS)

Whole-genome sequencing (WGS) using NGS technology has revolutionized genomics by providing insight into plant genomes and serves as the basis for backbone mapping. This sequencing technology opens the possibility for high-fidelity reference genome construction, which is important for the annotation of genes, regulatory domains, and structural variations, and all-purpose evolutionary analysis of the genome. WGS has been useful for model organisms, such as the reference specimen for functional genomics: *Arabidopsis thaliana*, and complex crops with big genomes that include many repeated sequences: polyploid wheat (*Triticum aestivum*), maize (*Zea mays*), and sugarcane (*Saccharum officinarum*) (Appels et al. 2018). Reference genomes help breeders and researchers identify key genes, understand important traits, and select target regions for molecular breeding and gene editing. In addition, the ability to study pan-genomes allows for understanding of intraspecific variation and diversity in adaptation (Golicz et al. 2016). In addition to breeding, the ability for WGS to address evolutionary biology questions plays a fundamental role.

3.2 Transcriptome Sequencing (RNA–Seq)

Transcriptome sequencing, or RNA–Seq, provides the capability to study the whole set of RNA transcripts from a cell or tissue under a specific condition. Wang et al. (2009) noted that RNA–Seq allows for the quantification of gene expression levels to an unprecedented degree, and enables the detection of new transcripts, alternative splicing, non-coding RNAs, and even different forms of a single transcript with great sensitivity and resolution. In plants, RNA–Seq is used broadly to study gene expression at different developmental stages, tissues, genotypes, and even stress situations like drought, salinity, or infection by pathogens (Kakumanu et al. 2012; Zhang et al. 2014). RNA–Seq provides an unparalleled advantages and flexibility when working with non-model species that lack genomic resources. In such species, the de novo transcriptome assembly method for building a transcriptome provides an efficient means of gene identification and annotation, which supports the improvement and conservation of specific traits (Martin and Wang 2011).

Additionally, RNA–Seq assists in the discovery of differentially expressed genes (DEGs), their associated co-expression networks, and offers perspectives into the complex trait regulatory systems. Combining RNA–Seq data with other omics dimensions, such as proteomics and metabolomics, deepens our insight into a plant's systems biology. From stress biology and developmental genetics to genotype–environment interactions, the applications of RNA–Seq are ubiquitous, highlighting its critical role in contemporary plant science.

3.3 Resequencing and SNP Discovery

The resequencing of plant genomes through NGS technologies has become a compelling method for examining variation within and across plant populations. They can identify SNPs, InDels, CNVs, and other structural variants. According to Huang et al. (2010) and Varshney et al. (2009), these genetic variations, together with phenotypic diversity, contribute to adaptive variation. It is achieved through the sequencing of several accession landraces or cultivars to a reference genome. SNPs are especially informative traits for population genetics, genome-wide association studies, QTL mapping, and marker-assisted selection due to their abundance. In combination with these methods, SNPs can also fulfill the requirements of GWAS and QTL.

Resequencing is also able to track high-accuracy mapping of complex agronomic traits such as yield, disease resistance, drought tolerance, and flowering time in crops like soybeans, wheat, maize, and rice (Chia et al. 2012; Zhou et al. 2015). Besides these uses, resequencing is also fundamental for the preliminary characterization of germplasm, uncovering genetic bottlenecks alongside studying domestication and evolutionary processes.

Integration of resequencing data with phenotypic and environmental datasets increases the predictability of genomic selection and enables specific breeding for climate-resilient varieties. With decreasing costs of sequencing and optimizing analytical pipelines, resequencing will increasingly speed crop improvement and conservation of biodiversity across the world.

3.4 Epigenomics

Epigenomics focuses on the study of heritable alterations in a biological organism's form or structure, which might include the changes in their physiology as a consequence of the activity of certain genes, which do not have any changes in the base DNA sequence. DNA methylation, histone modifications, and chromatin remodeling are some of the epigenetic traits that help in regulating gene activity in the development stage or in response to the surrounding environment in plants. With the advancement of NGS technologies, entire genomes can be subjected to epigenomic profiling at base-pair resolution via bisulfite sequencing (BS–Seq) for DNA methylation, or ChIP–Seq for histone modifications (Zhang et al. 2006).

NGS-based studies of epigenomics paved new ways in bisulfite sequencing. A lot of new variations have been recorded in several plant species, and they focused on crucial controlling structures that regulate the gene silencing, transposon suppression, and also the tissue-specific gene expression (Cokus et al. 2008). For instance, whole-genome bisulfite sequencing of *A. thaliana* showed distinctive patterns of CHG, CHH, and CG methylation in different functional genes of *A. thaliana*. Furthermore, transposable elements, such as the DNA methylation epigenomic variation, play a part in the phenotypic change of organisms and also contribute to stress adaptability in plants. Kawakatsu et al. claim it is increasingly viewed as a significant source of data applicable in developing breeding strategies aimed at enhancing crop climate change resilience (Kawakatsu et al. 2016).

The merging of epigenomic information with genomic and transcriptomic analysis offers a more complete picture of networks of gene regulation and developmental plasticity. Since epigenetic marks are occasionally reversible and environmentally responsive, epigenomics suggests promising lines for non-genetic control of traits and the preservation of rare and endangered plant species by illuminating environmentally regulated gene regulation.

3.5 Metagenomics and Microbiome Studies

The revelation of our understanding of microbial communities associated with plants arose due to the development of metagenomics and its application in NGS technology. Unlike traditional culture-dependent methods, this paradigm enables comprehensive microbiomics analysis through molecular techniques, as it utilizes DNA extracted from environmental samples. It makes it possible to characterize complex microbiomes in the rhizosphere, phyllosphere, endosphere, and soil, determining their diversity, functional capacity, and relationships with host plants (Bulgarelli et al. 2013; Turner et al. 2013). Researchers use shotgun metagenomic sequencing alongside 16S/ITS amplicon profiling to determine bacterial, fungal, archaeal, and viral constituents that govern plant health, nutrient absorption, growth, and resilience to biotic and abiotic stressors.

In-depth studies of plant microbiomes show that particular taxa and functional genes are closely integrated with a specific plant's genotype, developmental stage, and environment. Such information is vital for developing breeding programs integrated with biotechnological advances for sustainable agriculture and the design of effective beneficial microorganisms targeting biofertilization, biopesticide application, or stress mitigation (Mitter et al. 2016).

In addition, metagenomic analysis helps determine keystone microbial species and new functional genes in nitrogen fixation, phosphate solubilization, and pathogen suppression. Consequently, plant metagenomics is a promising avenuc to improve crop productivity and soil fertility while limiting chemical input dependence. Further merging of metagenomic with plant genomic and transcriptomic datasets will enhance our knowledge of the plant–microbe holobiont and its function in ecosystem services and crop improvement.

4 Integrating Next-Generation Sequencing with Omics Applications

The outreach of new NGS technology has opened a new chapter in plant biology with an emphasis on multi-omics. The ability to integrate genomic data from NGS with other omics technologies (transcriptomics, proteomics, metabolomics, phenomics, and epigenomics) allows us to investigate and understand all aspects of plant development, plant physiology, and plant interactions and responses to environmental stimuli in a system-based manner. This investigative approach also brings us to a new level of resolution in searching for gene-to-phenotype associations and facilitates the dissection of complex traits more precisely than ever.

NGS-based genomics provides the sequence of an organism's DNA, which allows for the discovery of genes, structural variants, and regulatory elements. Access to the transcriptome through RNA sequencing (RNA–Seq) gives researchers the ability to see changes in gene expression across developmental stages, tissues, and stress conditions, hence establishing a functional relevance to genomic regions (Wang et al. 2009; Zhang 2019). By utilizing proteomics in the same study, through mass spectrometry and mapping protein interactions, researchers gain access to the observable dynamics of proteins—their abundance, modifications, and interactions—which cannot be established from transcript data alone (Jorrín-Novo et al. 2009).

Metabolomics, which quantifies small molecules related to primary and secondary metabolism, provides a direct extension to the association of gene and protein functions to physiological phenotypes (Fukushima et al. 2011). For example, in tomato, transcriptome and metabolome data integration has clarified the metabolic transitions for fruit quality and ripening traits (Zhu et al. 2018). Phenomics, the high-throughput quantification of plant traits either in a controlled environment or in the field, provides another layer of observable, agronomically relevant endpoints for these layers of molecular data (Furbank and Tester 2011). NGS also extends its reach into epigenomic questions through the ability to identify DNA methylation and histone modification patterns. These epigenome layers, when integrated into gene expression profiles, can identify epigenetically regulated gene responses to stress or development (Zhang et al. 2006; Kawakatsu et al. 2016). Integration of these elements will be particularly important for crops with complex traits that are associated with genetics and the environment, including, drought tolerance, flowering time, and resistance to pests and diseases.

A widely seen example of multi-omics integration is incorporated in the dissection of drought stress responses in rice and maize. For instance, as a result of the unique application of genomic data acquired through next-generation sequencing (NGS) platforms, or transcriptomics and metabolomics approaches, it is possible to specify a number of valuable regulatory hubs, learn of candidate genes which are possibly linked to regulation of stress, and identify regulatory pathways leading to stress, that could equally be points for genetic enhancement (Obata and Fernie 2012; Lawas et al. 2018). Also, the integration, together with high-resolution expression quantitative trait *loci* (eQTL) and proteomic maps, better allows the identification of candidate genes for more complex traits, such as yield, and for quantifying traits associated with nutrient use efficiency (Huang et al. 2010; Pecchioni et al. 2012; Libbrecht and Noble 2015). This integrative approach will be important not only for basic research but importantly for practical research applications, including breeding, in which genomic selection (GS) models are increasingly incorporating omics-derived predictors to get more accurate forecasts of traits (Crossa et al. 2017).

The inclusion of NGS with other omics platforms not only enriches our knowledge of plant biology but also acts as a key stepping stone for its applied uses in agriculture and ecology. It helps researchers understand the complex plant system, enable them to find molecular determinants of important traits, and facilitate breeding for new crop varieties. As tools become more advanced and methods for the integration of diverse data become more sophisticated, multi-omics applications will continue to be a vital approach for addressing agriculture and sustainability-related global challenges. In the following sections, we define how this approach is being used for agronomic trait improvement, developing advanced disease diagnosis, addressing evolutionary relationships, supporting biodiversity conservation, and increasing resilience to both abiotic and biotic stressors.

4.1 Improving Agronomic Traits

The improvement of agronomic objectives includes yield, drought tolerance, nutrient use efficiency, time to flowering, disease resistance, and other traits relevant to crop improvement and agricultural production to ensure global food security. The introduction of NGS has changed the way we can begin to target the genetic basis of complex traits and launch breeding programs to include genomic and multi-omics data as part of the traits for crop improvement. NGS approaches also make the high-resolution mapping of quantitative trait *loci* (QTL), genome-wide association studies (GWASs), and rapid selection of molecular markers a more feasible task for modern plant breeding industries. Whole-genome sequencing (WGS) and resequencing of large germplasm collections have allowed for the identification of single-nucleotide polymorphisms (SNPs) and structural variants that are associated with desired agronomic traits. Various genome sequencing projects in rice (*O. sativa*) and maize (*Z. mays*) have established patterns of key allelic variants underlying traits such as grain size, stress tolerance, and flowering time, enabling a way to initiate marker-assisted selection (MAS) and genomic selection (GS) (Huang et al. 2010; Jiao et al. 2012). GS uses genome-wide markers to predict breeding values, and while not as widely adopted as MAS, it is becoming a suitable alternative to identify complex traits, particularly traits that have low heritability or strong environmental interaction (Crossa et al. 2017).

RNA–Seq expands upon genomic information by providing information on gene expression in various tissues, developmental time points, and environments. RNA–Seq allows for the identification of differentially expressed genes and regulatory networks that may underpin variation in phenotypic traits. For example, recent transcriptome analyses of wheat (*T. aestivum*) subjected to drought stress have identified genes related to abscisic acid signaling, osmotic adjustment, and antioxidant defense mechanisms, which could be used to identify potential targets for genetic improvement (Liu et al. 2015).

The integration of NGS with further omics approaches, such as proteomics and metabolomics, will advance understanding of how genetic variation momentarily manifests to produce phenotypic traits. Proteomics will identify post-translational modifications and protein–protein interactions that are essential for expressing a trait, and metabolomics analyses will provide biochemical pathways connected to physiological responses. A prime example of multi-omics can be illustrated with tomato (*Solanum lycopersicum*), where the integration of different omics approaches resulted in finding key insights into the regulation of fruit development, particularly regarding quality traits such as, for example, sugar content, acidity, and aroma (Tieman et al. 2017).

NGS has also greatly enhanced pan-genome analyses, which presents an entirely new approach towards trait discovery by capturing the entire range of genomic variation within a species or genotype. This data approach has proven particularly valuable in crops like *Brassica napus* and *Glycine max*, where analysis of presence/absence variation (PAV) and copy-number variation (CNV) has been shown to be associated with key agronomic traits (Song et al. 2020; Liu et al. 2020a, b).

NGS technologies have enabled rapid gene cloning and functional validation using MutMap and QTL–seq methods, which utilize bulked segregant analysis (BSA), allowing the fine mapping of genes controlling traits such as disease resistance and plant architecture in a fraction of the time of conventional methods. These technologies are expediting the development of climate-resilient, high-yield, resource-efficient crop varieties that will help address population growth and environmental change.

4.2 Disease Diagnostic Tools

Plant pathogens, such as fungi, bacteria, viruses, nematodes, and oomycetes, severely hamper crop production globally. To manage diseases and maintain sustainable agricultural systems, it is crucial that plant pathogens are identified early, accurately, and at high throughput. Metagenomics-based NGS has emerged as a useful diagnostic platform to detect a virtually infinite number of pathogens, including pathogens that have not been described or have not been detected, with no requirement for prior knowledge of genomes. Because NGS is an unbiased and sequence-based technology, it has the ability to profile the entire phytobiome, enabling precise diagnostics in symptomatic and asymptomatic plant tissue samples.

In conjunction with greater accuracy diagnostics of plant pathogens, metagenomic sequencing also provides the capacity to provide information on infection and disease-associated microbes that may be acting together or may additionally be acting through interference with each other. Some of the notable examples of NGS-based metagenomics studies include the documentation and identification of new viral species in fruit crops, a major root-rot pathogen in cereals, and bacterial blight in rice (Adams et al. 2009; Barzon et al. 2011). NGS offers an unbiased, sequence-independent approach that does not rely on predefined targets, thereby minimizing concerns related to concurrent infections or strain shifts. This capability has enabled it to overcome several limitations associated with conventional diagnostics, such as culturing, PCR, and ELISA. Tyler et al. (2006) provided a notable milestone study with the first description and identification of the causal agent of citrus greening disease, *Candidatus Liberibacter asiaticus*. Documented identification of it was undertaken using NGS metagenomic analysis with infected tissues without culture.

Amplicon sequencing is another emerging NGS-based technique referring to conserved genomic regions like the internal transcribed spacer (ITS) for fungi and 16S rRNA for bacteria, and pathogenic identification and classification of pathogens. This method has accurately detected fungal pathogens such as *Fusarium*,

Colletotrichum, and *Alternaria* from crops such as tomato and maize (Mohammed and Boyraz 2022). Similarly, NGS has improved the surveillance of viral populations with the use of small RNA sequencing (sRNA–seq), which detects virally derived RNAs (siRNAs) produced as a part of host responses to defense (Massart et al. 2014).

Beyond merely identifying the pathogen, NGS has the capability to monitor the evolution of pathogens, virulence genes, and new resistance-breaking strains. This will be invaluable in developing resistant cultivars and in deploying control strategies more efficiently. Whole-genome sequencing of pathogen isolates has the potential to reveal SNPs, insertion/deletion events, and gene gain/loss that contribute to pathogenicity and adaptation (Friesen et al. 2008). For example, comparative genomics has characterized *Magnaporthe oryzae* (rice blast fungus) and demonstrated the discovery of effector gene variants that are linked to host specificity and the breakdown of resistance.

The scalability of NGS has opened additional avenues for the application of the technology to both disease prediction and epidemiological surveillance. In particular, portable sequencing technologies such as the MinION from Oxford Nanopore provide opportunities for field-based diagnostics and allow for real-time disease surveillance and responses (Boykin et al. 2018). Further, NGS-based databases of pathogen sequences allow for the design of CRISPR-/Cas-based detection systems and DNA microarrays that can be utilized for rapid, affordable diagnostics. Although NGS has changed the game in terms of diagnostics, its uptake as a routine diagnostic tool for plants has been hampered by the cost, complexity of the data, and the need for standardized protocol and curated reference databases. Nevertheless, improvements in sequencing platforms, bioinformatics pipelines, and data-sharing infrastructures are removing these obstacles and continue to pave the way for an exciting future for NGS in plant disease diagnostics.

4.3 Evolutionary Analysis

Next-generation sequencing (NGS) has greatly improved our ability to conduct all-encompassing evolutionary studies in plants with high-resolution, genome-scale data. Unlike traditional molecular markers, where we would have a few loci to use for phylogenetic or population genetic studies, NGS has presented data of such large scale that we could use whole genomes, transcriptomes, and epigenomes to reconstruct evolutionary history, identify speciation events, and have clearer signatures of natural selection. Within phylogenomics, we can leverage whole-genome or transcriptome sequences across several species to make robust inferences about evolutionary relationships and divergence times. Under developmental genetics and molecular systematics, RAD–seq (Restriction-site-associated DNA sequencing), GBS (genotyping-by-sequencing), and genome skimming have proven useful for obtaining phylogenies at deep and relatively shallow taxonomic levels (Eaton and Ree 2013; Dodsworth 2015; Mishra et al. 2023). These techniques are of value in working with plants with complex genomes and species that hybridize frequently, where traditional markers or methods often fail to account for the complexity of evolutionary processes.

NGS has changed the study of domestication and crop evolution. By comparative genomics based on high-throughput resequencing of landraces, cultivars, and their wild relative, researchers can determine genomic regions targeted by selection during domestication. For example, genome-wide scans of rice and maize have uncovered selective sweeps based on traits such as seed shattering, flowering time, and plant architecture (Huang et al. 2012; Hufford et al. 2012). These outcomes provide insights related not only to evolutionary biology but also to crop improvement by identifying genes of agronomic interest. Population genomics is another important field of research, made possible by advances in NGS, which features the study of genetic variation, both within a population and across populations when using comparative genomics. Using whole-genome resequencing and identification of SNPs in a series of natural populations allows detection of *loci* under selection, gene flow patterns within populations, and adaptive divergence from environmental pressures (Narum et al. 2013). These data will also help to examine the impacts of biotic and abiotic stressors on plant adaptation and maintenance of genomic diversity in wild populations.

Moreover, NGS can aid in understanding polyploid evolution, which is widespread, especially in plants. Polyploidization can complicate phylogenetic studies through the phenomenon of genome duplication and gene redundancy, because they can be more difficult to resolve. However, long-read sequencing techniques (e.g., PacBio, Oxford Nanopore) allow much higher resolution to resolve homologous sequences and subgenomes. This can be even further clarified when a polyploid species has undergone a well-defined and straightforward evolution that regressed to diploidy (Renny-Byfield and Wendel 2014). NGS has been essential to elucidating both reticulate evolution and introgression.

In the genera *Helianthus* and *Populus*, gene flow between species was shown on a genome scale, which undermined the conventional tree model and lent itself more to network-based descriptions of evolution (Rieseberg et al. 2007; Suarez-Gonzalez et al. 2016). This demonstrates that evolution in plants is complex and reticulate. Finally, NGS has been applied, in conjunction with ecological and environmental metadata (climate variables, soil types, elevation), in what is known as landscape genomics. This discipline uses genetic variants that correlate with ecological gradients to inform adaptive evolution and conservation methods based on climate change (Fournier-Level et al. 2011). In this way, NGS is shaping the study of variation, divergence, and evolution and will be crucial to studying plants in evolutionary biology and modern studies of evolutionary biology.

4.4 Biodiversity Conservation

With the capacity to provide high-resolution, genome-wide coverage of genetic diversity, population structure, evolutionary relationships, and adaptive characteristics in plant species, NGS has become a fundamental feature in biodiversity conservation strategies. Conservation genetics traditionally relied on a limited number of molecular markers or approaches (microsatellites or AFLPs), which provided relatively low genomic resolution. By contrast, NGS has the capability of being able to assess thousands to millions of *loci* from individuals and populations together, and so develops a greater understanding of the processes behind plant biodiversity, informing management and conservation practices.

One of the major ways NGS plays a role in conservation biology is by determining levels of genetic diversity between and within populations. Advancements in techniques, namely, Restriction site-associated DNA sequencing (RAD–seq), genotyping-by-sequencing (GBS), and whole-genome resequencing, facilitate the ability of a researcher to characterize and represent how variation occurs across genome-wide DNA sequences (Andrews et al. 2016). This will help establish a baseline for detecting populations with low-genetic diversity, managing a genetic resource, and developing strategies to conserve or increase genetic variation that is fundamental for resilience and long-term maintenance.

Although NGS can build on species delimitation and phylogenetic reconstruction, helping to fundamentally resolve taxonomic ambiguities that may constrain conservation priorities. Some studies have used genome skimming, target-enrichment sequencing, and NGS methods to sort out difficult-to-identify cryptic species complexes and produce well-resolved phylogenies (Dodsworth 2015; Vargas et al. 2017). One example is the successful application of genome skimming, a low-coverage whole-genome sequencing approach used to retrieve plastid genomes, nuclear ribosomal DNA, and repetitive elements. This approach can inform identifications made at the level of evolutionarily significant units (ESUs) and help to develop informed conservation actions.

Another possible application is to develop DNA barcodes and genomic markers that can be used for species identification, monitoring, and forensic-type applications. It is becoming popular and increasingly applicable to use NGS metabarcoding techniques for ecological data collection as it has proven effective to identify multiple plant species in environmental matrices (e.g., soil, pollen, or herbivore gut contents) and has recently been championed for ecological monitoring and restoration projects (Hollingsworth et al. 2016). The genomic barcodes also have many applications in helping to mitigate illegal logging and trade in endangered plant species, given our genomic products can provide verifiable molecular evidence for species identification (Dormontt et al. 2015; Fitzpatrick and Keller 2015).

Moreover, NGS provides a means of tracking inbreeding, gene flow, and hybridization, all of which are components of conservation. High-throughput SNP genotyping is useful for tracking subtle patterns of gene flow among fragmented populations, assessing connectivity, and identifying introgression events with domesticated or invasive relatives. This information is incredibly important for maintaining genetic integrity and avoiding outbreeding depression in restoration programs (Allendorf et al. 2010). In summary, NGS has changed entirely how we assess, understand, and manage plant biodiversity in conservation genomics. With genome-scale high-resolution data obtained via NGS, we can make more informed decisions about conserving genetic resources in a time of rapid change.

4.5 Abiotic Stress and Biotic Stress

Plants are frequently impacted by a variety of abiotic and biotic stresses that adversely affect growth, development, and yield. Abiotic stresses include drought, salinity, temperature extremes, and heavy metals, and biotic stresses are mostly from pathogens, pests, and herbivory. To breed new varieties of crops that will be resilient to abiotic and biotic stress, it is critical to identify the molecular bases of plant responses to stressors. Interrogating stress biology at a high resolution and genome-wide scale has been greatly advanced by next-generation sequencing (NGS) because researchers can discover stress-responsive genes, regulatory elements, and epigenetic changes associated with stressors. Most notably, NGS has become the gold standard in the field of molecular biology for transcriptome profiling in stress conditions (especially with RNA–Seq). The RNA–Seq approach allows researchers to identify differential patterns of gene expression in plants subjected to various abiotic and biotic stresses, discover transcription factors that are activated by stressors, and identify known and novel non-coding RNAs playing roles in regulatory pathways associated with stress response in plants. For example, transcriptomic studies in major crops (rice, maize, wheat) have identified candidate genes and stress-inducible pathways that are associated with osmotic regulation, reactive oxygen species (ROS) scavenging, and heat shock responses under drought and heat stress (Kulski 2016).

NGS has also enabled the characterization of plant glycol in biotic stress contexts by analyzing gene expression changes in both host and pathogen through transcript sequencing. Because dual RNA–Seq (dual transcriptomics) allows full characterization of host defense responses and pathogen virulence responses (Westermann et al. 2012), there is now a greater opportunity to understand host–pathogen interactions in a more holistic fashion. Modern-day sequencing technologies are now also uncovering (R) genes, pathogen recognition receptors, and downstream signaling components that define immunity in plants (Liu et al. 2011), such as the NB–LRR proteins.

NGS has also helped uncover the role of small RNAs (sRNAs; e.g., miRNAs and siRNAs) in stress responses through NGS-based small RNA sequencing. The sRNA forms of regulatory RNAs are intermediaries of gene-silencing pathways that fine-tune the expression of stress-responsive genes. For instance, miR398 has been demonstrated to regulate the copper/zinc superoxide dismutases during oxidative stress, and miR393 has been shown to influence auxin signaling components during infection with bacteria (Sunkar et al. 2012).

Epigenetic modifications, specifically DNA methylation and histone modifications, are essential for the establishment of stress memory and acclimation. A deeper understanding of stress-induced epigenetic changes is provided by NGS applications such as bisulfite sequencing (BS–Seq) and chromatin immunoprecipitation sequencing (ChIP–Seq) (Secco et al. 2015). A remarkable finding from these studies was that there are dynamic methylation changes at stress-responsive loci and at transposable elements, demonstrating the epigenetic plasticity of plant genomes in coping with environmental challenges. Also, with the multi-omic approach, including transcriptomics, metabolomics, proteomics, and epigenomics, we have expanded our comprehension of complex molecular pathways involved in stress responses (Varshney et al. 2019).

In summary, NGS technologies have transformed our understanding of plant responses to abiotic and biotic stress, and facilitated a systems view of the complex molecular, genetic, and epigenetic networks involved in stress perception and adaptation. These systems will be crucial for engineering and breeding climate-resilient crops, an important goal for achieving global food security in changing environmental conditions.

5 Impact on Crop Improvement

NGS technologies have fundamentally changed the landscape of crop improvement through the provision of unique insights regarding plant genomes, the enhancement of the discovery and deployment of agriculturally functional traits. Genomics, along with NGS, enables high-throughput sequencing and genome-resolved bioinformatic analysis, which aids and/or facilitates the mapping of genetic variation, identification of beneficial alleles, and application of genomics-assisted breeding methods in a revolutionary way.

A fundamental advancement in NGS and crop improvement, but most importantly genomics-assisted breeding, is in genomic selection (GS). This contrasts with traditional breeding methods, which use phenotypic selection wherein selection decisions are made using growth and yield or related phenotypes. In GS, instead of using phenotypic data, breeders now make selective decisions based on genome-wide marker data that utilize the underlying genotypes of the breeding value prediction of individuals, which reduces breeding cycle time and increases selection accuracy (Meuwissen et al. 2001; Heffner et al. 2009). The availability of high-density single-nucleotide polymorphisms (SNP's) that are produced from processes of genotyping-by-sequencing (GBS) or whole-genome resequencing has enabled the development of genomic prediction models that are now commonly applied to major crops such as wheat, maize, and rice (Crossa et al. 2017).

Another area greatly influenced by NGS is marker-assisted selection (MAS). The identification of tightly linked molecular markers for targeted traits, such as disease resistance, drought tolerance, or grain quality, will enable rapid introgression of the desired traits into elite cultivars (Varshney et al. 2014). For example, the identification of quantitative trait loci (QTLs) using NGS has permitted improved rice varieties to be developed with submergence tolerance (e.g., Sub1A) and phosphorus uptake efficiency (e.g., Pup1 locus) (Kumar et al. 2014). Beyond improving selection accuracy, NGS also greatly assists de novo genome assembly and pan-genome analysis. By these efforts, it is possible to identify the genomic regions that we do not observe in reference genomes but could be important for trait expression in vast germplasm (Hirsch et al. 2014). These studies are also beneficial for the discovery of new alleles and structural variation that govern complex traits such as stress tolerance, yield components, and flowering time.

Another significant advancement is merging NGS and gene-editing methods, particularly CRISPR/Cas systems. The sequence information provided by NGS provides a best guess at the design of CRISPR constructs for targeted gene editing for genes influencing traits of importance. This has been completed for herbicide-tolerant, disease-resistant, and high-yielding cultivars for crops like tomato, rice, and soybean (Zhang et al. 2014), and other examples of success can be found for traits of importance to alternative crops. Exploration of the epigenetic landscape of crop genomes can also occur with NGS. Information about DNA methylation and histone modifications can be produced via bisulfite sequencing, and in addition towards gene expression studies (discussed below, and developed via high-output sequencing). Using sequencing data to develop an increased understanding of gene regulation under various environmental conditions enables avenues to develop breeding strategies that utilize epigenetic variation (Springer and Schmitz 2017). This will improve the likelihood of capitalizing on natural variation and environmental responses in crop improvement programs.

Integration of transcriptome profiling through RNA–Seq also allows for the discovery of important candidate genes and gene networks involved in agronomically important crop traits, particularly when RNA sequencing is performed under different stress or developmental conditions and is supported by physiological and agronomic characterizations and data. This can be critical to develop approaches for yield and the most productive phytonutrient and increasing adaptation to abiotic and biotic stressors (Mochida and Shinozaki 2010). Overall, we believe that the application of NGS data to areas of phenomics, environmental data, and machine learning will predictably accelerate the ability to breed climate-resilient and high-performing crop varieties going forward.

6 Challenges in Plant Genomics Using NGS

NGS has propelled plant genomics forward by enabling rapid, cheap, and high-throughput genome analysis. However, there are still hurdles to overcome that can limit the full use of NGS technologies. These hurdles arise from the nature of plant genomes, the technical constraints of sequencing instrumentation, and the data volume harvest requiring computational resources and bioinformatics expertise. Among the challenges, the complexity and size of plant genomes are first and foremost. Many of the crops of interest have large genomes filled with repetitive sequences, transposable elements, and polyploidy. For example, the wheat genome is approximately 17 gigabases and is made up of greater than 80% repetitive DNA (IWGSC 2018). Such complexity and size make genome assembly difficult to conduct accurately, resulting in fragmented assemblies and gaps. Although long-read sequencing technologies like PacBio and Oxford Nanopore have improved the quality of assembly, they come with additional cost, higher errors, and complementary methods are often needed (Jiao and Schneeberger 2017). The high complexity of plant genomics and the extensive amount of data generated require extensive computational resources and bioinformatics software that contribute to the barriers to entry for limited resource labs.

Polyploidy, or the existence of multiple sets of chromosomes, is common in many major crops, such as wheat, cotton, and potato. This provides further complications in distinguishing between homologous and homologous sequences and hampers variant calling, gene annotation, and functional analysis (Wendel 2015).

Resolving subgenomes of polyploid organisms remains a sizable bottleneck, necessitating bioinformatics tools and investigatory designs. Another serious challenge is the large volume and complexity of NGS data produced. Large datasets necessitate high-performance computing infrastructure for data storage, processing, and analysis. The agile management of raw sequence files, quality control, quality assessment, alignment, genome assembly, variant calling, and downstream functional annotation entails cumbersome pipeline establishment and the necessity of bioinformatics expertise and standardized, reproducible lab work/analysis (Wang 2023). Finally, many plant researchers also face roadblocks from a lack of adequate bioinformatics training or relevant computational tools/resources, resulting in setbacks and/or delays to their projects.

In addition, genomic variation among plants of the same species, including structural variants, copy number variants, and presence–absence variants, is an analytical challenge as well. Current short-read sequencing technologies may not capture large structural variants and cannot resolve complex genomic regions, which are important for our understanding of phenotypic diversity and adaptation (Alonge et al. 2020), and incorporating multiple sequencing techniques, such as long-read and optical mapping tools, is often an arduous, resource-intensive endeavor.

Epigenomic studies using NGS also face challenges since many aspects of methylation patterns and histone modifications are dynamic in response to environmental conditions as well as tissue-specific to particular groups of plants. Working around these layers of epigenetic regulation requires careful experimental design and interpretation to mitigate confounding results (Li et al. 2012). Finally, cost and sample preparation considerations can limit the scalability of NGS for some plant species, particularly those with large, complex genomes or that have limited genomic resources. The input quality, such as a sufficient yield of high molecular weight DNA or usable RNA extracted from the challenged plant tissue, was likely difficult due to the contamination of secondary metabolites and polysaccharides, which could affect sequencing success (Zhou et al. 2010).

In conclusion, while NGS provides powerful tools for tackling any issue in plant genomics and can take these studies to unprecedented levels, continuing to move forward and tackle these challenges will require continuous improvement of sequencing technologies along with spatial transcriptomics technologies, both sophisticated computational tools, increased bioinformatics training opportunities, and collaborative efforts to build reliable community resources and established best practices.

7 Future Prospects

The future of next-generation sequencing (NGS) in plant research lies in increasing incorporation with advanced computational biology, multi-omics analysis, and precision breeding. With ongoing advancements in sequencing technology, from Illumina short-read precision to Oxford Nanopore and ultra-long-read capabilities of PacBio, the practicality of full genome assemblies being available for even the most complex of plant species will grow increasingly rapidly. One of the key frontiers is the integration of NGS with other omics platforms, such as transcriptomics, proteomics, metabolomics, and epigenomics. This system biology strategy will enhance our insight into gene-to-phenotype correlations, which will enable accurate identification of regulatory networks governing traits such as drought tolerance, pest resistance, and yield potential. Moreover, machine learning and artificial intelligence technologies are becoming more prevalent to manage, analyze, and predict results from large-scale sequencing data. The technology has the potential to enhance genome annotation, gene discovery, and identification of new regulatory elements in other plant genomes.

Pangenomics, where the entire genetic variation of a species is considered, as opposed to one reference genome, is emerging as a central tool for studying intraspecies diversity and adaptive evolution. It has the potential to improve basic agronomic traits by utilizing extensive breeding programs that capture rare alleles and population-specific variations. Conservation biology shall benefit from the application of landscape genomics and eDNA technologies fueled by NGS in raising the monitoring of species, habitat assessment, and early genetic erosion detection. These technologies are specifically crucial for the conservation of vulnerable and endemic species, which fits with worldwide conservation objectives for the maintenance of biodiversity.

With the ongoing production of huge datasets in plant genome sequencing endeavors, reliable curated databases become a necessity for storage, retrieval, visualization, and comparative analysis of the data. There are a number of specialized genomic resources available that facilitate a broad range of functions ranging from gene annotation and variation analysis to synteny mapping and evolutionary investigations. A list of popularly used and reliable databases, their main applications, web addresses, and supported data types is presented to help researchers navigate tools for plant NGS data interpretation (Table 4).

Table 4 Essential plant NGS databases and resources

S.No.	Database/ Resource	Purpose	URL	Supported data types
1.	Phytozome	Plant genome browser and annotations	https://phytozome-next.jgi.doe.gov	Genomes, gene families
2.	Ensemble plants	Comparative plant genomics	https://plants.ensembl.org	Genome annotation, variation
3.	NCBI SRA	Raw sequencing read archive	https://www.ncbi.nlm.nih.gov/sra	RNA–Seq, WGS, ChIP–Seq
4.	TAIR	Arabidopsis gene function data	https://www.arabidopsis.org	Gene function, genome
5.	PLAZA	Comparative genomics of plants	https://bioinformatics.psb.ugent.be/plaza	Phylogeny, synteny, networks

Lastly, portability and miniaturization of sequencing platforms (e.g., MinION by Oxford Nanopore) open up the prospect for in-field diagnostics, real-time environmental and ecological monitoring, and citizen science, bridging the gap between laboratory research and application in agri- and ecological settings.

8 Conclusion

With its unparalleled capability to decipher the structural and functional intricacy of plant genomes, next-generation sequencing transformed plant biology. Ranging from first-generation Sanger sequencing to third-generation systems' real-time long-read technology, the method has evolved into a central part of contemporary genomics. Plant biology shifted from descriptive to predictive study with the integration of NGS and omics technologies. It has facilitated stable disease diagnosis, identification of adaptive traits, precise genetic improvement, and holistic conservation strategies. All these advancements are particularly important in regard to emerging challenges such as climate change, loss of biodiversity, and increasing demands for food security.

Despite challenges such as high data complexity, limited availability of reference genomes for non-model plant species, and bioinformatic bottlenecks, the adoption and advancement of NGS technologies continue to rise steadily. Ongoing improvement of sequencing chemistries, decreasing costs, and interdisciplinarity promise that NGS will continue to be at the center of the future of sustainable agriculture, forestry, and conservation genomics. Through embracing innovation and inclusivity in genomics, we stand to unleash the full potential of plant diversity, protecting ecosystems while nourishing and sustaining an expanding global population.

References

Adams IP, Glover RH, Monger WA, Mumford R, Jackeviciene E, Navalinskiene M, Samuitiene M, Boonham N (2009) Next-generation sequencing and metagenomic analysis: a universal diagnostic tool in plant virology. Mol Plant Pathol 10(4):537–545. https://doi.org/10.1111/j.1364-3703.2009.00545.x

Allendorf FW, Hohenlohe PA, Luikart G (2010) Genomics and the future of conservation genetics. Nat Rev Genet 11(10):697–709. https://doi.org/10.1038/nrg2844

Alonge M, Wang X, Benoit M, Soyk S, Pereira L, Zhang L, Suresh H, Ramakrishnan S, Maumus F, Ciren D, Levy Y (2020) Major impacts of widespread structural variation on gene expression and crop improvement in tomato. Cell 182(1):145–161. https://doi.org/10.1016/j.cell.2020.05.021

Altmann T (2005) Plant genomics. In: Encyclopedic reference of genomics and proteomics in molecular medicine. Springer, Berlin/Heidelberg, pp 1416–1420. https://doi.org/10.1007/3-540-29623-9_3390

Andrews KR, Good JM, Miller MR, Luikart G, Hohenlohe PA (2016) Harnessing the power of RADseq for ecological and evolutionary genomics. Nat Rev Genet 17(2):81–92. https://doi.org/10.1038/nrg.2015.28

Appels R, Eversole K, Stein N, Feuillet C, Keller B, Rogers J, Pozniak CJ, Choulet F, Distelfeld A, Poland J (2018) Shifting the limits in wheat research and breeding using a fully annotated reference genome. Science 361(6403):eaar7191. https://doi.org/10.1126/science.aar7191

Barba M, Czosnek H, Hadidi A (2014) Historical perspective, development and applications of next-generation sequencing in plant virology. Viruses 6(1):106–136

Barzon L, Lavezzo E, Militello V, Toppo S, Palù G (2011) Applications of next-generation sequencing technologies to diagnostic virology. Int J Mol Sci 12(11):7861–7884. https://doi.org/10.3390/ijms12117861

Berkman PJ, Lai K, Lorenc MT, Edwards D (2012) Next-generation sequencing applications for wheat crop improvement. Am J Bot 99(2):365–371

Bhat JA, Ali S, Salgotra RK, Mir ZA, Dutta S, Jadon V, Tyagi A, Mushtaq M, Jain N, Singh PK, Singh GP (2016) Genomic selection in the era of next generation sequencing for complex traits in plant breeding. Front Genet 7:221

Boykin LM, Ghalab A, De Marchi BR, Savill A, Wainaina JM, Kinene T, Lamb S, Rodrigues M, Kehoe MA, Ndunguru J, Tairo F (2018) Real time portable genome sequencing for global food security. bioRxiv:314526

Bulgarelli D, Schlaeppi K, Spaepen S, Van Themaat EVL, Schulze-Lefert P (2013) Structure and functions of the bacterial microbiota of plants. Annu Rev Plant Biol 64(1):807–838. https://doi.org/10.1146/annurev-arplant-050312-120106

Chen CY (2014) DNA polymerases drive DNA sequencing-by-synthesis technologies: both past and present. Front Microbiol 5:305

Chia JM, Song C, Bradbury PJ, Costich D, De Leon N, Doebley J, Elshire RJ, Gaut B, Geller L, Glaubitz JC, Gore M (2012) Maize HapMap2 identifies extant variation from a genome in flux. Nat Genet 44(7):803–807. https://doi.org/10.1038/ng.2313

Claros MG, Bautista R, Guerrero-Fernández D, Benzerki H, Seoane P, Fernández-Pozo N (2012) Why assembling plant genome sequences is so challenging. Biology 1(2):439–459

Cokus SJ, Feng S, Zhang X, Chen Z, Merriman B, Haudenschild CD, Pradhan S, Nelson SF, Pellegrini M, Jacobsen SE (2008) Shotgun bisulphite sequencing of the Arabidopsis genome reveals DNA methylation patterning. Nature 452(7184):215–219. https://doi.org/10.1038/nature06745

Crossa J, Pérez-Rodríguez P, Cuevas J, Montesinos-López O, Jarquín D, De Los Campos G, Burgueño J, González-Camacho JM, Pérez-Elizalde S, Beyene Y, Dreisigacker S (2017) Genomic selection in plant breeding: methods, models, and perspectives. Trends Plant Sci 22(11):961–975

Davis GL, McMullen MD, Baysdorfer C, Musket T, Grant D, Staebell M, Xu G, Polacco M, Koster L, Melia-Hancock S, Houchins K (1999) A maize map standard with sequenced core markers, grass genome reference points and 932 expressed sequence tagged sites (ESTs) in a 1736-locus map. Genetics 152(3):1137–1172

Deschamps S, Campbell MA (2010) Utilization of next-generation sequencing platforms in plant genomics and genetic variant discovery. Mol Breed 25:553–570

Dodsworth S (2015) Genome skimming for next-generation biodiversity analysis. Trends Plant Sci 20(9):525–527. https://doi.org/10.1016/j.tplants.2015.06.012

Dormontt EE, Boner M, Braun B, Breulmann G, Degen B, Espinoza E, Gardner S, Guillery P, Hermanson JC, Koch G, Lee SL (2015) Forensic timber identification: it's time to integrate disciplines to combat illegal logging. Biol Conserv 191:790–798

Eaton DA, Ree RH (2013) Inferring phylogeny and introgression using RADseq data: an example from flowering plants (*Pedicularis*: Orobanchaceae). Syst Biol 62(5):689–706. https://doi.org/10.1093/sysbio/syt032

Edwards D, Batley J, Snowdon RJ (2013) Accessing complex crop genomes with next-generation sequencing. Theor Appl Genet 126:1–11

Eren K, Taktakoğlu N, Pirim I (2022) DNA sequencing methods: from past to present. Eurasian J Med 54(Suppl 1):S47

Fitzpatrick MC, Keller SR (2015) Ecological genomics meets community-level modelling of biodiversity: mapping the genomic landscape of current and future environmental adaptation. Ecol Lett 18(1):1–16. https://doi.org/10.1111/ele.12376

Fournier-Level A, Korte A, Cooper MD, Nordborg M, Schmitt J, Wilczek AM (2011) A map of local adaptation in *Arabidopsis thaliana*. Science 334(6052):86–89. https://doi.org/10.1126/science.1209271

Friesen TL, Faris JD, Solomon PS, Oliver RP (2008) Host specific toxins: effectors of necrotrophic pathogenicity. Cell Microbiol 10(7):1421–1428

Fukushima A, Kusano M, Redestig H, Arita M, Saito K (2011) Metabolomic correlation-network modules in *Arabidopsis* based on a graph-clustering approach. BMC Syst Biol 5:1–12

Furbank RT, Tester M (2011) Phenomics–technologies to relieve the phenotyping bottleneck. Trends Plant Sci 16(12):635–644

Golicz AA, Batley J, Edwards D (2016) Towards plant pangenomics. Plant Biotechnol J 14(4):1099–1105. https://doi.org/10.1111/pbi.12499

Goodwin S, McPherson JD, McCombie WR (2016) Coming of age: ten years of next-generation sequencing technologies. Nat Rev Genet 17(6):333–351

Gupta N, Verma VK (2019) Next-generation sequencing and its application: empowering in public health beyond reality. In: Microbial technology for the welfare of society, pp 313–341

Hassan S, Bahar R, Johan MF, Mohamed Hashim EK, Abdullah WZ, Esa E, Abdul Hamid FS, Zulkafli Z (2023) Next-generation sequencing (NGS) and third-generation sequencing (TGS) for the diagnosis of thalassemia. Diagnostics 13(3):373

Heather JM, Chain B (2016) The sequence of sequencers: the history of sequencing DNA. Genomics 107(1):1–8

Heffner EL, Sorrells ME, Jannink JL (2009). Genomic selection for crop improvement. Crop Science 49(1):1–12. https://doi.org/10.2135/cropsci2008.08.0512

Hirsch CN, Foerster JM, Johnson JM, Sekhon RS, Muttoni G, Vaillancourt B, Peñagaricano F, Lindquist E, Pedraza MA, Barry K, de Leon N (2014) Insights into the maize pan-genome and pan-transcriptome. Plant Cell 26(1):121–135. https://doi.org/10.1105/tpc.113.119982

Hollingsworth PM, Li DZ, van der Bank M, Twyford AD (2016) Telling plant species apart with DNA: from barcodes to genomes. Philos Trans R Soc B 371(1702):20150338. https://doi.org/10.1098/rstb.2015.0338

Hu T, Chitnis N, Monos D, Dinh A (2021) Next-generation sequencing technologies: an overview. Hum Immunol 82(11):801–811

Huang X, Wei X, Sang T, Zhao Q, Feng Q, Zhao Y, Li C, Zhu C, Lu T, Zhang Z, Li M (2010) Genome-wide association studies of 14 agronomic traits in rice landraces. Nat Genet 42(11):961–967. https://doi.org/10.1038/ng.695

Huang X, Kurata N, Wei X, Wang ZX, Wang A, Zhao Q, Zhao Y, Liu K, Lu H, Li W, Guo Y (2012) A map of rice genome variation reveals the origin of cultivated rice. Nature 490(7421):497–501. https://doi.org/10.1038/nature11532
Hufford MB, Xu X, Van Heerwaarden J, Pyhäjärvi T, Chia JM, Cartwright RA, Elshire RJ, Glaubitz JC, Guill KE, Kaeppler SM, Lai J (2012) Comparative population genomics of maize domestication and improvement. Nat Genet 44(7):808–811. https://doi.org/10.1038/ng.2309
Iizuka R, Yamazaki H, Uemura S (2022) Zero-mode waveguides and nanopore-based sequencing technologies accelerate single-molecule studies. Biophys Physicobiol 19:e190032
International Wheat Genome Sequencing Consortium (IWGSC), Appels R, Eversole K, Stein N, Feuillet C, Keller B, Rogers J, Pozniak CJ, Choulet F, Distelfeld A, Poland J (2018) Shifting the limits in wheat research and breeding using a fully annotated reference genome. Science 361(6403):eaar7191
Jiao WB, Schneeberger K (2017) The impact of third generation genomic technologies on plant genome assembly. Curr Opin Plant Biol 36:64–70. https://doi.org/10.1016/j.pbi.2016.12.004
Jiao Y, Zhao H, Ren L, Song W, Zeng B, Guo J, Wang B, Liu Z, Chen J, Li W, Zhang M (2012) Genome-wide genetic changes during modern breeding of maize. Nat Genet 44(7):812–815. https://doi.org/10.1038/ng.2312
Jorrín-Novo JV, Maldonado AM, Echevarría-Zomeño S, Valledor L, Castillejo MA, Curto M, Valero J, Sghaier B, Donoso G, Redondo I (2009) Plant proteomics update (2007–2008): second-generation proteomic techniques, an appropriate experimental design, and data analysis to fulfill MIAPE standards, increase plant proteome coverage and expand biological knowledge. J Proteome 72(3):285–314
Kakumanu A, Ambavaram MM, Klumas C, Krishnan A, Batlang U, Myers E, Grene R, Pereira A (2012) Effects of drought on gene expression in maize reproductive and leaf meristem tissue revealed by RNA-Seq. Plant Physiol 160(2):846–867. https://doi.org/10.1104/pp.112.200444
Kawakatsu T, Huang SSC, Jupe F, Sasaki E, Schmitz RJ, Urich MA, Castanon R, Nery JR, Barragan C, He Y, Chen H (2016) Epigenomic diversity in a global collection of *Arabidopsis thaliana* accessions. Cell 166(2):492–505
Kochetkov N (2012) Organic chemistry of nucleic acids: part B. Springer Science & Business Media
Krehenwinkel H, Pomerantz A, Prost S (2019) Genetic biomonitoring and biodiversity assessment using portable sequencing technologies: current uses and future directions. Genes 10(11):858
Kulski J (2016) Next-generation sequencing an overview of the history, tools, and Omic applications. In: Next generation sequencing-advances, applications and challenges, pp 3–60
Kumar A, Dixit S, Ram T, Yadaw RB, Mishra KK, Mandal NP (2014) Breeding high-yielding drought-tolerant rice: genetic variations and conventional and molecular approaches. J Exp Bot 65(21):6265–6278
Kumar KR, Cowley MJ, Davis RL (2024) Next-generation sequencing and emerging technologies. Semin Thromb Hemost 50(7):1026–1038
Lang D, Zhang S, Ren P, Liang F, Sun Z, Meng G, Liu S (2020) Comparison of the two up-to-date sequencing technologies for genome assembly: HiFi reads of Pacific Biosciences Sequel II system and ultralong reads of Oxford Nanopore. Gigascience 9(12):giaa123. https://doi.org/10.1093/gigascience/giaa123
Lawas LMF, Zuther E, Jagadish SK, Hincha DK (2018) Molecular mechanisms of combined heat and drought stress resilience in cereals. Curr Opin Plant Biol 45:212–217
Lee LG, Connell CR, Woo SL, Cheng RD, McArdle BF, Fuller CW, Halloran ND, Wilson RK (1992) DNA sequencing with dye-labeled terminators and T7 DNA polymerase: effect of dyes and dNTPs on incorporation of dye-terminators and probability analysis of termination fragments. Nucleic Acids Res 20(10):2471–2483
Levinson G, Gutman GA (1987) Slipped-strand mispairing: a major mechanism for DNA sequence evolution. Mol Biol Evol 4(3):203–221
Li X, Zhu J, Hu F, Ge S, Ye M, Xiang H, Zhang G, Zheng X, Zhang H, Zhang S, Li Q (2012) Single-base resolution maps of cultivated and wild rice methylomes and regulatory roles of DNA methylation in plant gene expression. BMC Genomics 13:1–15. https://doi.org/10.1038/nbt.2468

Li C, Lin F, An D, Wang W, Huang R (2017) Genome sequencing and assembly by long reads in plants. Genes 9(1):6

Li H, Li L, ShangGuan G, Jia C, Deng S, Noman M, Liu Y, Guo Y, Han L, Zhang X, Dong Y (2020) Genome-wide identification and expression analysis of bZIP gene family in *Carthamus tinctorius* L. Sci Rep 10(1):15521. https://doi.org/10.1038/s41598-020-72390-z

Libbrecht MW, Noble WS (2015) Machine learning applications in genetics and genomics. Nat Rev Genet 16(6):321–332

Liu J, Elmore JM, Lin ZJD, Coaker G (2011) A receptor-like cytoplasmic kinase phosphorylates the host target RIN4, leading to the activation of a plant innate immune receptor. Cell Host Microbe 9(2):137–146

Liu Z, Xin M, Qin J, Peng H, Ni Z, Yao Y, Sun Q (2015) Temporal transcriptome profiling reveals expression partitioning of homeologous genes contributing to heat and drought acclimation in wheat (*Triticum aestivum* L.). BMC Plant Biol 15:1–20

Liu Y, Du H, Li P, Shen Y, Peng H, Liu S, Zhou GA, Zhang H, Liu Z, Shi M, Huang X (2020a) Pan-genome of wild and cultivated soybeans. Cell 182(1):162–176. https://doi.org/10.1016/j.cell.2020.05.023

Liu Y, Pang D, Tian Y, Li Y, Jiang H, Sun Y, Xia L, Chen L (2020b) Comparative transcriptomic analysis of the tea plant (*Camellia sinensis*) reveals key genes involved in pistil deletion. Hereditas 157:1–11. https://doi.org/10.1186/s41065-020-00153-x

Mahmoud M, Gobet N, Cruz-Dávalos DI, Mounier N, Dessimoz C, Sedlazeck FJ (2019) Structural variant calling: the long and the short of it. Genome Biol 20:1–14

Mardis ER (2008) Next-generation DNA sequencing methods. Annu Rev Genomics Hum Genet 9(1):387–402

Martin JA, Wang Z (2011) Next-generation transcriptome assembly. Nat Rev Genet 12(10):671–682. https://doi.org/10.1038/nrg3068

Massart S, Olmos A, Jijakli H, Candresse T (2014) Current impact and future directions of high throughput sequencing in plant virus diagnostics. Virus Res 188:90–96. https://doi.org/10.1016/j.virusres.2017.05.021

Maxam AM, Gilbert W (1977) A new method for sequencing DNA. Proc Natl Acad Sci 74(2):560–564

Maxam AM, Gilbert W (1980) [57] sequencing end-labeled DNA with base-specific chemical cleavages. In: Methods in enzymology, vol 65. Academic, pp 499–560

Meinke DW, Cherry JM, Dean C, Rounsley SD, Koornneef M (1998) *Arabidopsis thaliana*: a model plant for genome analysis. Science 282(5389):662–682

Meuwissen TH, Hayes BJ, Goddard M (2001) Prediction of total genetic value using genome-wide dense marker maps. Genetics 157(4):1819–1829. https://doi.org/10.1093/genetics/157.4.1819

Michael TP, VanBuren R (2020) Building near-complete plant genomes. Curr Opin Plant Biol 54:26–33. https://doi.org/10.1016/j.pbi.2019.12.009

Mishra G, Meena RK, Kant R, Pandey S, Ginwal HS, Bhandari MS (2023) Genome-wide characterization leading to simple sequence repeat (SSR) markers development in *Shorea robusta*. Funct Integr Genomics 23(1):51

Mitter B, Pfaffenbichler N, Sessitsch A (2016) Plant–microbe partnerships in 2020. Microb Biotechnol 9(5):635–640

Mochida K, Shinozaki K (2010) Genomics and bioinformatics resources for crop improvement. Plant Cell Physiol 51(4):497–523

Mohammed TA, Boyraz N (2022) Molecular detection of fungal plant pathogens: emphasizing recent advancement. In: *2nd international conference on sustainable ecological agriculture 2. UluslararasıSürdürülebilirEkolojikTarımKongresi*, p 422

Myburg AA, Grattapaglia D, Tuskan GA, Hellsten U, Hayes RD, Grimwood J, Jenkins J, Lindquist E, Tice H, Bauer D, Goodstein DM (2014) The genome of *Eucalyptus grandis*. Nature 510(7505):356–362. https://doi.org/10.1038/nature13308

Narum SR, Buerkle CA, Davey JW, Miller MR, Hohenlohe PA (2013) Genotyping-by-sequencing in ecological and conservation genomics. Mol Ecol 22(11):2841. https://doi.org/10.1111/mec.12350

Obata T, Fernie AR (2012) The use of metabolomics to dissect plant responses to abiotic stresses. Cell Mol Life Sci 69:3225–3243
Okubo K, Hori N, Matoba R, Niiyama T, Fukushima A, Kojima Y, Matsubara K (1992) Large scale cDNA sequencing for analysis of quantitative and qualitative aspects of gene expression. Nat Genet 2(3):173–179
Pareek CS, Smoczynski R, Tretyn A (2011) Sequencing technologies and genome sequencing. J Appl Genet 52:413–435
Pecchioni N, Milc JA, Pasquariello M, Francia E (2012) Barley: omics approaches for abiotic stress tolerance. In: Improving crop resistance to abiotic stress, pp 779–884
Peng JH, Lapitan NL (2005) Characterization of EST-derived microsatellites in the wheat genome and development of eSSR markers. Funct Integr Genomics 5:80–96
Plomion C, Aury JM, Amselem J, Leroy T, Murat F, Duplessis S, Faye S, Francillonne N, Labadie K, Le Provost G, Lesur I (2018) Oak genome reveals facets of long lifespan. Nat Plants 4(7):440–452. https://doi.org/10.1038/s41477-018-0172-3
Renny-Byfield S, Wendel JF (2014) Doubling down on genomes: polyploidy and crop plants. Am J Bot 101(10):1711–1725
Rieseberg LH, Kim SC, Randell RA, Whitney KD, Gross BL, Lexer C, Clay K (2007) Hybridization and the colonization of novel habitats by annual sunflowers. Genetica 129:149–165. https://doi.org/10.1126/science.1086887
Salk JJ, Schmitt MW, Loeb LA (2018) Enhancing the accuracy of next-generation sequencing for detecting rare and subclonal mutations. Nat Rev Genet 19(5):269–285
Sanger F (1981) Determination of nucleotide sequences in DNA. Science 214(4526):1205–1210
Sanger F, Nicklen S, Coulson AR (1977) DNA sequencing with chain-terminating inhibitors. Proc Natl Acad Sci 74(12):5463–5467
Schadt EE, Turner S, Kasarskis A (2010) A window into third-generation sequencing. Hum Mol Genet 19(R2):R227–R240
Searle B, Müller M, Carell T, Kellett A (2023) Third-generation sequencing of epigenetic DNA. AngewandteChemie 135(14):e202215704
Secco D, Wang C, Shou H, Schultz MD, Chiarenza S, Nussaume L, Ecker JR, Whelan J, Lister R (2015) Stress induced gene expression drives transient DNA methylation changes at adjacent repetitive elements. elife 4:e09343. https://doi.org/10.7554/eLife.09343
Sentmanat MF, Peters ST, Florian CP, Connelly JP, Pruett-Miller SM (2018) A survey of validation strategies for CRISPR-Cas9 editing. Sci Rep 8(1):888
Shendure JA, Porreca GJ, Church GM (2008) Overview of DNA sequencing strategies. Curr Protoc Mol Biol 81(1):7–1
Shoemaker R, Keim P, Vodkin L, Retzel E, Clifton SW, Waterston R, Smoller D, Coryell V, Khanna A, Erpelding J, Gai X (2002) A compilation of soybean ESTs: generation and analysis. Genome 45(2):329–338
Song JM, Guan Z, Hu J, Guo C, Yang Z, Wang S, Liu D, Wang B, Lu S, Zhou R, Xie WZ (2020) Eight high-quality genomes reveal pan-genome architecture and ecotype differentiation of *Brassica napus*. Nat Plants 6(1):34–45. https://doi.org/10.1038/s41477-019-0577-7
Song B, Ning W, Wei D, Jiang M, Zhu K, Wang X, Edwards D, Odeny DA, Cheng S (2023) Plant genome resequencing and population genomics: current status and future prospects. Mol Plant 16(8):1252–1268. https://doi.org/10.1016/j.molp.2023.07.009
Springer NM, Schmitz RJ (2017) Exploiting induced and natural epigenetic variation for crop improvement. Nat Rev Genet 18(9):563–575. https://doi.org/10.1038/nrg.2017.45
Suarez-Gonzalez A, Hefer CA, Christe C, Corea O, Lexer C, Cronk QC, Douglas CJ (2016) Genomic and functional approaches reveal a case of adaptive introgression from Populus balsamifera (*balsam poplar*) in *P. átrichocarpa* (black cottonwood). Mol Ecol 25(11):2427–2442
Sunkar R, Li YF, Jagadeeswaran G (2012) Functions of microRNAs in plant stress responses. Trends Plant Sci 17(4):196–203. https://doi.org/10.1016/j.tplants.2012.01.010
Supple MA, Shapiro B (2018) Conservation of biodiversity in the genomics era. Genome Biol 19:1–12. https://doi.org/10.1186/s13059-018-1520-3

Tieman D, Zhu G, Resende MF Jr, Lin T, Nguyen C, Bies D, Rambla JL, Beltran KSO, Taylor M, Zhang B, Ikeda H (2017) A chemical genetic roadmap to improved tomato flavor. Science 355(6323):391–394. https://doi.org/10.1126/science.aal1556
Turner TR, James EK, Poole PS (2013) The plant microbiome. Genome Biol 14:1–10. https://doi.org/10.1186/gb-2013-14-6-209
Tuskan GA, Difazio S, Jansson S, Bohlmann J, Grigoriev I, Hellsten U, Putnam N, Ralph S, Rombauts S, Salamov A, Schein J (2006) The genome of black cottonwood, *Populus trichocarpa* (Torr & Gray). Science 313(5793):1596–1604. https://doi.org/10.1126/science.1128691
Tyler BM, Tripathy S, Zhang X, Dehal P, Jiang RH, Aerts A, Arredondo FD, Baxter L, Bensasson D, Beynon JL, Chapman J (2006) Phytophthora genome sequences uncover evolutionary origins and mechanisms of pathogenesis. Science 313(5791):1261–1266
Van Dijk EL, Jaszczyszyn Y, Naquin D, Thermes C (2018) The third revolution in sequencing technology. Trends Genet 34(9):666–681
Vargas OM, Ortiz EM, Simpson BB (2017) Conflicting phylogenomic signals reveal a pattern of reticulate evolution in a recent high-Andean diversification (Asteraceae: Astereae: Diplostephium). New Phytol 214(4):1736–1750. https://doi.org/10.1111/nph.14530
Varshney RK, Nayak SN, May GD, Jackson SA (2009) Next-generation sequencing technologies and their implications for crop genetics and breeding. Trends Biotechnol 27(9):522–530. https://doi.org/10.1016/j.tibtech.2009.05.006
Varshney RK, Terauchi R, McCouch SR (2014) Harvesting the promising fruits of genomics: applying genome sequencing technologies to crop breeding. PLoS Biol 12(6):e1001883. https://doi.org/10.1371/journal.pbio.1001883
Varshney RK, Pandey MK, Bohra A, Singh VK, Thudi M, Saxena RK (2019) Toward the sequence-based breeding in legumes in the post-genome sequencing era. Theor Appl Genet 132(3):797–816. https://doi.org/10.1007/s00122-018-3272-8
Verma M, Kulshrestha S, Puri A (2016) Genome sequencing. In: Bioinformatics: volume I: *data, sequence analysis, and evolution*, pp 3–33
Wang M (2021) Next-generation sequencing (NGS). In: Clinical molecular diagnostics, pp 305–327
Wang X (2023) Next-generation sequencing data analysis. CRC Press
Wang Z, Gerstein M, Snyder M (2009) RNA-Seq: a revolutionary tool for transcriptomics. Nat Rev Genet 10(1):57–63. https://doi.org/10.1038/nrg2484
Weber AP, Weber KL, Carr K, Wilkerson C, Ohlrogge JB (2007) Sampling the Arabidopsis transcriptome with massively parallel pyrosequencing. Plant Physiol 144(1):32–42
Wee Y, Bhyan SB, Liu Y, Lu J, Li X, Zhao M (2019) The bioinformatics tools for the genome assembly and analysis based on third-generation sequencing. Brief Funct Genomics 18(1):1–12
Wendel JF (2015) The wondrous cycles of polyploidy in plants. Am J Bot 102(11). https://doi.org/10.3732/ajb.1500293
Westermann AJ, Gorski SA, Vogel J (2012) Dual RNA-seq of pathogen and host. Nat Rev Microbiol 10(9):618–630. https://doi.org/10.1038/nrmicro2852
Wicke S, Schneeweiss GM (2015) Next-generation organellar genomics: potentials and pitfalls of high-throughput technologies for molecular evolutionary studies and plant systematics. In: Hörandl E, Appelhans MS (eds) Next-generation sequencing in plant systematics, vol 158. Koeltz Scientific Books, Schmitten, pp 9–50
Yamamoto K, Sasaki T (1997) Large-scale EST sequencing in rice. Plant Mol Biol 35:135–144
Ye CY, Fan L (2021) Orphan crops and their wild relatives in the genomic era. Mol Plant 14(1):27–39
Ying YL, Hu ZL, Zhang S, Qing Y, Fragasso A, Maglia G, Meller A, Bayley H, Dekker C, Long YT (2022) Nanopore-based technologies beyond DNA sequencing. Nat Nanotechnol 17(11):1136–1146
Zhang, H., 2019. The review of transcriptome sequencing: principles, history and advances. In IOP conference series: earth and environmental science (332, 4, p. 042003). IOP Publishing

Zhang X, Yazaki J, Sundaresan A, Cokus S, Chan SWL, Chen H, Henderson IR, Shinn P, Pellegrini M, Jacobsen SE, Ecker JR (2006) Genome-wide high-resolution mapping and functional analysis of DNA methylation in *Arabidopsis*. Cell 126(6):1189–1201. https://doi.org/10.1016/j.cell.2006.08.003

Zhang H, Zhang J, Wei P, Zhang B, Gou F, Feng Z, Mao Y, Yang L, Zhang H, Xu N, Zhu JK (2014) The CRISPR/C as9 system produces specific and homozygous targeted gene editing in rice in one generation. Plant Biotechnol J 12(6):797–807. https://doi.org/10.1038/nbt.2833

Zhou X, Ren L, Li Y, Zhang M, Yu Y, Yu J (2010) The next-generation sequencing technology: a technology review and future perspective. Sci China Life Sci 53:44–57

Zhou Z, Jiang Y, Wang Z, Gou Z, Lyu J, Li W, Yu Y, Shu L, Zhao Y, Ma Y, Fang C (2015) Resequencing 302 wild and cultivated accessions identifies genes related to domestication and improvement in soybean. Nat Biotechnol 33(4):408–414. https://doi.org/10.1038/nbt.3096

Zhu G, Wang S, Huang Z, Zhang S, Liao Q, Zhang C, Lin T, Qin M, Peng M, Yang C, Cao X (2018) Rewiring of the fruit metabolome in tomato breeding. Cell 172(1):249–261

Zischewski J, Fischer R, Bortesi L (2017) Detection of on-target and off-target mutations generated by CRISPR/Cas9 and other sequence-specific nucleases. Biotechnol Adv 35(1):95–104

Genome-Wide Association Studies (GWAS) in Plants

Narkhede Gopal Wasudeo, G. Harish Kumar, Manchikatla Arun Kumar, and Shabir H. Wani

Abstract Genome-wide association studies (GWASs) have been a crucial and powerful method for the elucidation of plant traits by studying natural genetic variation and linkage disequilibrium in different genotype populations. The chapter covers all the historical events, the basic theories, and all the methodological developments that have led to GWAS becoming a dominant tool in modern crop breeding. It discusses the main reasons for the success of GWAS, such as population structure, the choice of genotyping technology, statistical methods, and the increasing influence of multi-omics integration, machine learning, and high-throughput phenotypic evaluation on raising detection power and improving biological interpretation, all of which are already well acknowledged. This chapter also discusses the methods used to deal with population stratification, polyploid genomes, and the detection of rare alleles, which pose the main challenges. The authors present their vision of the future and the latest breakthroughs, where GWAS is still a fundamental instrument in the speedy production of climate-resilient and high-yielding varieties.

Keywords Genome-wide association studies · SNP · Linkage disequilibrium · Trait mapping · Genomic selection · Abiotic stress · Biotic stress · Polyploid crops · Phenotyping · Multi-omics

N. G. Wasudeo · G. Harish Kumar · M. A. Kumar
Department of Genetics and Plant Breeding, School of Agriculture, SR University, Warangal, Telangana, India

S. H. Wani (✉)
National Seed Project Laboratory, Sher-e-Kashmir University of Agricultural Sciences and Technology of Kashmir, Srinagar, Jammu and Kashmir, India
e-mail: shabirhwani@skuastkashmir.ac.in

L. M. Vaschetto (ed.), *Omics Technologies in Plant Science*, Interdisciplinary Biotechnological Advances, https://doi.org/10.1007/978-981-92-0533-2_2

Abbreviations

BLINK	Bayesian information and linkage disequilibrium iteratively nested keyway
CRISPR	Clustered regularly interspaced short palindromic repeats
FarmCPU	Fixed and random model circulating probability unification
G-BLUP	Genomic best linear unbiased prediction
GEBV	Genomic estimated breeding value
GLM	General linear model
GS	Genomic selection
GWAS	Genome-wide association studies
LASSO	Least absolute shrinkage and selection operator
LD	Linkage disequilibrium
MAGIC	Multi-parent advanced generation inter-cross
MAS	Marker-assisted selection
MLM	Mixed linear model
MLMM	Multi-locus mixed model
NAM	Nested association mapping
NGS	Next-generation sequencing
PCA	Principal component analysis
QTL	Quantitative trait loci
RNAi	RNA interference
RR-BLUP	Ridge regression best linear unbiased prediction
SNP	Single nucleotide polymorphism
TF	Transcription factor

1 Introduction

Genome-wide association studies (GWASs) give a clear and detailed view of the different genomic regions related to the complex traits through the investigation of the correlation between the genetic markers, mostly being single-nucleotide polymorphisms (SNPs) and the phenotypic variation of different plant populations (Marees et al. 2018; Zhang et al. 2010). By exploiting natural genetic variation and linkage disequilibrium (LD), GWAS has been the main method for unraveling yield, stress tolerance, and disease resistance, which are the most important agronomic traits (Bhandari et al. 2020; Hu et al. 2021). Its precision mapping makes it possible not only to discern the candidate genes and the genomic regions that are responsible for the variation of traits in the plant germplasm but also to provide the insights that go beyond the application of traditional marker-assisted selection and conventional breeding programs (Varshney et al. 2021).

1.1 History of GWAS

GWAS entered plant research in the early 2000s, mainly due to the progress in genotyping technologies and increasing the quality and availability of reference genomes, especially for *Arabidopsis thaliana* (Marees et al. 2018). By promising a lot in the linking between natural genetic variation and phenotypic diversity, the early plant GWAS attempts were able to demonstrate the power of the method even in germplasm panels structured and unstructured (Zhang et al. 2010). One of the first and major milestones in using GWAS in Arabidopsis could help detect flowering time and other quantitative traits reliably (Zhang et al. 2010). The rapid development of high-throughput SNP genotyping and next-generation sequencing (NGS) was the basis of wider application across crops like maize, rice, wheat, and soybean (An et al. 2020; Gao et al. 2021; Amadu et al. 2025). One example of such an application is the maize study conducted by Liu et al. (2016) that led to the identification of many loci for agronomically important traits, and at the same time, the method was being used for breeding. The continued development of population structure and kinship models into the statistical models that were able to account for these factors has so far strengthened the accuracy of association mapping (Zhang et al. 2010; Bose et al. 2023). Currently, GWAS is one of the major tools in plant genomics, and the fields supported candidate gene discovery, genetic architecture elucidation, and molecular breeding facilitation (Varshney et al. 2021). The merging of GWAS with multi-omics datasets and machine learning is still ongoing and is one of the main reasons for the constant growth of the method's power to uncover complex traits in various and polyploid crops (Ma et al. 2021; Mathew et al. 2022).

1.2 Role of GWAS in Modern Plant Breeding and Biotechnology

GWAS have become an essential genomic approach in contemporary plant breeding and biotechnology (Varshney et al. 2021). This approach employs the natural genetic variation present in diverse germplasm panels along with high-resolution genotyping, thus identifying marker–trait associations for complex quantitative traits faster than the traditional linkage mapping method (Bhandari et al. 2020; Pavan et al. 2020). The advantages that GWAS has brought to the different crop species in terms of trait dissection, marker discovery, and precision breeding are enormous. One of its main benefits is the power to reveal the genetic basis of multifactorial traits, for instance, yield, abiotic and biotic stress tolerance, and quality attributes, through the finding of a large number of small-effect loci (Hu et al. 2021; Amadu et al. 2025). The high mapping resolution provided by GWAS not only allows the identification of causal regions with better precision but also gives breeders strong SNP markers for marker-assisted selection (MAS) (Bhandari et al. 2020). Such markers not only enhance genomic prediction models but also lead to the rise of the accuracy of genomic selection (GS), which in turn shortens breeding cycles (Sehgal et al. 2020; Crossa et al. 2022).

GWAS is not only about statistical associations but also acts as a gateway to identify genes and regulatory networks that cause phenotypic variation. If the GWAS technique is combined with datasets from transcriptomics, epigenomics, and metabolomics, it can not only support functional annotation but also reveal the molecular basis of trait expression (Ma et al. 2021; Yuan et al. 2023). The increasing focus on the widening of the genetic base of crops has drawn attention more to landraces, wild relatives, and germplasm collections. The application of GWAS on these different resources leads to the discovery of novel alleles and beneficial haplotypes that have been extensively recombined throughout history (Zeng et al. 2022; Gao et al. 2021). Such findings make it easier for pre-breeding, allele mining, and developing resilient cultivars. Fine-scale mapping of loci allows for the selective introgression of preferred alleles into elite backgrounds by means of backcrossing or using advanced breeding pipelines (Varshney et al. 2021). The insights gained from GWAS are now used immediately in molecular breeding, wherein genomic data direct the selection of parental combinations and predict offspring performance through simulation tools (Lee et al. 2024). After candidate genes or haplotypes are discovered, the functional validation can then be performed using gene-editing platforms like CRISPR/Cas9, RNA interference, or transgenic overexpression (Shelake et al. 2022).

This continuum—from association mapping to gene validation—provides a bridge between breeding and molecular biotechnology, allowing precise and accurate trait manipulation. As the global agricultural landscape changes toward climate-resilient crops, GWAS remains a core method for discovering the genetic bases of adaptation traits, including early maturity, water-use efficiency, and heat tolerance (Chang-Brahim et al. 2024. The machine learning and AI techniques are applied more and more to decode intricate GWAS data, enhance the signal detection, and make predictions regarding breeding strategies more powerful (Mathew et al. 2022; Yang et al. 2024).

2 Basic Concepts of GWAS

The use of genome-wide association studies (GWASs) has become a mainstay in modern plant breeding strategies for the identification of genetic variants responsible for complex traits (Zhang et al. 2010; Marees et al. 2018). GWAS achieves higher resolution than classical QTL mapping through drawing on the natural recombination patterns and by analyzing the whole-genome SNP variation in a wide variety of germplasm (Bhandari et al. 2020; Varshney et al. 2021). Through statistical association models, the variation in phenotypes is connected to particular genes in the genome, hence allowing the identification of genes and regions associated with yield, stress tolerance, and quality (Hu et al. 2021; Amadu et al. 2025). The obtained knowledge instantly endorses the use of marker-assisted selection, genomic selection, and gene discovery activities in crop enhancement programs (Sehgal et al. 2020; Crossa et al. 2022).

2.1 *The Genetic Basis of Trait Variation*

Plant trait variation includes visible distinctions in morphology, physiology, biochemistry, and agriculture among individuals within a population (Marees et al. 2018). Knowledge of the genetic factors responsible for this variation is of paramount importance in contemporary plant breeding, since it is not only the foundation of recognizing the best phenotypes but also the means of deploying appropriate selection methods, and consequently, the source of genetic progress that lasts over time (Varshney et al. 2021; Zeng et al. 2022).

2.2 *Study Design and Population Structure*

The success of the GWAS technique is highly dependent on the design of the study and the genetic composition of the population (Pavan et al. 2020; Bose et al. 2023). A well-designed study makes use of a large variety of genotypes to support the work, together with accurate phenotyping and high-density marker data (Sehgal et al. 2020). Different types of populations—natural panels, biparental populations, MAGIC lines, or NAM populations—offer different levels of mapping power and resolution (Liu et al. 2016; Gao et al. 2021). Because population structure and relatedness can cause misleading associations, some statistical models like the mixed linear models that combine kinship matrices and structure covariates (e.g., principal components or Q-matrices) are employed to significantly reduce the chances of false positives (Zhang et al. 2010; Pavan et al. 2020). The criteria for population size, genetic diversity, marker density, and appropriate statistical controls need to be robust and logically reasoned if marker-trait associations are to be detected with a high level of confidence (Marees et al. 2018; Gerard 2021).

2.3 *Linkage Disequilibrium (LD)*

Linkage disequilibrium (LD) is the term used for the combination of alleles at different loci that are non-randomly and unfavorably associated (Marees et al. 2018). Through GWAS, LD points out those genetic markers that are not causal but are just substitutes for adjacently located causal variants (Zhang et al. 2010). Notably, LD patterns are different in every human population, and they are one of the major factors that determine the quality and the ability of association mapping to be effective (An et al. 2020; Liu et al. 2016). The decreasing pattern of LD—meaning the correlation between alleles decreases with a rise in the physical or genetic distance between loci—is among the many factors that directly affect the power of GWAS in assigning trait-associating areas to specific genes or variations (Gao et al. 2021). Consequently, a thorough understanding of LD structure and its decay rate within the target population is essential. This knowledge ensures optimal genotyping density selection, effective population panel planning, and accurate interpretation of association signals (Amadu et al. 2025; Varshney et al. 2021).

2.4 Statistical Models in GWAS

The statistical models that are employed in GWAS play a major role in the detection of real marker-trait associations and in the reduction of false positives (Zhang et al. 2010). General linear model (GLM), which is one of the early methods, uses population structure and thus yields fewer false discoveries (Marees et al. 2018). Mixed linear models (MLMs) further include population structure and kinship, and therefore their ability to eliminate confounding effects is better (Liu et al. 2016; Pavan et al. 2020). More sophisticated methods that fall under the category of FarmCPU, BLINK, and MLMM are to conduct waves of getting rid of fixed and random effects, or they are to upgrade multi-locus testing procedures even further in terms of statistical power and computer efficiency (Ma et al. 2021; Mathew et al. 2022; Bose et al. 2023). Since these models portray different ways of dealing with confounding variables, effect sizes, and multiple testing burdens, the choice of a particular model will consequently impact GWAS results in terms of robustness and resolution (Crossa et al. 2022; Yang et al. 2024) (Table 1). The criteria for model selection will thus take into account the genetic window of the trait, how big the

Table 1 Key components influencing successful GWAS in plants

Component	Description	References
Population structure	Determines the presence of stratification; requires correction using PCA or Q-matrix to avoid false positives	Zhang et al. (2010), Pavan et al. (2020), Bose et al. (2023)
Genetic diversity	Broader diversity increases mapping resolution and discovery of novel alleles	Gao et al. (2021), Zeng et al. (2022), Varshney et al. (2021)
Phenotyping quality	Accurate, consistent, multi-environment phenotyping improves trait–marker association power	Sehgal et al. (2020), Crossa et al. (2022), Mathew et al. (2022)
Genotyping platform	Includes SNP arrays, GBS, WGR; marker density governs power and resolution	Alonge et al. (2020), An et al. (2020), Gao et al. (2021)
Statistical model choice	GLM, MLM, FarmCPU, BLINK, and MLMM differ in handling confounding structures and multi-locus testing	Liu et al. (2016), Pavan et al. (2020), Ma et al. (2021), Mathew et al. (2022)
Linkage disequilibrium (LD)	Species-specific LD decay affects mapping resolution; dictates marker density requirements	Marees et al. (2018), Zhang et al. (2010), Amadu et al. (2025)
Sample size	Larger panels improve the detection of small-effect loci and rare variants	Gerard (2021), Varshney et al. (2021)

Summary of major biological, statistical, and design factors affecting the reliability, resolution, and interpretability of genome-wide association studies across plant species

Table 2 Major statistical models used in GWAS

Model	Key features	Advantages	Limitations	References
GLM—General linear model	Uses population structure covariates	Simple, fast	Does not correct kinship → more false positives	Zhang et al. (2010), Marees et al. (2018)
MLM—Mixed linear model	Accounts for population structure + kinship	Reduces spurious associations	Computationally heavier	Liu et al. (2016), Pavan et al. (2020)
FarmCPU	Iterative fixed and random effect modeling	High power, fewer false positives	Requires careful parameterization	Ma et al. (2021), Bose et al. (2023)
BLINK	Bayesian & LD-based marker grouping	Very fast; improved power	May miss complex interactions	Mathew et al. (2022), Crossa et al. (2022)
MLMM	Multi-locus mixed model	Models several loci simultaneously	Increased complexity; may overfit	Gerard (2021), Pavan et al. (2020)

Overview of commonly used statistical models in plant GWAS, highlighting their key features, strengths, methodological limitations, and appropriate use cases

population structure is, and the characteristics of the study population (Gerard 2021; Varshney et al. 2021) (Table 2).

3 Methods and Technologies in GWAS

GWAS marked the beginning of a new era in genomics, combining large-scale trait measurements with rigorous statistical approaches to establish trustworthy marker–trait associations (Zhang et al. 2010; Marees et al. 2018). Advances in high-throughput genotyping technologies such as SNP arrays, sequencing-based methods, and whole-genome resequencing not only cover the required marker density of various crop genomes but also make it feasible to study the trait mapping of the crop (Alonge et al. 2020; An et al. 2020; Gao et al. 2021). Statistical methods used in this scenario comprise GLM and MLM together with more contemporary FarmCPU and BLINK that play a role in mitigating the spurious effects from population structure, relatedness, and environment (Liu et al. 2016; Pavan et al. 2020; Ma et al. 2021; Mathew et al. 2022). Furthermore, the integration of phenotypic data, environmental factors, and structural covariates leads to better detection of associations (Crossa et al. 2022; Sehgal et al. 2020). GWAS, which is usually viewed as a one-size-fits-all tool, has actually become more dependent on the choice of the genotyping technique and the analytical model influenced by species, genome architecture, and research goals, thus rendering it a powerful and extensively applied tool for revealing complex traits in plant breeding (Varshney et al. 2021; Amadu et al. 2025) (Table 3).

Table 3 Genotyping platforms used in plant GWAS

Platform	Description	Advantages	Limitations	References
SNP arrays	Fixed-content arrays with high SNP density	High accuracy, standardized data	Limited to known variation; costly for large species	An et al. (2020), Varshney et al. (2021)
GBS—Genotyping-by-sequencing	Reduced-representation sequencing capturing thousands of SNPs	Cost-effective, high throughput	More missing data; lower depth	Alonge et al. (2020), Gao et al. (2021)
Whole-genome resequencing (WGR)	Comprehensive sequencing of full genomes	Maximum resolution; detects rare alleles	Expensive; requires strong computing resources	Amadu et al. (2025), Zeng et al. (2022)
KASP assays	Allele-specific PCR-based SNP genotyping	Ideal for validation and breeding	Not suitable for discovery; low throughput	Shelake et al. (2022), Lee et al. (2024)

Comparison of major genotyping technologies supporting genome-wide association studies, including their advantages, limitations, and suitability for different crop systems

4 Applications of GWAS in Plant Research

GWAS have played a significant role in the area of plant science by allowing the systematic and direct identification of genomic loci responsible for complex traits such as yield, stress tolerance, disease resistance, and several other quality attributes (Bhandari et al. 2020; Hu et al. 2021). Apart from just detecting variation linked to traits, GWAS has opened up learning about the genetic structure of quantitative traits and the identification of biologically plausible candidate genes (Varshney et al. 2021; Zeng et al. 2022) (Table 4). All these developments significantly boost both marker-assisted and genomic selection methods by pinpointing superior alleles that are still present in the different repositories of germplasm (Sehgal et al. 2020; Crossa et al. 2022). Additionally, GWAS brings to light the existing knowledge of evolutionary processes, domestication paths, and gene–environment interactions that give rise to different phenotypes (Gao et al. 2021; Amadu et al. 2025; Chang-Brahim et al. 2024). Furthermore, when GWAS is combined with multi-omics data and high-throughput phenotyping, it provides "the" way for biological complexity to be expressed in a comprehensive manner (Ma et al. 2021; Mathew et al. 2022; Yuan et al. 2023) and thus reinforces its central position in modern plant genetics and crop breeding programs (Varshney et al. 2021).

Table 4 Applications of GWAS in plant breeding

Application	GWAS contribution	References
Yield improvement	Identification of loci with small and large effects contributing to yield architecture	Bhandari et al. (2020), Hu et al. (2021), Amadu et al. (2025)
Abiotic stress tolerance	Discovery of alleles for drought, heat, and salinity tolerance; integration with transcriptomics	Gao et al. (2021), Ma et al. (2021), Chang-Brahim et al. (2024)
Biotic stress resistance	Mapping of resistance genes for diseases and pests in diverse germplasm	Varshney et al. (2021), Zeng et al. (2022)
Quality traits	Identification of loci influencing grain quality, nutritional composition, and product attributes	An et al. (2020), Amadu et al. (2025)
Genomic selection enhancement	Incorporation of GWAS SNPs improves GP model accuracy and prediction of breeding values	Sehgal et al. (2020), Crossa et al. (2022)
Candidate gene discovery	Functional annotation and integration with multi-omics for regulatory pathway identification	Ma et al. (2021), Yuan et al. (2023)
Breeding pipeline acceleration	Supports marker-assisted selection, haplotype analysis, and targeted introgression	Varshney et al. (2021), Lee et al. (2024)

Summary of key application areas where GWAS contributes to trait dissection, allele mining, genomic selection, and breeding pipeline acceleration across crops

5 Challenges and Limitations of GWAS in Plants

Genome-wide association studies have to deal with various methodological and biological constraints that can restrict their interpretation and use in plant breeding (Marees et al. 2018; Zhang et al. 2010). One of the major issues is population structure along with cryptic relatedness, which can lead to false associations if not properly controlled (Pavan et al. 2020; Bose et al. 2023). The ability to discover rare alleles remains very limited, especially in the case of traits controlled by a large number of loci with small effects (Gerard 2021; Ma et al. 2021). It is common for phenotypic datasets to be affected by environmental variation, inconsistent measurements, and low throughput, all of which contribute to the low accuracy of marker–trait associations (Sehgal et al. 2020; Crossa et al. 2022). In some crops, the presence of high linkage disequilibrium brings to the surface not the causal variants but genotyping errors and missing data, which together lead to a further reduction in the precision of mapping (Liu et al. 2016; Alonge et al. 2020; Gao et al. 2021). These problems are aggravated in polyploid species, where the complex genome organization makes allele dosage estimation and LD modeling difficult (LaPierre et al. 2021; Amadu et al. 2025). Therefore, GWAS-identified loci have to pass a functional validation of extremely rigorous levels before their use in breeding programs can be considered reliable (Shelake et al. 2022; Lee et al. 2024).

6 Recent Advances and Future Directions

The latest in plant GWAS is a result of more advanced and interconnected, data-rich, discovery pipelines (Varshney et al. 2021; Crossa et al. 2022). The use of high-throughput phenotyping, better genome assemblies that come along with pangenome resources, and polyploid-aware analytical tools enables a more precise identification of genotype–phenotype relationships (Mathew et al. 2022; Zeng et al. 2022; LaPierre et al. 2021). Concomitantly, multi-omics datasets comprising transcriptomic, metabolomic, and epigenomic data are becoming more common to support causal inference and uncover regulatory networks involved in complex traits (Ma et al. 2021; Yuan et al. 2023). Furthermore, predictive frameworks are maturing: a combination of machine-learning models, genomic prediction, and environmentally informed G × E analyses leads to a more accurate prediction of traits across different environments (Crossa et al. 2022; Yang et al. 2024; Chang-Brahim et al. 2024). Among these, haplotype-based GWAS is emerging as a potent method to identify functional variation that may be overlooked by single-locus analyses (Gao et al. 2021; Amadu et al. 2025).

Connecting these advancements to disturbances in a natural way is a matter of course. A combination of CRISPR-based functional validation, GWAS-informed genomic selection, and speed breeding offers a smooth transition from the detection of a locus to the development of a new variety (Shelake et al. 2022; Lee et al. 2024). All in all, the integration of these advancements positions GWAS as the driving force behind the speeding up of crop improvement through the enhancement of yield, quality, and climate resilience (Varshney et al. 2021; Chang-Brahim et al. 2024).

7 Conclusion

Genomic-wide association studies (GWASs) have turned out to be the main strategy for revealing the genetic origin of complex plant traits such as agronomy, quality, and stress resistance. Natural genetic variation and high-density genotyping have enabled GWAS to identify genomic regions and candidate genes underlying important phenotypes. This supports the advancement of molecular breeding, genomic selection, and functional genomics in both model species and major crops. But still challenges such as population structure, low power to find rare alleles, and multicollinearity in polygenomic genomes have not been completely overcome. The progress achieved in statistical modeling, multi-omics integration, high-throughput phenotyping, and pangenome resources is breaking down these limitations. The creation of polyploid-aware GWAS frameworks and AI-driven prediction tools is additionally opening up new areas for application in the case of diverse and complicated plant genomes. In the future, the combination of GWAS with gene editing, genomic selection, and breeding in climate-resilient varieties will certainly lead to the faster realization of genetic gains. As more and more collaborative platforms and open-access databases are established, GWAS will always be a key player in unlocking plant genetic diversity and promoting sustainable agricultural innovation.

References

Alonge M, Wang X, Benoit M, Soyk S, Pereira L, Zhang L et al (2020) Major impacts of widespread structural variation on gene expression and crop improvement in tomato. Cell 182:145–161

Amadu MK, Beyene Y, Chaikam V, Tongoona PB, Danquah EY, Ifie BE et al (2025) Genome-wide association mapping and genomic prediction analyses reveal the genetic architecture of grain yield under drought and optimal conditions in maize. BMC Plant Biol 25:135

An Y, Chen L, Li Y, Li C, Shi Y, Zhang D et al (2020) Genome-wide association studies and whole-genome prediction reveal the genetic architecture of kernel row number in maize. BMC Plant Biol 20:1–11

Bhandari A, Sandhu N, Bartholome J, Cao-Hamadoun TV, Ahmadi N, Kumari N, Kumar A (2020) Genome-wide association study for yield and related traits under reproductive-stage drought in rice. Rice 13:1–22

Bose A, Burch M, Chowdhury A, Paschou P, Drineas P (2023) Structure-informed clustering for population stratification in association studies. BMC Bioinf 24:411

Chang-Brahim I, Koppensteiner LJ, Beltrame L, Bodner G, Saranti A, Salzinger J et al (2024) Essential roles of remote phenotyping, GWAS, and explainable AI in drought-tolerant winter wheat breeding. Front Plant Sci 15:1319938

Crossa J, Montesinos-Lopez OA, Pérez-Rodríguez P, Costa-Neto G, Fritsche-Neto R, Ortiz R et al (2022) Genome- and environment-based prediction of complex traits incorporating genotype × environment interaction. In: Genomic prediction of complex traits. Springer, pp 245–283

Gao L, Meng C, Yi T, Xu K, Cao H, Zhang S et al (2021) Genome-wide association study reveals the genetic basis of yield and quality traits in wheat. BMC Plant Biol 21:1–11

Gerard D (2021) Pairwise linkage disequilibrium estimation for polyploids. Mol Ecol Resour 21:1230–1242

Hu P, Zheng Q, Luo Q, Teng W, Li H, Li B, Li Z (2021) Genome-wide association study of yield and related traits in wheat under salt stress. BMC Plant Biol 21:1–20

LaPierre N, Taraszka K, Huang H, He R, Hormozdiari F, Eskin E (2021) Identifying causal variants by fine-mapping across multiple studies. PLoS Genet 17:e1009733

Lee AMJ, Foong MYM, Song BK, Chew FT (2024) Genomic selection for crop improvement in fruits and vegetables: a systematic scoping review. Mol Breed 44:60

Liu X, Huang M, Fan B, Buckler ES, Zhang Z (2016) Iterative usage of fixed and random effect models for efficient genome-wide association studies. PLoS Genet 12:e1005767

Ma Y, Min L, Wang J, Li Y, Wu Y, Hu Q et al (2021) Multi-omic dissection of high-temperature-induced male sterility in cotton. New Phytol 231:165–181

Marees AT, De Kluiver H, Stringer S, Vorspan F, Curis E, Marie-Claire C, Derks EM (2018) A tutorial on conducting genome-wide association studies. Int J Methods Psychiatr Res 27:e1608

Mathew B, Hauptmann A, Léon J, Sillanpää MJ (2022) NeuralLasso: neural networks meet LASSO in genomic prediction. Front Plant Sci 13:800161

Pavan S, Delvento C, Ricciardi L, Lotti C, Ciani E, D'Agostino N (2020) Recommendations for selecting genotyping methods and quality control in crop GWAS. Front Genet 11:447

Sehgal D, Rosyara U, Mondal S, Singh R, Poland J, Dreisigacker S (2020) Incorporating GWAS results into genomic prediction models in wheat. Front Plant Sci 11:197

Shelake RM, Kadam US, Kumar R, Pramanik D, Singh AK, Kim JY (2022) Engineering drought and salinity tolerance traits in crops through CRISPR. Plant Commun 3:100417

Varshney RK, Bohra A, Yu J, Graner A, Zhang Q, Sorrells ME (2021) Designing future crops: genomics-assisted breeding comes of age. Trends Plant Sci 26:631–649

Yang Z, Cieza B, Reyes-Dumeyer D, Montesinos R, Soto-Añari M, Custodio N, Tosto G (2024) Benchmark study on GWAS models in admixed populations. Brief Bioinform 25:bbad437

Yuan H, Song M, Liu Y, Xie Q, Cao W, Zhu Y et al (2023) Field phenotyping monitoring systems for high-throughput applications. Agronomy 13:2832

Zeng T, Meng Z, Yue R, Lu S, Li W, Sun Q (2022) Genome-wide association analysis for yield traits in maize. BMC Plant Biol 22:449

Zhang Z, Ersoz E, Lai C-Q, Todhunter RJ, Tiwari HK, Gore MA et al (2010) Mixed linear model approach adapted for genome-wide association studies. Nat Genet 42:355–360

Transcriptomics and Related Omics: The Role of Regulatory Noncoding RNAs in Plant Development and Stress Responses

Shashidhar B. Reddappa, Pavan K. Naik, Gyan P. Mishra, Nandakumar Shekharappa, Muralidhar S. Aski, Manju Kohli, Bidwan Rath, Pream Kumar, Rohit Sarkar, Mallikarjun Biradar, K. N. Kiran, and Harsh K. Dikshit

Abstract Regulatory noncoding RNAs (rncRNAs) have emerged as important modulators of gene expression, influencing plant development and adaptation to diverse stresses. This chapter presents a comprehensive overview of current knowledge on the biogenesis, structural features, and modes of action of major rncRNA classes, including RNA-directed DNA methylation, chromatin remodeling, translational repression, and mRNA cleavage. We discuss how rncRNAs orchestrate developmental transitions from germination to senescence and mediate plant responses to biotic and abiotic stresses. Advances in high-throughput sequencing, transcriptomic profiling, and machine-learning-based target prediction have greatly accelerated the discovery of trait-associated rncRNAs, while modern genome-editing tools such as CRISPR/Cas enable their precise manipulation with minimal off-target effects. Despite these rapid advances, critical knowledge gaps remain, particularly regarding rncRNA cross talk, long-distance mobility, and organelle-specific functions. Future research should focus on expanding and functionally validating ncRNA repertoires linked to yield, stress resilience, and immunity; unraveling mechanisms

S. B. Reddappa
Krishi Vigyan Kendra, ICAR-Vivekananda Parvatiya Krishi Anusandhan Sansthan, Almora, Uttarakhand, India

P. K. Naik · M. S. Aski · M. Kohli · B. Rath · P. Kumar · R. Sarkar · H. K. Dikshit
Division of Genetics, Indian Agricultural Research Institute, New Delhi, India

G. P. Mishra (✉)
Division of Seed Science and Technology, Indian Agricultural Research Institute, New Delhi, India

N. Shekharappa
ICAR-Central Arid Zone Research Institute, Jodhpur, Rajasthan, India

M. Biradar
ICAR-National Bureau of Plant Genetic Resources, New Delhi, India

K. N. Kiran
ICAR-Indian Institute of Horticultural Research, Bengaluru, Karnataka, India

L. M. Vaschetto (ed.), *Omics Technologies in Plant Science*, Interdisciplinary Biotechnological Advances, https://doi.org/10.1007/978-981-92-0533-2_3

governing ncRNA movement across cells and tissues; and integrating ncRNA-based regulatory modules into crop improvement pipelines to develop smart and climate-resilient cultivars. Harnessing the full potential of the plant ncRNome will be pivotal for advancing sustainable agriculture and ensuring global food security.

Keywords Abiotic · Biotic · Genomics · Genome · ncRNA

Abbreviations

ABA	Abscisic Acid
circRNA	Circular RNA
CK	Cytokinin (contextual abbreviation)
DNA	Deoxyribonucleic Acid
dsRNA	Double-Stranded RNA
ET	Ethylene
hc-siRNA	Heterochromatic siRNA
IAA	Indole-3-acetic acid (Auxin)
JA	Jasmonic Acid
lncRNA	Long Noncoding RNA
miRNA	MicroRNA
mRNA	Messenger RNA
nat-siRNA	Natural Antisense siRNA
ncRNA	Noncoding RNA
NGS	Next-Generation Sequencing
ORF	Open Reading Frame
phasiRNA	Phased Secondary siRNA
piRNA	PIWI-Interacting RNA
PTGS	Post-Transcriptional Gene Silencing
RdDM	RNA-Directed DNA Methylation
RNA	Ribonucleic Acid
rncRNA	Regulatory Noncoding RNA
rRNA	Ribosomal RNA
siRNA	Small Interfering RNA
sncRNA	Small Noncoding RNA
snoRNA	Small Nucleolar RNA
snRNA	Small Nuclear RNA
ssRNA	Single-Stranded RNA
tasiRNA	Trans-Acting siRNA
TEs	Transposable Elements
TGS	Transcriptional Gene Silencing
tRNA	Transfer RNA
UTR	Untranslated Region
VIGS	Virus-Induced Gene Silencing
vsiRNA	Virus-Derived siRNA

1 Introduction

The intensifying demands of an ever-growing human population and rapid industrialization continue to accelerate climate change, creating critical challenges for agricultural sustainability and food security (Raza et al. 2023; Sarkar et al. 2025). This environmental shift has simultaneously exacerbated both abiotic stresses (such as drought, salinity, heat, cold, and extreme temperature fluctuations) and biotic stresses (from pathogens, pests, and viruses) (Zorb et al. 2019; Mantri et al. 2012). Rising temperatures and reduced frost days due to climate change will elevate wildfire risks, extend cultivation periods, and alter growing seasons, necessitating greater irrigation (Crimmins et al. 2016). Further, warmer/colder (variation in the temperature) climates and altered precipitation patterns facilitate the spread of weeds, insect pests, and pathogens into new agro-ecological regions, thereby amplifying pest and disease pressure. Heat stress causes pronounced reductions in plant growth, shifts in flowering times, decreases in flower production, and disrupts plant–pollinator synchrony, reducing fruit set and seed set (Crimmins et al. 2016). The cumulative effect of these stressors can reduce crop yields by up to 70%, sometimes leading to complete losses (100%) under severe viral infestations (Mantri et al. 2012; Zorb et al. 2019; Biradar et al. 2025). As sessile organisms, plants must evolve intricate molecular mechanisms to perceive, respond to, and adapt to these external stimuli, as both abiotic and biotic stressors, acting individually or synergistically, disrupt fundamental physiological, biochemical, and molecular processes essential for growth and development (Suzuki et al. 2014; Zandalinas et al. 2016; Kohli et al. 2024).

For decades, the central dogma of molecular biology: the linear flow of genetic information from DNA to RNA to protein, served as the framework for understanding gene expression. However, this view has been significantly expanded by the discovery that the vast majority (approximately 98%) of the eukaryotic transcriptome consists of noncoding RNAs (ncRNAs) (Yu et al. 2019; Zhang et al. 2025). Transcribed from traditionally "silent" regions, including intergenic regions, transposons, repetitive sequences, and pseudogenes, these molecules were once dismissed as transcriptional noise or junk. Today, ncRNAs are recognized as crucial regulators of gene expression in diverse biological systems, including plant development, physiological homeostasis, and adaptation to environmental stresses (Chekanova 2015; Zhang et al. 2025). The ncRNAs are functional RNA molecules that, unlike messenger RNAs (mRNAs) which carry the genetic instructions from DNA for translation into a functional protein, lack an open reading frame (ORF) and instead serve as regulatory elements through their unique RNA sequence and structure modulating gene expression at the transcriptional, post-transcriptional, and epigenetic levels via specific RNA–DNA, RNA–RNA, or RNA–protein interactions. Although ncRNAs often exhibit lower sequence conservation compared to protein-coding genes, they maintain conserved structural motifs that are essential for their regulatory functions (Table 1) (Tian et al. 2025).

The ncRNAs are broadly categorized into two main groups: structural ncRNAs and regulatory ncRNAs (rncRNA). Structural ncRNAs include essential cellular

Table 1 ncRNA classes, their functions, and conservation level

ncRNA type	Size	Primary functions	Conservation level
miRNAs	~22 nt	Posttranscriptional gene silencing via mRNA degradation/repression	Highly conserved
siRNAs	20–25 nt	Defense against viruses/transposons; RNA interference (RNAi), and can also mediate DNA methylation	Moderately conserved
lncRNAs	>200 nt	Chromatin remodeling, transcriptional activation/ repression, nuclear organization, and serving as miRNA sponges.	Lineage specific
circRNAs	Variable	Covalently closed, single-stranded RNA loops formed by back-splicing and function as miRNA sponges, protein scaffolds, and disease biomarkers	Moderately conserved

miRNA micro-RNA, *siRNA* small interfering RNA, *lncRNA* long noncoding RNA, *circRNA* circular RNA

components like ribosomal RNA (rRNA), transfer RNA (tRNA), small-nucleolar RNA (snoRNA), and small-nuclear RNA (snRNA) (Zhu et al. 2018).The rncRNAs are sub-classified based on size and structure into long ncRNAs (lncRNAs, >200 nt), medium-sized ncRNAs, small ncRNAs (sncRNAs, 18–31 nt, miRNA, siRNA, and piRNA), and circular RNAs (circRNA, 100–4000 nt) (Zhu et al. 2018; Bhogireddy et al. 2021). These rncRNAs function through various mechanisms, including RNA-directed DNA methylation (RdDM), translational repression, mRNA cleavage, chromatin remodeling, and interaction with transcription factors to regulate plant development and stress responses (Ariel et al. 2015; Wang and Chekanova 2016).

Crop improvement has traditionally focused on selecting traits such as high yield, pest resistance, or climate adaptability. However, the integration of one beneficial trait often comes at the compromise of others (Varshney et al. 2020). The most promising solution lies in manipulating regulatory genes and their upstream control mechanisms, offering precision without detrimental trade-offs. This direction has become increasingly accessible with the advent of genomics tools such as high-throughput microarrays, next-generation sequencing (NGS), and transcriptomic profiling (Varshney et al. 2020; Raza et al. 2023). The understanding of the cross talk among different classes of ncRNAs, where lncRNAs can act as miRNA sponges or scaffolds, while miRNAs and siRNAs co-regulate overlapping gene sets in feedback loops. Such multilayered regulatory networks confer plants with the plasticity needed to survive dynamic environmental conditions (Bhogireddy et al. 2021; Raza et al. 2023). The exploration of the ncRNome remains one of the most dynamic frontiers in modern biology and plant science, which is the central focus of this chapter, detailing the roles of rncRNAs in plant development, stress responses, and their associated regulatory networks.

2 Historical Perspective: From Junk to Key Regulators

The study on junk DNA began in the mid-twentieth century with the identification of rRNA and tRNA as essential components of the translational machinery. In the 1950s, George Palade characterized ribosomes as the site of protein synthesis via electron microscopy, and Paul Zamecnik discovered tRNA: the first functional ncRNA, initially called "soluble RNA," which acts as an adaptor linking mRNA to amino acids (Zamecnik 2005; Mattick and Amaral 2022). The structural complexity of RNA was revealed in 1965 when Robert Holley sequenced the 80-nucleotide alanine tRNA from yeast, evolving RNA's capacity to form defined cloverleaf secondary structures without encoding proteins (Holley 1965). The 1980s marked a paradigm shift with the discovery of the first regulatory ncRNAs unconnected to protein synthesis. In 1984, the *micF* RNA in *Escherichia coli* was characterized as an independent gene encoding a small RNA that base-pairs with *ompF* mRNA to repress porin expression during osmotic stress. This work, led by Mizuno and Inouye, established the concept of *trans*-acting regulatory RNA genes and RNA-RNA interactions as a gene regulatory mechanism (Mizuno et al. 1984; Delihas 2015). Simultaneously, early long noncoding RNAs (lncRNAs) like *H19* and *Xist* (X-inactivation specific transcript) were identified in mammals (Jarroux et al. 2017). The lncRNA *Xist*, discovered by Brockdorff and Willard, was later shown to have a role in X-chromosome inactivation by wrapping around the chromosome from which it was produced and causing heterochromatin formation, a landmark finding in epigenetic regulation (Brockdorff et al. 1992).

The importance of ncRNA was established in 1993 with the discovery of the first microRNA, *lin-4*, in *Caenorhabditis elegans* by the teams of Victor Ambros, Gary Ruvkun, and Rosalind Lee. They demonstrated that *lin-4* encodes ~22-nt RNAs that repress LIN-14 protein via antisense complementarity to its 3' UTR, revealing a novel post-transcriptional gene-silencing mechanism (Ambros and Ruvkun 2018). The repression of LIN-14 by *lin-4* controls the developmental timings in *C. elegans*. In 2000, Ruvkun's group identified *let-7* as the first evolutionarily conserved microRNA, with homologs from flies to humans, proving that RNA-mediated regulation was a universal feature of metazoan genomes (Ambros and Ruvkun 2018). These discoveries sparked global interest in small regulatory RNAs, leading to the classification of miRNAs, siRNAs, and piRNAs as central components of RNA interference, a breakthrough that was recognized with the 2006 Nobel Prize awarded to Andrew Fire and Craig Mello (Mello 2007). The early 2000s witnessed an explosion in ncRNA discovery due to high-throughput technologies such as microarrays and RNA and DNA sequencing that revealed pervasive transcription across eukaryotic genomes. The ENCODE Consortium project indicated that 70–80% of the human genome is transcribed, uncovering thousands of lncRNAs (Jarroux et al. 2017; Mattick and Amaral 2022). Databases such as Rfam, launched in 2003 with 25 families, now catalog >4000 ncRNA families (Mattick and Amaral 2022).

3 Structure and Biogenesis of rncRNA

3.1 MicroRNAs (miRNAs)

MicroRNAs (miRNAs) are endogenous, single-stranded noncoding RNAs of 21–24 nucleotides (nt) in length, characterized by stem-loop-structured precursor transcripts (Bhogireddy et al. 2021; Bashir and Husaini 2025). Plant genomes encode hundreds of *MIR* genes, many organized into evolutionarily conserved families, classified as either intergenic (transcribed independently by RNA Polymerase II, Pol II) or intronic (processed from host gene introns). The primary miRNA transcripts, called pri-miRNAs, exhibit similar Pol II transcribed mRNA features like 5′ capping, 3′ polyadenylation, and splicing and fold into complex hairpin structures (60–500 nt) comprising a terminal loop, upper/lower stems, and flanking arms (Bartel 2004; Song et al. 2021). This structure undergoes sequential cleavage in the nucleus by the DCL1-HYL1-SE complex: Dicer-Like 1 (DCL1), an RNase III enzyme, performs two endo-nucleolytic cleavages; first liberating a 70–300 nt precursor miRNA (pre-miRNA) from the pri-miRNA, then excising the miRNA/miRNA duplex with 2-nt 3′ overhangs (21–24 nt) (Li et al. 2023). The accessory proteins Hyponastic Leaves 1 (HYL1, dsRNA-binding protein) and zinc-finger protein serrate (SE) stabilize the complex and ensure accurate cleavage, while the resulting duplex is protected from degradation by 3′-end 2′-O-methylation via HUA Enhancer 1 (HEN1). The mature miRNA strand (guide strand) is loaded onto ARGONAUTE 1 (AGO1) to form the core of the RNA-induced silencing complex (RISC), which is subsequently exported to the cytoplasm by EXPORTIN1 (EXPO1) or via HASTY (HST) (Wang et al. 2019a; Li et al. 2023). The guide strand directs the RISC complex to target messenger RNAs (mRNAs) through near-perfect complementarity, leading to its final inhibitory effects: the PIWI domain of AGO1 catalyzes mRNA cleavage (the dominant mechanism in plants), followed by rapid transcript destabilization through deadenylation (by the CCR4-NOT2/3/5 complex), decapping (VARICOSE, DCP1/2), and 5′ to 3′ decay (XRN4 exonucleases); miRNAs can also initiate translational repression through ribosomal stalling (Li et al. 2023).

3.2 Small Interfering RNAs (siRNAs)

Small interfering RNAs (siRNAs) are a diverse class of ncRNAs (20–25 nt) that play fundamental roles in genome defense and developmental programming, differing from miRNAs by originating from dsRNA precursors (Carthew and Sontheimer 2009; Sanan Mishra et al. 2021). These dsRNA precursors are generated from viral replication, transposon activity, convergent transcription, or the action of RNA-dependent RNA polymerases (RDRs) on single-stranded RNA templates, before being processed by Dicer-like (DCL) enzymes into siRNA duplexes that are loaded into specific Argonaute (AGO) proteins (Sanan Mishra et al. 2021).

Based on siRNA biogenesis and function, they are classified into four main types: heterochromatic siRNAs (hc-siRNAs), phased secondary siRNAs (phasiRNAs), natural antisense siRNAs (nat-siRNAs), and virus-derived siRNAs (vsiRNAs) (Sanan Mishra et al. 2021). The 24-nt hc-siRNAs are produced by DCL3 and RDR2 and require RNA Polymerase IV (Pol IV) for transcription from repetitive elements and transposable elements (TEs); loaded into AGO4, they guide the RdDM pathway for TGS and genome stability (Vazquez and Hohn 2013). Phased secondary siRNAs (phasiRNAs) and *trans*-acting siRNAs (tasiRNAs) are 21–22 nt in length and generated through a unique amplification loop termed transitivity. This process requires an initial miRNA or siRNA that cleaves a target RNA transcript. The cleaved fragment is then converted into dsRNA by RDR6, and subsequently diced by DCL4 (or DCL2) into a series of siRNAs phased in 21- or 22-nt fragments from the cleavage site (Fei et al. 2013; Vaucheret and Voinnet 2024). PhasiRNAs can arise from noncoding transcripts (e.g., *TAS* genes) or protein-coding mRNAs. Biogenesis follows either a one-hit model (using a single 22-nt miRNA trigger) or a two-hit model (requiring two miRNA target sites) (Fahlgren et al. 2006). Functionally, phasiRNAs act *in trans* to silence genes distinct from their origin. A classic example involves *miR390* triggering the production of tasiR–ARFs from *TAS3* transcripts via the two-hit mechanism. These tasiR–ARFs are loaded into AGO1 or AGO7 and target *AUXIN RESPONSE FACTOR 3* (*ARF3*/*ETT*) and *ARF4* mRNAs for creating adaxial–abaxial (dorsoventral) leaf polarity (Fahlgren et al. 2006). The *TAS3* transcripts are expressed adaxially, generating tasiR–ARFs that move to the abaxial side to restrict *ARF3*/*ARF4* expression and promote proper leaf patterning (Fahlgren et al. 2006; Garcia et al. 2006; Nogueira and Timmermans 2007). PhasiRNAs also regulate large gene families like *Nucleotide-Binding Leucine-Rich Repeat (NB–LRR)* disease resistance genes and *Pentatricopeptide Repeat (PPR)* genes, reducing their expression to prevent autoimmunity or coordinate stress responses (Fei et al. 2013; Vaucheret and Voinnet 2024).

Natural antisense siRNAs are derived from dsRNA formed by overlapping sense/antisense transcript pairs and typically function *in cis* (Vazquez and Hohn 2013). They are synthesized by RDR6, which amplifies the overlapping dsRNA, followed by dicing through DCL1 (yielding 21-nt siRNAs) or DCL2 (yielding 22-nt siRNAs). A nat-siRNA, SRO5-P5CDH pair induced during salt stress in *Arabidopsis* leads to the accumulation of proline, a compatible osmolyte for osmotic adjustment to enhance salt tolerance (Vazquez and Hohn 2013). Virus-derived siRNAs (vsiRNAs) are important components of siRNAs the plant's antiviral RNAi defense (Mohamed et al. 2022). They are formed from dsRNA replication intermediates of RNA viruses, highly structured regions of viral single-stranded RNAs, or bidirectional transcription of DNA viral genomes (Sanan Mishra et al. 2021; Mohamed et al. 2022). Cytoplasmically, vsiRNAs are generated by DCL4 (21-nt) and DCL2 (22-nt), while nuclear-replicating viruses (e.g., *Caulimoviruses*, *Geminiviruses*) also engage DCL3 (24-nt). These vsiRNAs are loaded into AGO1- or

AGO2-containing RISC complexes to direct the cleavage of complementary viral RNAs in a sequence-specific manner (Kong et al. 2022). The vsiRNA-mediated silencing can spread systemically through the following mechanism: primary vsiRNAs trigger RDR6-dependent amplification, generating secondary vsiRNAs that move through plasmodesmata and confer resistance throughout the plant. This amplification and systemic movement are the basis of virus-induced gene silencing (VIGS) (Liu et al. 2021; Kong et al. 2022; Vaucheret and Voinnet 2024).

The siRNAs can induce both post-transcriptional gene silencing (PTGS) and transcriptional gene silencing (TGS) (Dana et al. 2017). In PTGS, guided by perfect or near-perfect complementarity between the siRNA guide strand and its mRNA target, the PIWI domain of AGO proteins catalyzes endonucleolytic cleavage of the target mRNA between nucleotides complementary to positions 10 and 11 of the siRNAs and cleaved by exonucleases similar to miRNA (Li et al. 2023). Some siRNAs can repress translation without cleavage, although less common than with miRNAs. In TGS, which is mediated by 24-nt hc-siRNAs in complex with AGO4, siRNAs guide DNA methylation and histone modifications at homologous genomic loci, leading to chromatin compaction and suppression of transcription. Certain siRNAs (e.g., tasiR–ARFs, secondary vsiRNAs) act as systemic silencing signals across tissues and exhibit high plasticity and redundancy (Dana et al. 2017; Li et al. 2023).

3.3 Long Noncoding RNAs (lncRNAs)

Long noncoding RNAs (lncRNAs) are defined as non-protein-coding RNA molecules greater than 200 nt (Statello et al. 2021; Bousios et al. 2025). Primarily transcribed by Pol II, with 5′-capped, polyadenylated transcripts, these transcripts are classified by genomic origin as: *intergenic (lincRNAs)*, *antisense (lncNATs)* overlapping coding genes inversely, and *intronic* RNAs (Engreitz et al. 2016). Some lncRNAs (e.g., *MALAT1*, *NEAT1*) are Pol II-transcribed but are processed in a noncanonical, non-polyadenylated manner, and their transcription is regulated by specialized promoters. They are exported to the cytoplasm via NXF1/TAP or CRM1 pathways, and their stability is regulated by degradation mechanisms such as the RNA exosome complex, nonsense-mediated decay (NMD), and XRN1/XRN2 exoribonucleases (Statello et al. 2021). Structurally, lncRNAs adopt complex secondary/tertiary conformations (e.g., stem-loops, R-loops) that interact with nucleic acids and proteins (Statello et al. 2021). They regulate gene expression through (1) chromatin remodeling: *COLDAIR* binds PRC2 to deposit H3K27me3 at *FLC*, repressing flowering (Heo and Sung 2011); (2) Transcriptional interference: *COOLAIR* antisense lncRNA evicts activators (e.g., H3K4me3) from *FLC* (Csorba et al. 2014); (3) miRNA sponging: *IPS1* sequesters *miR399* to protect *PHO2* mRNA during phosphate stress (Franco-Zorrilla et al. 2007); (4) Splicing modulation: *ASCO* lncRNA binds PRP8 to alter spliceosome activity (Rigo et al. 2020) and (5) R-loop formation: *SEP3* circRNA stabilizes RNA:DNA hybrids to regulate parental gene splicing (Wang and Chang 2011; Conn et al. 2017).

3.4 Circular RNAs (circRNAs) and Piwi-Interacting RNAs (piRNAs)

Circular RNAs (circRNAs) are a unique class of ncRNAs, typically are 100 nt to several kb (commonly 100–4000 nt), single-stranded, covalently closed, and form stable loops lacking 5′ caps or 3′ poly-A tails (Lu 2020; Zhang et al. 2024; Srinivasan et al. 2025). They are formed through back-splicing, where a downstream 5′ splice donor joins an upstream 3′ splice acceptor, facilitated by three key mechanisms: (1) Exon skipping, where lariat intermediates circularize after intron removal; (2) Intron pairing, by complementary sequences (e.g., *Miniature Inverted-repeat Transposable Elements*, MITEs) in flanking introns; and (3) RNA-binding protein (RBP) mediated circularization, where proteins like SR45 dimerize to bridge splice sites (Liu et al. 2023; Zhang et al. 2024). Unlike animals, plant circRNAs rarely rely on *Alu*-like repeats for circularization and have unique evolutionary adaptations (Zhang et al. 2020). They are classified by composition: exonic circRNAs (ecircRNAs, cytoplasm-localized), exon–intron circRNAs (EIciRNAs, nuclear), and circular intronic RNAs (ciRNAs, nuclear), each with distinct functional properties (Zhang et al. 2020, 2024). Their closed structure confers the highest stability against exonucleases (e.g., RNase R), hence prolonged regulatory activity (Zhang et al. 2020; Liu et al. 2023). CircRNAs exhibit tissue-specific expression and act through miRNA sponging, R-loop formation, protein interaction and rarely translate into protein (Greene et al. 2017; Zhang & Dai 2022, 2024). In contrast, piRNAs are 24–31 nt ncRNAs, prominent in animal germline cells, where they silence TEs and maintain genomic integrity through association with Piwi-clade Argonaute and ping-pong cycle (Iwasaki et al. 2015). Unlike siRNAs and miRNAs, piRNA biogenesis is Dicer-independent, originates from genomic clusters enriched in TE fragments, and is absent in plants (Siomi et al. 2011; Iwasaki et al. 2015).

4 Mode of Action of Plant rncRNA

4.1 Epigenetic Regulation

Epigenetic regulations are heritable changes achieved without altering the underlying DNA sequence. Regulatory noncoding RNAs (rncRNAs) in plants regulate genes at multiple levels, including epigenetic, transcriptional, and post-transcriptional control, as well as through interactions with phytohormone signaling pathways (Haag and Pikaard 2011; Matzke and Mosher 2014). Instead,

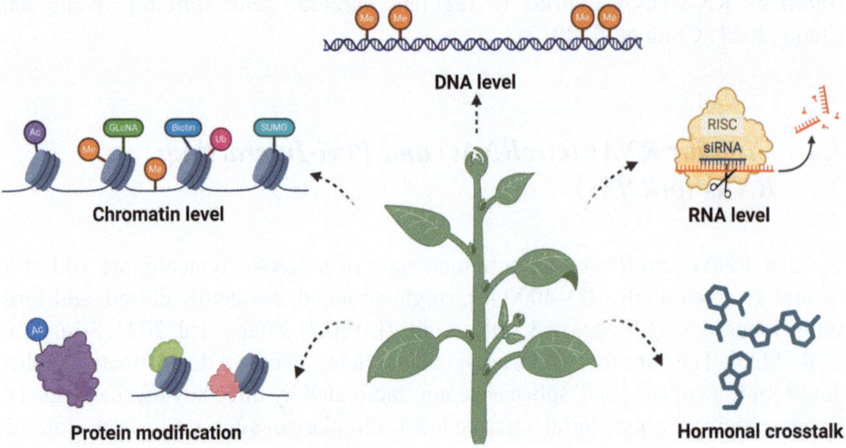

Fig. 1 Regulatory ncRNA mechanisms on plant developmental processes and stress at different levels, such as DNA, RNA, protein, and hormonal cross talk. (Figure created using Biorender)

these changes occur through chemical modifications of DNA (such as DNA methylation), histone modifications (such as acetylation or methylation), RNA splicing, and chromatin remodeling (Fig. 1) (Haag and Pikaard 2011; Matzke and Mosher 2014; Wierzbicki 2012). These chemical and structural modifications influence the cellular growth and development, stress response, and other metabolism.

4.1.1 RNA-Directed DNA Methylation (RdDM)

DNA methylation in plants is mediated by the RdDM pathway. This process consists of two sequential steps: (a) biogenesis of 24-nucleotide (nt) siRNAs, which requires Pol IV, RDR2 (RNA-dependent RNA polymerase 2), and DCL3 (Dicer-like 3), and (b) de novo methylation, which depends on Pol-V-dependent scaffold RNAs, AGO4-bound 24-nt siRNAs, and the de-novo DNA methyltransferase, DRM2 (Domains Rearranged Methyltransferase 2), as well as a number of specific proteins whose functions are not fully understood (Law and Jacobsen 2010; Zhang and Zhu 2011; Pikaard et al. 2012; Saze et al. 2012). In 1994, Wassenegger et al. first reported that the potato spindle tuber viroid (PSTVd) transgene in tobacco was methylated, a process later identified to occur via siRNA intermediates (Wassenegger et al. 1994; Wierzbicki et al. 2008). These 24-nt siRNAs are generated by DICER-LIKE 3 (DCL3) from long double-stranded RNA (dsRNA) precursors, and then the small RNAs are loaded into AGO4 (Law and Jacobsen 2010). The small RNA–AGO4 complex is recruited to the RdDM target loci through sequence complementarity between the siRNA and the scaffold RNA, leading to the recruitment of the DNA methyltransferase DRM2 to establish methylation. The siRNAs are produced from both exogenous and endogenous loci. Endogenously, transposable elements produce numerous siRNAs, and they mediate the maintenance of CHH (where H is any nucleotide but C) DNA methylation via the RdDM pathway following DNA replication (Onodera et al. 2005). Exogenous RNAs such as transcripts from viruses or transgenes can be processed into siRNAs by the plant RdDM pathway to silence these foreign sequences at the transcriptional level (Onodera et al. 2005). Along with the 24-nt siRNAs, the longer noncoding RNAs, i.e., the scaffold RNA, play a critical role in guiding the methyltransferase to target loci (Wassenegger et al. 1994; Jones et al. 1999).

4.1.2 Histone Modification

Small interfering RNAs (siRNAs) are also involved in histone modification (Zhou et al. 2018). Lin et al. reported that overexpression of the mutated forms of histone H3 (H3.1K36M and H3.3K36M) causes serious developmental defects (Lin et al. 2018). The AGO4/siRNAs recruit histone methyltransferases such as SUVH4/KYP, which deposit H3K9me2, thereby reinforcing gene suppression in tandem with RdDM for TE silencing, where siRNAs direct SUVH4/5/6 to create H3K9me2-marked heterochromatin (Law and Jacobsen 2010). Similarly, some lncRNAs, like *ASL*, engage histone acetyltransferases (HATs) to enhance the activation mark H3K9ac (Roule et al. 2022), while the lncRNA APOLO interacts with LIKE HETEROCHROMATIN PROTEIN 1 (LHP1) and PRC1/PRC2 to modify H3K27me3 at auxin-responsive genes (Ariel et al. 2014). In *Arabidopsis thaliana*, the epigenetic silencing of the *Flowering Locus C* (FLC) gene during vernalization is an early example of histone modification, by deposition of H3K27me3, a repressive histone mark. This activity is due to the lncRNA, *COLDAIR*, which is transcribed from the first intron of the FLC gene during cold exposure. Unlike the antisense RNA *COOLAIR*, which indirectly contributes to silencing, *COLDAIR* directly recruits the Polycomb Repressive Complex 2 (PRC2) to the *FLC* locus. PRC2 catalyzes the trimethylation of histone H3 at lysine 27 (H3K27me3), which is a stable and heritable repressive chromatin state that silences FLC expression. This modification regulates long-term repression of *FLC*, allowing the plant to transition to flowering after winter, sensing environmental signals (Heo and Sung 2011).

4.1.3 Chromatin Remodeling

Chromatin remodeling, the dynamic modification of chromatin structure to regulate DNA accessibility, is a major regulatory target for ncRNAs, particularly lncRNAs. Basic mechanisms include competitive inhibition, where lncRNAs bind directly to chromatin remodelers, blocking their interactions with genomic targets. They recruit repressive complexes to specific loci, as in the case of *COLDAIR* and *COOLAIR* systems in *Arabidopsis*. Additionally, scaffold lncRNAs (*APOLO*) interact with transcription factors like WRKY42 to loop chromatin and activate genes like PIN7 during root development (Ariel et al. 2020), while stress-responsive lncRNAs (*DANA1*) recruit histone deacetylases to open chromatin at drought-responsive genes (Wierzbicki et al. 2021). In Arabidopsis, Pol V-derived lncRNAs act as scaffolds to recruit RNA-binding proteins like IDN2, essential for RNA-directed DNA methylation (RdDM) (Ausin et al. 2009; Zheng et al. 2010). IDN2 interacts with SWI3B, a subunit of the SWI/SNF chromatin remodeling complex, guiding it to Pol V-transcribed loci (Bohmdorfer and Wierzbicki 2015).

4.2 Role of siRNAs in Transposon Silencing and Genome Stability

Mobilization of TE can be mutagenic to the plant, so the host genomes have evolved mechanisms to suppress their activities in somatic and gametic cells (Matzke and Mosher 2014). In *Arabidopsis*, RNA Polymerase IV-dependent small interfering RNAs (p4-siRNAs) are produced from thousands of genomic regions, covering over 1% of the genome, which largely include TEs and repetitive sequences and can accumulate in specific tissues such as endosperm (Ito 2012). These siRNAs maintain genome stability via two major pathways: post-transcriptional gene silencing (PTGS), where siRNAs target TE transcripts for degradation, and transcriptional gene silencing (TGS), where siRNAs guide chromatin-modifying complexes to induce heterochromatin formation through histone modifications and DNA methylation (Matzke et al. 2015). In *Arabidopsis thaliana*, AGO4/siRNAs target nascent RNA transcripts from RNA Pol V and recruit SUVH4/5/6 and DRM1/2, which mediate repressive histone modification (H3K9me2) and DNA methylation, respectively, reinforcing the silencing of TE chromatin (Zilberman et al. 2004; Tran et al. 2005; Zhong et al. 2014). The RDR6-RdDM pathway recognizes TEs that have escaped silencing or have just entered the genome and post-transcriptionally breaks down TE mRNAs (Zhong et al. 2014).

4.3 Transcriptional and Post-transcriptional Regulation

Transcription is a coordinated process involving RNA polymerase, transcription factors (TFs), and various regulatory complexes. One key player is the MEDIATOR complex, which facilitates communication between TFs and RNA Polymerase II, and influences transcriptional activity. In *Arabidopsis*, the long noncoding RNA *ELENA1* (*ELF18-INDUCED LONG-NONCODING RNA1*) has been shown to interact with MED19a, a subunit of the MEDIATOR complex, and regulate its recruitment to specific gene promoters (Seo et al. 2017). The process of lncRNA transcription itself can play a regulatory role by directly influencing RNA polymerase II (RNAPII) activity (Tremblay and Qüesta 2023). LncRNAs can also function as negative regulators of gene transcription. In *Arabidopsis*, the lncRNA

HIDDEN TREASURE-1 (*HID1*) suppresses the transcriptional activity of PHYTOCHROME-INTERACTING FACTOR 3 (PIF3) by binding to its first intron, leading to enhanced hypocotyl elongation (Tremblay and Qüesta 2023). Natural antisense transcripts (NATs), which are lncRNAs, form RNA pairs (dsRNA) with complementary mRNAs (cis-NATs or trans-NATs), leading to RNA degradation or recruiting protein partners to modulate transcriptional activity (Lapidot and Pilpel 2006, Britto-Kido et al. 2013). In rice, the NAT-lncRNA *TWISTED LEAF* (*TL*) is transcribed from the strand opposite to that of the *OsMYB60* gene, which encodes an R2R3 MYB transcription factor. When TL expression is suppressed using RNA interference, *OsMYB60* expression increases significantly, resulting in twisted leaf blades (Liu et al. 2018a).

4.4 lncRNAs as Scaffolds, Decoys, or Guides for Protein Complexes

Long noncoding RNAs (lncRNAs) are essential molecular platforms for assembling and guiding multi-protein complexes, controlling chromatin dynamics and gene expression. As scaffolds, lncRNAs assemble multiple proteins into functional complexes, facilitating coordinated regulatory actions. They also act as decoys by binding and sequestering proteins or miRNAs, avoiding them from interacting with their natural targets. This mechanism allows lncRNAs to act as competitive inhibitors. The lncRNA, *IPS1*, functions as a target mimic, or molecular sponge, for *miR399* in *Arabidopsis*. Under phosphate-deficient conditions, plants induce *IPS1*, which contains a motif with partial complementarity to *miR399*. Normally, *miR399* directs the degradation of PHO2, a negative regulator of phosphate uptake, thereby enhancing phosphate acquisition. However, *IPS1* sequesters *miR399* by binding to it without perfect complementarity, preventing *miR399*-mediated cleavage. This interaction protects PHO2 mRNA from degradation and fine-tunes phosphate homeostasis (Franco-Zorilla et al. 2007). Similarly, the *ASCO*-lncRNA acts as a decoy to regulate root development. It competes with the natural targets of nuclear speckle RNA-binding proteins (NSRs), which are important regulators of alternative splicing. This sequestration of NSRs alters the splicing patterns of their target mRNAs, resulting in the formation of alternative splice isoforms and causing a shift in the developmental outcomes of plant roots (Bardou et al. 2014).

4.5 Cross Talk with Phytohormones

Phytohormones, also known as plant hormones, are naturally occurring chemical messengers that regulate various aspects of plant growth and development. The ncRNAs are essential for controlling phytohormone signaling pathways, which affect how plants grow, develop, and react to stress. They interact with phytohormone signaling pathways such as auxin, cytokinin, abscisic acid (ABA), and jasmonic acid (JA), forming complex regulatory networks.

The auxin signaling pathway includes auxin receptor F-box proteins TIR1/AFBs, AUX/IAA (auxin/indole-3-acetic acid) proteins, and ARFs. A miRNA, *miR393*, affects plant sensitivity to auxin signaling by negatively regulating the expression of the target genes TIR1/AFBs and modulates salt stress adaptation of *Arabidopsis* through the redox system and osmotic pressure (Chen et al. 2015). In the cytokinin pathway, *miR156* and *miR172* control age-dependent transitions by targeting *SQUAMOSA PROMOTER BINDING PROTEIN-LIKE* (*SPL*) and *APETALA2* (*AP2*) transcription factors, respectively, which are also influenced by cytokinin levels (Wang et al. 2014). The ABA pathway is similarly regulated by ncRNAs, where *miR159* targets *MYB33/101* to control ABA-responsive seed germination and *miR393* influences drought responses by suppressing *TIR1*, an auxin receptor that cross talks with ABA signaling (Jung et al. 2007). A soybean miRNA, *miR5678*, was predicted to target an ethylene receptor 1-related gene (*Glyma.12G241700*) and its sequence is reverse-complementary to a circRNA, *Gma-circ_0000352*, suggesting that *Gma-circ_0000352* may affect nodulation through the interaction with miRNA (Wang et al. 2020). Ethylene production and lycopene accumulation were largely repressed in lncRNA1459 mutants of tomato. The genes related to ethylene and carotenoid biosynthesis were distinctly downregulated in lncRNA1459 mutants compared with wild-type fruits, and expression of numerous ripening-related genes was changed significantly when lncRNA1459 was knocked out (Li et al. 2018a). The *miR319d* indirectly downregulates the jasmonic acid (JA) biosynthetic gene *lipoxygenase 2* (*LOX2*) by directly targeting and inhibiting the transcription factor *TEOSINTE BRANCHED/CYCLOIDEA/PCF10* (*TCP10*) (Martín-Rodríguez et al. 2018). Under drought stress, the expression of *miR396* is downregulated, leading to the activation of JA and ABA biosynthesis. This activation enhances the accumulation of JA-Ile in vivo, which helps protect chloroplasts and photosynthetic signaling systems. Additionally, the reduced expression of *miR396* contributes to decreased stomatal density and size by maintaining a balanced level of JA and ABA hormones (Fracasso et al. 2021).

5 ncRNAs in Plant Developmental Processes

Noncoding RNAs (ncRNAs) have emerged as crucial regulators throughout the entire life cycle of a plant, from embryogenesis to senescence. Their involvement is so pervasive and essential that there are virtually no major developmental or physiological processes in plants that occur without the participation of ncRNAs. These molecules have a vast array of regulatory mechanisms; their versatility and complexity have made them one of the most rapidly expanding areas of plant molecular research (Table 2).

Table 2 Regulation of plant developmental processes via ncRNA

Type	ncRNA	Crop/Species	Function of ncRNA	References
lncRNA	*TPSI1*	*Solanum lycopersicum*	Phosphate homeostasis	Liu et al. (1997)
	GmENOD40	*Glycine max*	Root nodules formation	Rohrig et al. (2002)
	BcMF11 and *HID1*	*Brassica campestris*	Flowering regulation and seedling photomorphogenesis	Song et al. (2021); Wang et al. (2014)
	APOLO and *ASL*	*Arabidopsis thaliana*	Auxin response; lateral root development; flowering	Ariel et al. (2014); Shin & Chekanova (2014)
	LAIR *TL*	*Oryza sativa*	Rice grain yield and leaf shape remodeling	Wang et al. (2018a); Liu et al. (2018a)
	MISSEN	*Oryza sativa*	Seed development	Zhou et al. (2020)
	lncRNA39026	*Lycopersicon esculentum*	Disease resistance	Hou et al. (2020)
	TPSI1	*Solanum lycopersicum*	Phosphate homeostasis	Liu et al. (1997)
miRNA	*miR156* targeting *SPL* family	*Arabidopsis* and *Zea mays*	Plastochron length, promoting flowering; leaf development, root development, secondary metabolism, and abiotic stress; tillering and corn development	Aukerman and Sakai (2003); Chuck et al. (2007, 2010); Wang et al. (2008); Xu et al. (2016a); Dai et al. (2019)
	miR165, *miR166*, and *miR167* targeting ARFs	*Arabidopsis* and *Oryza*	Development of the male organ, roots, stems, leaves, and flowers, flowering time, embryonic development, seed development, and stress	Williams et al. (2005); Jia et al. (2015); Merelo et al. (2016); Yan et al. (2016)
	mir397 on *OsLAC*	*Oryza*	Grain yield and panicle branches	Zhang et al. (2013)
	miR171 on *SCL*	*Arabidopsis* and barley	Chlorophyll biosynthesis, phase transitions, and floral meristem determinacy	Curaba et al. (2013); Ma et al. (2014); Li et al. (2020)
	miR159 on *GAMYB*	*Arabidopsis*	Vegetative and reproductive development	Allen et al. (2007); Millar et al. (2019)

5.1 Embryogenesis and Seed Development

MicroRNAs (miRNAs) are essential regulators of plant embryogenesis and seed development, mainly involved in establishing and maintaining embryo patterning and controlling developmental timing. Among the several miRNAs involved are *miR156* and *miR172*, which function within a conserved, spatiotemporally regulated network. The miRNA *miR172* controls *APETALA2* (*AP2*)-like genes, which affect flowering time and floral organ identity, whereas *miR156* targets *SQUAMOSA*

PROMOTER BINDING PROTEIN-LIKE (*SPL*) transcription factors, which are crucial for fostering the juvenile-to-adult phase transition (Wu et al. 2009). The suppression of *SPL* genes by *miR156* results in the maintenance of the embryonic state and delays the transition to the later developmental phase. While *miR172* targets *APETALA2*-like transcription factors and becomes more active later in development, it boosts the maturation of embryo structures and encourages the transition from vegetative to reproductive phases. The antagonistic interaction between *miR156* and *miR172* ensures proper timing of developmental transitions (Xu et al. 2016a). Other miRNAs, including miR160, miR165/166, and miR167, directly regulate auxin response factors and HD–ZIP III genes, contributing to embryo patterning and organ polarity (Nodine and Bartel 2010). In rice, the lncRNA *LAIR* (a reverse antisense transcript of *LRK1*) acts as a scaffold, binding to the *LRK1* genomic region to recruit *OsMOF* and *OsWDR5*. This complex mediates the deposition of activating histone marks (H4K16ac and H3K4me3), resulting in the upregulation of *LRK1* expression and an associated increase in grain yield (Wang et al. 2018a).

5.2 *Root and Shoot Development*

The molecular regulation of root growth and lateral root (LR) development is more characterized in the eudicot *Arabidopsis thaliana* than in monocot species. In this species, the lncRNA *APOLO* (*Auxin Regulated Promoter Loop RNA*) promotes the formation of a chromatin loop with the promoter region of the *PINOID* gene, which encodes a regulatory kinase responsible for the polar positioning of an auxin transporter. The *PINOID* gene is situated 5248 base pairs downstream of the *APOLO* gene locus. Both *APOLO* and *PINOID* are induced by auxin, and their expression levels have been observed to decline simultaneously after 12 hours of auxin exposure. Silencing *APOLO* through RNAi also suppresses *PINOID* expression and delays the plant's gravitropic response, an effect resembling the phenotype of *pinoid* mutants (Ariel et al. 2014). *APOLO* also activates RHD6 (Root Hair Defective 6) by promoting a chromatin loop involving PRC1/PRC2 (Polycomb Repressive Complexes) complexes and interacts with *WRKY42* to enhance root hair growth under cold stress (Moison et al. 2021). Two nuclear speckle RNA-binding proteins, NSRA and NSRB, regulate the alternative splicing of 85 genes, and the *nsra/nsrb* double mutant shows fewer and shorter roots. These proteins interact with lncRNAs and modulate auxin-dependent splicing events important for lateral root (LR) formation (Bardou et al. 2014).

In *Arabidopsis*, LR initiation is primarily triggered by the auxin-mediated degradation of the repressor protein IAA14, which otherwise inhibits downstream signaling. Upon IAA14 degradation, the transcription factors *ARF7* and *ARF19* are activated and subsequently induce the expression of *LBD/ASL* genes (Xie et al. 2000; Guo et al. 2005). The *LBD/ASL* transcription factors activate components of the cell cycle machinery and genes governing cellular patterning, ultimately facilitating the emergence of a lateral root primordium. The transcription factor *NAC1* also contributes to auxin signal transduction, promoting LR development; however, its expression is negatively regulated by *miR164*, which directs *NAC1* mRNA cleavage and thereby dampens auxin responses (Li et al. 2012). Adventitious root (AR) formation shares developmental features with LR development, as both originate from pericycle-like cell layers (Sundararajan et al. 2023). AR formation is regulated by several plant-specific transcription factors belonging to the *WUSCHEL*-related homeobox (*WOX*) family. In particular, *WOX11*, *LATERAL ORGAN BOUNDARIES DOMAIN 16*, and *SMALL AUXIN-UP RNA 36* (*SAUR36*) function as crucial regulators of AR initiation and development within the auxin-signaling framework (Liu et al. 2022a).

Shoot development is similarly influenced by small RNAs that govern the establishment and maintenance of the shoot apical meristem (SAM). Several miRNAs act within the STM–WUS–CLV signaling pathway to maintain SAM homeostasis. Among these, miR394 is produced in the epidermal L1 layer of the SAM and moves downward to the Organizing Center (OC), where it represses the expression of leaf curling responsiveness (LCR). LCR repression results in the downregulation of WUS, a central regulator of stem cell maintenance in the SAM. Although *miR394* accumulates predominantly in the L1 layer, its functional activity is restricted to the OC, highlighting the importance of precise miRNA concentration gradients for developmental regulation (Knauer et al. 2013). ARGONAUTE10 (AGO10) also has a role in maintaining SAM by specifically binding and sequestering *miR165/166*. These microRNAs normally target the Class III homeodomain leucine zipper transcription factors (HD–ZIP III TFs), which are essential for regulating SAM formation and maintenance. By sequestering *miR165/166*, AGO10 prevents the degradation or translational repression of HD–ZIP III transcripts, thereby promoting their expression. This upregulation of HD–ZIP III TFs ensures proper development and function of the SAM (Zhu and Helliwell 2011).

5.3 Alternative Splicing and Root Nodule Formation

Alternative splicing (AS) regulation involving circRNAs has also been observed in *Arabidopsis thaliana*. One such circRNA, derived entirely from exon 6 of the *SEPALLATA 3* (*SEP3*) gene, was shown to enhance the levels of a naturally occurring splice variant, *SEP3.3*, which lacks exon 6. *SEP3* belongs to the MADS-box transcription factor family, involved in flower development. Changes in *SEP3* splicing can lead to floral homeotic alterations. Interestingly, the exon 6 circRNA of *SEP3* is capable of binding to its own DNA locus, forming an RNA: DNA hybrid known as an R-loop. This interaction causes transcriptional pausing and is associated with the recruitment of splicing factors, thereby influencing AS (Conn et al. 2017). A link between *miR390*-mediated signaling and nodulation regulation has been identified. Specifically, the *miR390*/TAS3 pathway was shown to interact with two key nodulation transcription factors, NSP1 and NSP2, which are essential components of the nodulation signaling cascade. Overexpression of *miR390* suppressed rhizobia-induced expression of *NSP1*/*NSP2*, while MIM390 lines and ago7 mutants showed elevated levels of these genes, suggesting that *miR390* negatively regulates them. Overexpression of *miR171h*, which targets *NSP2*, also reduces nodule formation (Hobecker et al. 2017).

5.4 Floral and Reproduction

Numerous factors have been identified that influence the timing of flowering in plants. Recent studies reveal that sex determination and floral transition are controlled by a complex genetic network initiated during differentiation in addition to environmental cues and phytohormones (Teotia and Tang 2015; Zhao et al. 2019). The transition from the vegetative to the reproductive phase is driven by a group of plant-specific transcription factors known as SBP/SPL proteins. During the juvenile stage, their expression is suppressed by the microRNAs (miR156 and miR157). As the levels of *miR156/miR157* decline, SBP/SPL protein levels rise, triggering the shift from the vegetative to the reproductive phase (Xu et al. 2016b; He et al. 2018; Fouracre et al. 2021). The *miR156–SPL3* directly regulates the expression of *FLOWERING LOCUS T* (*FT*) in leaves, influencing flowering response to ambient temperature. Overexpression of *miR156* results in delayed flowering at lower temperatures (16 °C), which is linked to reduced expression of both *FT* and *FRUITFULL*. Among the *miR156* target genes, *SPL3* mRNA levels showed the most significant reduction, likely due to increased *miR156*-mediated cleavage at 16 °C (Xie et al. 2000; Kim et al. 2012). When a *miR156*-resistant form of *SPL3* [*SPL3(−)*] was overexpressed, plants exhibited early flowering regardless of temperature, accompanied by increased expression of *FT* and *FRUITFULL*. Similarly, reducing *miR156* activity through target mimicry produced a phenotype comparable to that of *SUC2::rSPL3* plants. In *GVG-rSPL3* plants, dexamethasone treatment led to an upregulation of *FT*. Further, altering *FT* expression in leaves, either through misexpression or suppression using artificial microRNAs, impacted the temperature-responsive flowering pattern of plants overexpressing *miR156* and *SPL3(−)*. The role of *miR156* and *SPLs* in flower development has been reported in several crops (Xie et al. 2000, Kim et al. 2012).

Elevated expression of the long noncoding RNA known *as long-day-specific male-fertility-associated RNA* (*LDMAR*) is essential for pollen development under long-day conditions. A single-nucleotide polymorphism (SNP) at the *LDMAR* locus enhances RdDM at its promoter region, leading to reduced *LDMAR* transcription (Ding et al. 2012; Bhogireddy et al. 2021). MicroRNAs also regulate flower and seed development in monocots. In rice, Zhu et al. reported that overexpression of *miR172* leads to spikelet loss, abnormalities in floral organ development, and reduced fertility (Zhu et al. 2009). *OsmiR397*, another miRNA, is highly expressed in young panicles and grains, and contributes to increased grain yield by downregulating its target gene *OsLAC*. Overexpression of OsmiR397 enhances grain size and stimulates panicle branching (Zhang et al. 2013).

5.5 Leaf Development and Plant Growth

Leaf is a fundamental organ, and its initiation and maturation occur through coordinated changes in gene expression, tightly regulated by microRNAs and trans-acting siRNAs (Pulido and Laufs 2010). The *miR164* negatively regulates the *CUP-SHAPED COTYLEDON* (*CUC*) genes, *CUC1* and *CUC2*, which encode a pair of paralogous *NAC* transcription factors involved in leaf serration (Nikovics et al. 2006). Likewise, *miR165/166* targets class III homeodomain-leucine zipper (HD–ZIP III) transcription factors, which are essential for regulating key processes such as meristem initiation, lateral organ polarity, and vascular development (Zhou et al. 2007; Zhang and Zhang 2012; Zhou et al., 2015). Similarly, *miR319* and its TCP targets *(TEOSINTE BRANCHED1, CYCLOIDEA*, and *PCNA-binding factor*) also regulate leaf morphology. Mutations in *MIR319* or gain-of-function mutations in *TCP4* that block *miR319* regulation disrupt cotyledon boundary formation and leaf serration, acting through the CUC pathway. Conversely, loss-of-function mutations in both *miR319*-targeted and non-targeted TCP genes promote serrated leaves and alter downstream gene expression (Koyama et al. 2017). As plants reach the maturation stage, leaves undergo organ-level senescence, which is a critical process for optimal offspring production and plant survival. Age-related (AR) lncRNAs *AT4G36648*, *AT5G23410*, and *AT1G26208* interact with *miR164*, *miR169*, and *miR156*, respectively, to regulate *ORESARA1, NUCLEAR FACTOR Y (NF-Y),* and *SPL* genes (Kim et al. 2022). Plant architecture, a major determinant of productivity, is also influenced by the conserved *miR319* family (Zhang et al. 2016; Koyama et al. 2017; Sun et al. 2017; Fan et al. 2020). In common wheat, the *miR319/TaGAMYB3* regulatory module has been reported to control plant architecture and grain yield. Suppressing *tae-miR319* through short tandem target mimics led to desirable architectural traits such as increased plant height, fewer tillers, larger spikes and flag leaves, thicker culms, and improved grain yield in field trials. In contrast, overexpression of *tae-miR319* resulted in opposite effects on both plant structure and yield. *TaPCF8* and *TaGAMYB3* were identified as targets of miR319, and genetic complementation experiments revealed that only the *miR319*-resistant form of *TaGAMYB3* (*rTaGAMYB3*) could reverse the *miR319*-induced growth suppression of flag leaves and spikes. TaGAMYB3 functions as a transcriptional activator by binding to the promoters of downstream genes such as *TaPSKR1, TaXTH23, TaMADS5*, and *TaMADS51*. Moreover, *TaGAMYB3* physically interacts with *TaBA1*, a key regulator of spike development, to synergistically enhance the transcription of downstream targets like *TaMADS5* (Wang et al. 2021; Jian et al. 2022a).

6 ncRNAs in Plant Stress Responses

Climate change has significantly increased the vulnerability of agriculture, leading to substantial yield losses from both biotic and abiotic stresses. Major abiotic challenges include drought, heat, salinity, cold, mineral toxicity, nutrient deficiencies, and flooding, whereas rising temperatures and shifting climatic patterns also accelerate pest infestations and disease outbreaks in many crops (Zhang et al. 2023). As discussed in earlier sections, ncRNAs regulate a wide range of cellular and developmental processes that are essential for plant survival. Because plants are sessile and cannot escape environmental fluctuations, they have evolved sophisticated defense and adaptation mechanisms that fine-tune gene expression under stress conditions (Table 3) (Zhang et al. 2022).

Table 3 Regulation of abiotic and biotic stress responses through rncRNAs

Stress	miRNA	Crop	Function of ncRNA	References
Abiotic stress	*miR166*	Maize	*STTM166* knockdown improves stress tolerance, alters hormones	Yadav et al. (2024)
Alkaline soil	*L0760, L6247*, etc. (lncRNAs)	Wheat	lncRNA-miRNA-mRNA networks regulate alkaline stress responses	Wei et al. (2022)
White Fly	*miR170, miR171, miR164, miR166*	Tomato	Regulates transcripts of *GRAS* family and *NAC* domain proteins involved in defense	Wang et al. (2018b)
Cold	*miR1320*	Rice	Targets *PHD17*, affects ROS homeostasis, JA biosynthesis	Wang et al. (2024)
	miR156	Rice	Targets *OsSPL3* and *OsWRKY71*, regulates *OsMYB2* and stress genes	Zhou and Tang (2019)
	Tae-miR398, lncR9A, lncR117, lncR616	Wheat	ceRNA network controls *CSD1* expression and cold response	Lu et al. (2020)
Cold and heat	*miR396*	Poplar	Increases photosynthesis and heat tolerance	Zhao et al. (2021)
Cold and heat stress	*miR156, miR159, miR393*	Banana	Involved in auxin signaling, redox homeostasis	Zhu et al. (2019)
Disease resistance	*ALEX1* (lncRNA)	Rice	Activates JA pathway, enhances resistance to bacterial blight	Jian et al. (2022b)

(continued)

Table 3 (continued)

Stress	miRNA	Crop	Function of ncRNA	References
Drought	*miR164, miR160*	Maize	Regulates ABA and downstream metabolism	Liu et al. (2019)
	mdm-miR156, miR166, etc.	Apple	Regulates ABA biosynthesis (via *MdNCED3*)	Li et al. (2020)
	miR408	Rice	Regulates growth and drought tolerance via copper protein targets	Balyan et al. (2023)
	gma-miR398c	Soyabean	Negative regulator; overexpression reduces drought tolerance	Zhou et al. (2020)
	miR166	Soyabean	Regulates *ATHB14-LIKE* and ABA signaling, enhances tolerance	Zhao et al. (2024)
	hvu-miR172b-3p/5p	Barley	Coordinate flowering and sugar regulation under drought	Swida-Barteczka et al. (2023)
Drought, salinity, dehydration, PEG, ABA	*OsmiR535*	Rice	Negative regulator of stress tolerance; knockout enhances tolerance	Yue et al. (2020)
Fusarium oxysporum f. sp. cucumerinum	*miR319a, miR6300*	Cucumber	Regulates immune-related genes (*BEE1, DAHP1, PERK2*) for fungal resistance	Xu et al. (2021)
	miR164a, miR168a, miR393	Garlic	Enhances fungal resistance by reducing pathogen growth and upregulating defense genes	Zhang et al. (2017)
Heat stress	*miR397*	Banana	Improves high-temp tolerance without biomass loss	Patel et al. (2019)
Heavy metal (cd)	*miR390*	Rice	Reduces MDA level and seedling growth inhibition	Ding et al. (2016)
	miR166	Rice	Restricts cd movement from root to shoot	Ding et al. (2018)
Hypoxia/anoxia	*miR172a, miR160f-5p*	Maize	Controls shift from aerobic to anaerobic respiration	Azahar et al. (2020)
Hypoxia/ Waterlogging	*miR394a, miR396a-3p*	Cucumber	Adventitious root formation	Xu et al. (2019)

(continued)

Table 3 (continued)

Stress	miRNA	Crop	Function of ncRNA	References
Brown planthopper	*miR396*	Rice	Regulates *OsGRF8* to modulate responsive pathways under insect stress	Dai et al. (2019)
	miR398b, miR531, miR167a-5p, miR156	Rice	Modulates MAPK cascade and transcriptional stress pathways	Wu et al. (2017)
Nitrogen starvation	*TaMIR444a*	Wheat	Regulates N acquisition, antioxidant enzymes, and ROS detox	Gao et al. (2016)
Nutrient imbalance	*miR160a-5p, miR166g-3p*	Barley	Enhances photosynthesis under low K	Ye et al. (2021)
	SlmiR168	Tomato	Stimulates root growth, modifies CK/ABA signaling	Liu et al. (2020)
	miR399, miR398	Alfalfa	Enhances Pi uptake, metabolic adjustments	Li et al. (2018b)
	miR1139	Wheat	Improves Pi acquisition	Liu et al. (2018b)
Phosphate starvation, salt	*TaemiR408*	Wheat	Enhances biomass, photosynthesis, and stress tolerance	Bai et al. (2018)
Phosphorus deficiency	*miR399, PILNCR2*	Maize	Regulate *ZmPHT1* and Pi homeostasis	Wang et al. (2023)
Plectosphaerella cucumerina	*miR773*	Arabidopsis	Mediates PAMP-triggered immunity by targeting *MET2*	Salvador-Guirao et al. (2018)
Ralstonia solanacearum	*osa-miR169, ppt-miR1223, aly-miR398, gra-miR482, gma-miR4415, mtr-miR5261, ath-miR167, ath-miR1886*	Rice	Upregulates 62 defense-related genes, including NBS-LRR, ABC transporters, and heat shock proteins	Snigdha and Prasath (2021)

(continued)

Table 3 (continued)

Stress	miRNA	Crop	Function of ncRNA	References
Salinity	*miR164*	Maize	Modifies leaf morphology and upregulates salt-responsive genes	Shan et al. (2020)
	miR414c	Cotton	Enhances root growth and ROS scavenging via *GhFSD1*	Wang et al. (2019b)
	miR167	Tobacco	Promotes auxin transport and organogenesis under salt stress	Makkar et al. (2022)
	miR396b / GRF6	Rice	Enhances salt tolerance via ROS-scavenging, increases survival	Yuan et al. (2024)
	Gm-miR396a	Soyabean	CRISPR-edited lines show enhanced salinity tolerance	Chen et al. (2024)
Salinity, copper, stripe rust	*tae-miR408*	Wheat	Targets *TaCLP1* for biotic and abiotic stress responses	Feng et al. (2013)
Salinity, drought	*OsmiR535*	Rice	Regulates *SPL19*, improves root and shoot traits	Yue et al. (2020)
UV-B radiation	*miR395, miR408, miR165*	Saltwater cress	Enhances phenolic/ anthocyanin production and energy dissipation	Pashkovskiy et al. (2021)
	miR165, miR166, miR393	Wheat	Enhances UV-B tolerance via oxidative stress detox enzymes	Casati (2013)

6.1 Abiotic Stress Responses

6.1.1 Drought and Salinity

The role of ncRNAs in drought and salinity stress has been extensively studied across plant species. Both lncRNAs and miRNAs respond to water-deficit conditions by modulating hormonal signaling, chromatin remodeling, and translational regulation (Sanan Mishra et al. 2021). These ncRNA-mediated pathways coordinate the expression of drought-responsive genes to promote resistance or tolerance. In switchgrass, lncRNAs regulate drought responses through ABA and trehalose biosynthesis. Higher expression of *XLOC_053020*, *XLOC_014465*, and *XLOC_033252* upregulates *Pavir.Ia01153*, *Pavir.Bb00347*, and *Pavir.Eb01847*, genes associated with ABA accumulation (Zhong et al. 2014). In *Arabidopsis*, the drought-responsive lncRNA *DANA1* interacts with *DIP1* and promotes histone deacetylation of

CYP707A1 and *CYP707A2*, enhancing ABA-mediated drought tolerance (Cai et al. 2024). Another *Arabidopsis* lncRNA, *DRIR*, is drought-induced and contributes to both drought and salt tolerance (Liu et al. 2022b). In cassava, the drought-induced intergenic lncRNA *DIR* enhances stress tolerance by increasing proline accumulation through mRNA metabolism and has been cloned (Dong et al. 2022). Similar drought-responsive lncRNAs have also been identified in rice, maize, pearl millet, and mustard (Orek et al. 2023).

Changes in miRNA expression regulate drought-responsive genes, hormones, and transcription factors across crops such as rice, maize, wheat, sorghum, Arabidopsis, and soybean. In rice, drought induces *miR171f*, which accumulates in roots and floral tissues and negatively regulates its target *Os03g0828701-00*, supporting deeper root growth for improved water uptake (Chung et al. 2016). In maize, downregulation of *miR167* leads to increased expression of Phospholipase D, a key regulator of ABA signaling and stomatal movement during drought (Chung et al. 2016). Both lncRNAs and miRNAs also play critical roles in salinity tolerance. In rice and Arabidopsis, higher expression of osa-*MIR396c* and *osa-MIR393* modulates growth regulatory genes during salt stress (Gao et al. 2010). In cotton, *lncRNA973* is strongly induced under salinity and regulates genes associated with ROS scavenging, transcription factors, and salt-responsive pathways. Its overexpression enhances germination rate, root elongation, and biomass accumulation. Another cotton lncRNA, *TRABA*, suppresses *BGLU24* (β-glucosidase) expression and contributes to salt-stress responses (Zhang et al. 2019; Cui et al. 2024).

6.1.2 Extreme Temperatures

Noncoding RNAs, particularly lncRNAs, miRNAs, and siRNAs, play vital roles in plant responses to both heat and cold stress. Their major functions include regulating ROS scavenging, modulating hormone levels, and coordinating epistatic interactions that enhance stress tolerance.

Heat Stress Tolerance

High temperatures adversely affect plant growth, commonly leading to pollen sterility, growth retardation, and increased susceptibility to diseases. A study by Guan et al. (2013) shows that miR398, produced in response to heat stress in *Arabidopsis*, targets *CSD1* and *CSD2* (closely related copper/zinc superoxide dismutases), *Cox5b-1* and *CCS1* (a copper chaperone for SOD) (Guan et al. 2013). These genes are responsible for regulating the accumulation of ROS positively and negatively. Downregulation of these genes by *miR398* induces heat stress resistance response in *Arabidopsis*. There are other reports of the possible existence of regulatory genes, such as *miR156* and *miR172,* that induce heat stress resistance. The *miR172* is also reported in wheat, sunflower, and other crops. In crops like wheat and sunflower, *miR172* is heat-inducible, while *miR396* in sunflower targets *HaWRKY6*; reduced *HaWRKY6* expression triggers heat-stress tolerance (Mantri et al. 2012).

Cold-Induced lncRNAs and miRNAs in Thermotolerance

Cold stress is a major constraint on global crop productivity, where the temperature drops below 0 °C, called freezing, and 0–15 °C, called chilling. Numerous lncRNAs have been identified as regulators of cold tolerance in plants, which function through chromatin remodeling, splicing, and gene silencing. In *Arabidopsis*, the novel cold-induced lncRNA *CIL1* (*COLD INDUCED lncRNA 1*) functions as a positive regulator of cold tolerance. The *CIL1* expression increases sharply under cold conditions, whereas *CIL1*-silenced mutants show heightened sensitivity, marked by elevated ROS accumulation and reduced levels of osmoregulatory metabolites. *CIL1* modulates the expression of multiple cold-responsive genes during the seedling stage (Liu et al. 2022b). In cotton, cold stress also triggers significant changes in lncRNA expression. The lncRNA *XH123* has been shown to confer cold tolerance, as *XH123*-silenced plants display chloroplast damage and increased ROS levels under cold stress. Transcriptome analysis revealed similarity between *XH123*-silenced seedlings and *PIF3*-silenced cold-stressed plants, indicating that *XH123* functions in the same regulatory pathway involving *PIF3* (Cao et al. 2021). There are many other reports of lncRNA-mediated cold stress tolerance, namely, *MtCIR2* in *Medicago truncatula* for freeze tolerance, *CsLncRNA94* in cucumber, *dglncTCP1* in Chrysanthemum, and NAT–lncRNAs in Arabidopsis (Wielogórska et al. 2025).

6.1.3 Biotic Stress Responses

lncRNAs play a crucial role in controlling plant defense responses. They can act as endogenous target mimics by binding to miRNAs, thereby preventing these miRNAs from silencing defense-related messenger RNAs. Through such interactions, lncRNAs modulate both PAMP-triggered immunity (PTI) and effector-triggered immunity (ETI), influencing the activation of defense signaling pathways and defense-associated gene expression. Additionally, several lncRNAs regulate the biosynthesis and signaling of key plant defense hormones, including salicylic acid (SA), jasmonic acid (JA), and ethylene, the central components of resistance against a wide spectrum of pathogens (Liu et al. 2022c; Moghadam et al. 2024). Some lncRNAs also function through epigenetic mechanisms by recruiting chromatin-modifying complexes, such as the Polycomb Repressive Complex 2 (PRC2), to specific gene promoters, altering chromatin states and modulating immunity-related gene expression (Faust et al. 2012; Xu and Xu 2013).

A well-characterized lncRNA, *ELENA1*, is strongly induced by pathogen-associated molecular patterns (PAMPs) such as *elf18* and flagellin (*flg22*). ELENA1 acts as a positive regulator of defense, enhancing resistance to *Pseudomonas syringae* pv. tomato DC3000 by promoting the expression of *PR1* and *PR2*, which encode β-1,3-glucanases involved in callose deposition. Overexpression of *ELENA1* activates several SA-responsive genes, strengthening plant defense, whereas *ELENA1*-deficient plants show enhanced susceptibility (Mach 2017). Another important lncRNA, *Salicylic Acid Biogenesis Controller 1 (SABC1)*, modulates immunity by regulating SA biosynthesis. *SABC1* recruits *PRC2* to repress its neighboring gene

NAC3, a transcription factor that activates *ICS1*, a key enzyme in SA biosynthesis. By suppressing *NAC3* and *ICS1* expression, *SABC1* reduces SA levels and defense responses in favor of growth under non-stress conditions (Liu et al. 2022c). Furthermore, several lncRNAs regulate early immune events such as reactive oxygen species (ROSs) accumulation and calcium influx, highlighting the diverse layers of lncRNA-mediated defense regulation (Song et al. 2021). In addition to lncRNAs, miRNAs play pivotal roles in modulating plant defense mechanisms. For instance, *Osa-miR393-OsTIR1*-mediated auxin signaling in rice interacts with the JA pathway during RBSDV infection. RBSDV triggers JA signaling-related genes, while overexpression of *Osa-miR393* reduces *OsTIR1* levels, attenuating JA signaling and revealing a complex regulatory interplay between auxin and JA pathways (Shen et al. 2024). Similarly, the targets of *Osa-miR156*, namely, *OsSPL7* and *OsSPL14* (*squamosa promoter binding protein-like*), enhance resistance to bacterial blight (*Xanthomonas oryzae* pv. oryzae) when overexpressed. These SPL proteins interact with the JA biosynthesis gene *OsAOS2* and the SA receptor gene *OsNPR1*, activating their transcription and strengthening disease resistance (Zhang et al. 2022; Jia et al. 2025). However, miRNA-mediated defense is often accompanied by growth penalties; for example, reducing *Osa-miR156* enhances resistance but negatively affects panicle development and grain yield (Liu et al. 2019). Balancing defense and growth through miRNA manipulation remains challenging. Nonetheless, the suppression of *Osa-miR168* in rice demonstrates the potential to achieve both improved immunity and enhanced agronomic traits. Reduced *Osa-miR168* elevates *OsAGO1* levels, influencing downstream miRNAs such as *Osa-miR1320, Osa-miR535*, and *Osa-miR164*. This regulatory network enhances immunity to rice blast, increases grain yield, and promotes early flowering, showcasing the multifunctional nature of miRNA-guided regulatory modules (Wang et al. 2021).

7 Conclusion

Plant noncoding RNAs have emerged as critical players in developmental stages and adaptive responses. Among these, miRNAs are the most extensively studied and modulate plant growth and developmental transitions from germination to senescence through various gene silencing pathways. However, the functions of other ncRNA classes, especially lncRNAs and circRNAs, remain largely unexplored. Advances in sequencing, machine learning, and genome editing technologies have opened new possibilities to uncover the biogenesis, mobility, and targets of diverse ncRNAs. Despite this progress, many challenges persist, including understanding ncRNA cross talk, their movement between organs, and functional roles within organelles such as the nucleus, mitochondria, and chloroplasts. Research on ncRNA should prioritize the identification and validation of trait-specific ncRNAs, their target genes, and their utilization. Developing a comprehensive catalog of ncRNAs involved in growth, stress tolerance, and immunity will help in designing smart crops tailored for future climatic challenges. DNA-free editing can transform functional genomics by minimizing off-target effects and enabling precise ncRNA manipulation. The integration of ncRNA-based strategies in crop improvement programs holds tremendous potential for achieving sustainable agriculture and enhancing food and nutritional security.

Acknowledgement First author acknowledges Gyan P. Mishra for the opportunity and the support received from ICAR-VPKAS, Almora.

Conflict of Interest Disclosure The authors declare that they have no conflict of interest regarding the content, authorship, or publication of this book chapter.

References

Allen RS, Li J, Stahle MI, Dubroué A, Gubler F, Millar AA (2007) Genetic analysis reveals functional redundancy and the major target genes of the Arabidopsis miR159 family. Proc Natl Acad Sci 104(41):16371–16376

Ambros V, Ruvkun G (2018) Recent molecular genetic explorations of *Caenorhabditis elegans* microRNAs. Genetics 209(3):651–673

Ariel F, Christ A, Latrasse D, Romero-Barrios N, Benhamed M, Crespi M (2014) Noncoding transcription by alternative RNA polymerases dynamically regulates an auxin-driven chromatin loop. Mol Cell 55(3):383–3965

Ariel F, Romero-Barrios N, Jégu T, Benhamed M, Crespi M (2015) Battles and hijacks: noncoding transcription in plants. Trends Plant Sci 20(6):362–371

Ariel F, Lucero L, Christ A, Mammarella MF, Jegu T, Veluchamy A et al (2020) R-loop mediated trans action of the APOLO long noncoding RNA. Mol Cell 77(5):1055–1065

Aukerman MJ, Sakai H (2003) Regulation of flowering time and floral organ identity by a microRNA and its APETALA2-like target genes. Plant Cell 15(11):2730–2741

Ausin I, Mockler TC, Chory J, Jacobsen SE (2009) IDN1 and IDN2 are required for de novo DNA methylation in *Arabidopsis thaliana*. Nat Struct Mol Biol 16(12):1325–1327

Azahar I, Ghosh S, Adhikari A, Adhikari S, Roy D, Shaw AK, Singh K, Hossain Z (2020) Comparative analysis of maize root sRNA transcriptome unveils the regulatory roles of miRNAs in submergence stress response mechanism. Environ Exp Bot 171:103924

Bai Q, Li Y, Zhao J, Liu C, Geng S (2018) Wheat miRNA TaemiR408 acts as an essential mediator in plant tolerance to Pi deprivation and salt stress via modulating stress-associated physiological processes. Front Plant Sci 9:499. https://doi.org/10.3389/fpls.2018.00499

Balyan S, Kumar M, Mutum RD, Raghuvanshi S (2023) Delineating the tissue-mediated drought stress governed tuning of conserved miR408 and its targets in rice. Funct Integr Genomics 23(2):187. https://doi.org/10.1007/s10142-023-00960-1

Bardou F, Ariel F, Simpson CG, Romero-Barrios N, Laporte P, Balzergue S et al (2014) Long noncoding RNA modulates alternative splicing regulators in *Arabidopsis*. Dev Cell 30(2):166–1765

Bartel DP (2004) MicroRNAs: genomics, biogenesis, mechanism, and function. Cell 116(2):281–297

Bashir T, Husaini AM (2025) Non-coding RNAs and their role in plants: prospective omics-tools for improving growth, development and stress tolerance in field crops. Mol Biol Rep 52(1) Springer Nature

Bhogireddy S, Mangrauthia SK, Kumar R, Pandey AK, Singh S, Jain A, Budak H, Varshney RK, Kudapa H (2021) Regulatory non-coding RNAs: a new frontier in regulation of plant biology. Funct Integr Genomics 21(3):313–330

Biradar M, Tripathi K, Meena VS, Basavaraj PS, Kumari J, Reddappa SB et al (2025) Morpho-physiological traits underpinning high-temperature stress tolerance in cowpea (*Vigna unguiculata* L. Walp.). Genet Resour Crop Evol:1–14

Bohmdorfer G, Wierzbicki AT (2015) Control of chromatin structure by long noncoding RNA. Trends Cell Biol 25(10):623–632. https://doi.org/10.1016/j.tcb.2015.07.002

Bousios A, Kakutani T, Henderson IR (2025) Centrophilic retrotransposons of plant genomes. Annu Rev Plant Biol 76:579

Britto-Kido SDA, Ferreira Neto JRC, Pandolfi V, Marcelino-Guimarães FC, Nepomuceno AL, Vilela Abdelnoor R, Benko-Iseppon AM, Kido EA (2013) Natural antisense transcripts in plants: a review and identification in soybean infected with Phakopsora pachyrhizi SuperSAGE library. Sci World J

Brockdorff N, Ashworth A, Kay GF, McCabe VM, Norris DP, Cooper PJ et al (1992) The product of the mouse *Xist* gene is a 15 kb inactive X-specific transcript containing no conserved ORF and located in the nucleus. Cell 71(3):515–526

Cai J, Zhang Y, He R, Jiang L, Qu Z, Gu J, Wang D (2024) LncRNA DANA1 promotes drought tolerance and histone deacetylation of drought responsive genes in Arabidopsis. EMBO Rep 25(2):796

Cao Z, Zhao T, Wang L et al (2021) The lincRNA XH123 is involved in cotton cold-stress regulation. Plant Mol Biol 106:521–531. https://doi.org/10.1007/s11103-021-01169-1

Carthew RW, Sontheimer EJ (2009) Origins and mechanisms of miRNAs and siRNAs. Cell 136(4):642–655

Casati P (2013) Analysis of UV-B regulated miRNAs and their targets in maize leaves. Plant Signal Behav 8:e26758

Chekanova JA (2015) Long non-coding RNAs and their functions in plants. Curr Opin Plant Biol 27:207–216

Chen Z, Hu L, Han N, Hu J, Yang Y, Xiang T et al (2015) Overexpression of a miR393-resistant form of transport inhibitor response protein 1 (mTIR1) enhances salt tolerance by increased osmoregulation and Na^+ exclusion in *Arabidopsis thaliana*. Plant Cell Physiol 56:73–83

Chen X, Zhang X, Wang Y et al (2024) Gene editing and overexpression of soybean miR396a reveals its role in salinity tolerance and development. Crop J. https://doi.org/10.1016/j.cj.2024.08.003

Chuck G, Cigan AM, Saeteurn K, Hake S (2007) The heterochronic maize mutant Corngrass1 results from overexpression of a tandem microRNA. Nat Genet 39(4):544–549

Chuck G, Whipple C, Jackson D, Hake S (2010) The maize SBP-box transcription factor encoded by tasselsheath4 regulates bract development and the establishment of meristem boundaries. Development 137(8):1243–1250

Chung PJ, Jung H, Jeong DH, Ha SH, Choi YD, Kim JK (2016) Transcriptome profiling of drought responsive noncoding RNAs and their target genes in rice. BMC Genomics 17(1):563

Conn VM, Hugouvieux V, Nayak A, Conos SA, Capovilla G, Cildir G et al (2017) A circRNA from *SEPALLATA3* regulates splicing of its cognate mRNA through R-loop formation. Nat Plants 3(5):1–5

Crimmins A, Balbus J, Gamble JL, Beard CB, Bell JE, Dodgen D et al (2016) The impacts of climate change on human health in the United States: a scientific assessment. In: The impacts of climate change on human health in the United States: a scientific assessment

Csorba T, Questa JI, Sun Q, Dean C (2014) Antisense COOLAIR mediates the coordinated switching of chromatin states at *FLC* during vernalization. Proc Natl Acad Sci 111(45):16160–16165

Cui C, Wan H, Li Z, Ai N, Zhou B (2024) Long noncoding RNA TRABA suppresses β-glucosidase-encoding *BGLU24* to promote salt tolerance in cotton. Plant Physiol 194(2):1120–1138

Curaba J, Talbot M, Li Z, Helliwell C (2013) Over-expression of microRNA171 affects phase transitions and floral meristem determinancy in barley. BMC Plant Biol 13(1):6

Dai Z, Tan J, Zhou C, Yang X, Yang F, Zhang S, Sun S, Miao X, Shi Z (2019) The OsmiR396–OsGRF8–OsF3H-flavonoid pathway mediates resistance to the brown planthopper in rice (*Oryza sativa*). Plant Biotechnol J 17:1657–1669

Dana H, Chalbatani GM, Mahmoodzadeh H, Karimloo R, Rezaiean O, Moradzadeh A, Mehmandoost N, Moazzen F, Mazraeh A, Marmari V, Ebrahimi M, Rashno MM, Abadi SJ, Gharagouzlo E (2017) Molecular mechanisms and biological functions of siRNA. Int J Biomed Sci 13(2):48–57

Delihas N (2015) Discovery and characterization of the first non-coding RNA that regulates gene expression, micF RNA: a historical perspective. World J Biol Chem 6(4):272

Ding J, Shen J, Mao H, Xie W, Li X, Zhang Q (2012) RNA-directed DNA methylation is involved in regulating photoperiod-sensitive male sterility in rice. Mol Plant 5:1210–1216

Ding Y, Ye Y, Jiang Z, Wang Y, Zhu C (2016) MicroRNA390 is involved in cadmium tolerance and accumulation in rice. Front Plant Sci 7:235

Ding Y, Gong S, Wang Y, Wang F, Bao H, Sun J, Cai C, Yi K, Chen Z, Zhu C (2018) MicroRNA166 modulates cadmium tolerance and accumulation in rice. Plant Physiol 177:1691–1703

Dong S, Liang X, Li Z, Sun J, Yang H et al (2022) A novel long non-coding RNA, DIR, increases drought tolerance in cassava by modifying stress-related gene expression. J Integr Agric 21(9):2588–2602

Engreitz JM, Haines JE, Perez EM, Munson G, Chen J, Kane M et al (2016) Local regulation of gene expression by lncRNA promoters, transcription and splicing. Nature 539(7629):452–455

Fahlgren N, Montgomery TA, Howell MD, Allen E, Dvorak SK, Alexander AL, Carrington JC (2006) Regulation of *AUXIN RESPONSE FACTOR3* by TAS3 ta-siRNA affects developmental timing and patterning in *Arabidopsis*. Curr Biol 16(9):939–944

Fan D, Ran L, Hu J, Ye X, Xu D, Li J, Su H, Wang X, Ren S, Luo K (2020) miR319a/TCP module and DELLA protein regulate trichome initiation synergistically and improve insect defenses in *Populus tomentosa*. New Phytol 227:867–883

Faust T, Frankel A, D'Orso I (2012) Transcription control by long non-coding RNAs. Transcription 3(2):78–86

Fei Q, Xia R, Meyers BC (2013) Phased, secondary, small interfering RNAs in posttranscriptional regulatory networks. Plant Cell 25(7):2400–2415

Feng H, Zhang Q, Wang Q, Wang X, Liu J (2013) Target of tae-miR408, a chemocyanin-like protein gene (TaCLP1), plays positive roles in wheat response to high-salinity, heavy cupric stress and stripe rust. Plant Mol Biol 83:433–443. https://doi.org/10.1007/s11103-013-0097-3

Fouracre JP, He J, Chen VJ, Sidoli S, Poethig RS (2021) VAL genes regulate vegetative phase change via miR156-dependent and independent mechanisms. PLoS Genet 17:e1009626

Fracasso A, Vallino M, Staropoli A, Vinale F, Amaducci S, Carra A (2021) Increased water use efficiency in miR396-downregulated tomato plants. Plant Sci 303:110729

Franco-Zorrilla JM, Valli A, Todesco M, Mateos I, Puga MI, Rubio-Somoza I et al (2007) Target mimicry provides a new mechanism for regulation of microRNA activity. Nat Genet 39(8):1033–1037

Gao P, Bai X, Yang L, Lv D, Li Y, Cai H et al (2010) Over-expression of osa-MIR396c decreases salt and alkali stress tolerance. Planta 231(5):991–1001

Gao S, Wang X, Tian Y (2016) Wheat microRNA member TaMIR444a is nitrogen deprivation-responsive and involves plant adaptation to the nitrogen-starvation stress. Plant Mol Biol Report 34(5):931–946. https://doi.org/10.1007/s11105-016-0973-3

Garcia D, Collier SA, Byrne ME, Martienssen RA (2006) Specification of leaf polarity in *Arabidopsis* via the trans-acting siRNA pathway. Curr Biol 16(9):933–938

Greene J, Baird AM, Brady L, Lim M, Gray SG, McDermott R, Finn SP (2017) Circular RNAs: biogenesis, function and role in human diseases. Front Mol Biosci 4:38

Guan Q, Lu X, Zeng H, Zhang Y, Zhu J (2013) Heat stress induction of mi R 398 triggers a regulatory loop that is critical for thermotolerance in Arabidopsis. Plant J 74(5):840–851

Guo HS, Xie Q, Fei JF (2005) MicroRNA directs mRNA cleavage of the transcription factor NAC1 to downregulate auxin signals for *Arabidopsis* lateral root development. Plant Cell 17:1376–1386

Haag JR, Pikaard CS (2011) Multisubunit RNA polymerases IV and V: purveyors of non-coding RNA for plant gene silencing. Nat Rev Mol Cell Biol 12:483–492

He J, Xu M, Willmann MR, McCormick K, Hu T (2018) Threshold-dependent repression of SPL gene expression by miR156/miR157 controls vegetative phase change in *Arabidopsis thaliana*. PLoS Genet 14:e1007337

Heo JB, Sung S (2011) Vernalization-mediated epigenetic silencing by a long intronic noncoding RNA. Science 331(6013):76–79

Hobecker KV, Reynoso MA, Bustos-Sanmamed P, Wen J, Mysore KS, Crespi M et al (2017) The microRNA390/TAS3 pathway mediates symbiotic nodulation and lateral root growth. Plant Physiol 174(4):2469–2486

Holley RW (1965) Structure of an alanine transfer ribonucleic acid. JAMA 194(8):868–871

Hou X, Cui J, Liu W, Jiang N, Zhou X, Qi H, Luan Y (2020) LncRNA39026 enhances tomato resistance to Phytophthora infestans by decoying miR168a and inducing PR gene expression. Phytopathology 110(4):873–880

Ito H (2012) Small RNAs and transposon silencing in plants. Develop Growth Differ 54(1):100–107

Iwasaki YW, Siomi MC, Siomi H (2015) PIWI-interacting RNA: its biogenesis and functions. Annu Rev Biochem 84(1):405–433

Jarroux J, Morillon A, Pinskaya M (2017) History, discovery, and classification of lncRNAs. In: Long non coding RNA biology, pp 1–46

Jia X, Ding N, Fan W, Yan J, Gu Y, Tang X, Tang G (2015) Functional plasticity of miR165/166 in plant development revealed by small tandem target mimic. Plant Sci 233:11–21

Jia Y, Wei K, Qin J, Zhai W, Li Q, Li Y (2025) The roles of microRNAs in the regulation of rice–pathogen interactions. Plants 14(1):136

Jian C, Hao P, Hao C, Liu S, Mao H, Song Q et al (2022a) The miR319/TaGAMYB3 module regulates plant architecture and improves grain yield in common wheat (*Triticum aestivum*). New Phytol 235(4):1515–1530

Jian L, Wang Y, Xiong H, Wang G (2022b) Mutation of OsMYB60 reduces rice resilience to drought stress by attenuating cuticular wax biosynthesis. Plant J 112(2):339–351. https://doi.org/10.1111/tpj.15947

Jones L, Hamilton AJ, Voinnet O, Thomas CL, Maule AJ, Baulcombe DC (1999) RNA-DNA interactions and DNA methylation in post-transcriptional gene silencing. Plant Cell 11:2291–2301

Jung JH, Seo YH, Seo PJ, Reyes JL, Yun J, Chua NH, Park CM (2007) The *GIGANTEA*-regulated microRNA172 mediates photoperiodic flowering independent of *CONSTANS* in *Arabidopsis*. Plant Cell 19(9):2736–2748

Kim JJ, Lee JH, Kim W, Jung HS, Huijser P, Ahn JH (2012) The microRNA156-SQUAMOSA PROMOTER BINDING PROTEIN-LIKE3 module regulates ambient temperature-responsive flowering via FLOWERING LOCUS T in *Arabidopsis*. Plant Physiol 159(1):461–478. https://doi.org/10.1104/pp.111.192369

Kim JY, Lee J, Kang MH, Trang TTM, Lee J, Lee H, Jeong H, Lim PO (2022) Dynamic landscape of long noncoding RNAs during leaf aging in *Arabidopsis*. Front Plant Sci 13:1068163

Knauer S, Holt AL, Rubio-Somoza I, Tucker EJ, Hinze A (2013) A protodermal miR394 signal defines a region of stem cell competence in the *Arabidopsis* shoot meristem. Dev Cell 24:125–132. https://doi.org/10.1016/j.devcel.2012.12.009

Kohli M, Bansal H, Aski M, Mishra GP, Shashidhar BR, Roy A et al (2024) Genome-wide association mapping of biochemical traits and its correlation with MYMIV resistance in mungbean (*Vigna radiata* L. Wilczek). Sci Rep 14(1):31805

Kong X, Yang M, Le BH, He W, Hou Y (2022) The master role of siRNAs in plant immunity. Mol Plant Pathol 23(10):1565–1574

Koyama T, Sato F, Ohme-Takagi M (2017) Roles of miR319 and TCP transcription factors in leaf development. Plant Physiol 175:874–885

Lapidot M, Pilpel Y (2006) Genome-wide natural antisense transcription: coupling its regulation to its different regulatory mechanisms. EMBO Rep 7:1216–1222

Law JA, Jacobsen SE (2010) Establishing, maintaining and modifying DNA methylation patterns in plants and animals. Nat Rev Genet 11:204–220

Li J, Guo G, Guo W, Guo G, Tong D, Ni Z et al (2012) miRNA164-directed cleavage of *ZmNAC1* confers lateral root development in maize (*Zea mays* L.). BMC Plant Biol 12(1):220

Li R, Fu D, Zhu B, Luo Y, Zhu H (2018a) CRISPR/Cas9-mediated mutagenesis of lncRNA1459 alters tomato fruit ripening. Plant J 94(3):513–524. https://doi.org/10.1111/tpj.13872

Li Z, Xu H, Li Y, Wan X, Ma Z, Cao J, Li Z, He F, Wang Y, Wan L (2018b) Analysis of physiological and miRNA responses to Pi deficiency in alfalfa (*Medicago sativa* L.). Plant Mol Biol 96:473–492

Li X, Chen P, Xie Y, Yan Y, Wang L, Dang H, Zhang J, Xu L, Ma F, Guan Q (2020) Apple SERRATE negatively mediates drought resistance by regulating MdMYB88 and MdMYB124 and microRNA biogenesis. Hortic Res 7:98

Li Q, Shen H, Yuan S, Dai X, Yang C (2023) miRNAs and lncRNAs in tomato: roles in biotic and abiotic stress responses. Front Plant Sci 13:1094459

Lin G, Zhou Y, Li M, Fang Y (2018) Histone 3 lysine 36 to methionine mutations stably interact with and sequester SDG8 in *Arabidopsis thaliana*. Sci China Life Sci 61:225–234

Liu C, Muchhal US, Raghothama KG (1997) Differential expression of TPS11, a phosphate starvation-induced gene in tomato. Plant Mol Biol 33(5):867–874

Liu X, Li D, Zhang D, Yin D, Zhao Y, Ji C, Zhao X, Li X, He Q, Chen R et al (2018a) A novel antisense long noncoding RNA, TWISTED LEAF, maintains LEAF blade flattening by regulating its associated sense R2R3-MYB gene in rice. New Phytol 218:774–788

Liu Z, Wang X, Chen X, Shi G, Bai Q, Xiao K (2018b) TaMIR1139: a wheat miRNA responsive to Pi-starvation, acts as a critical mediator in modulating plant tolerance to Pi deprivation. Plant Cell Rep 37:1293–1309

Liu X, Zhang X, Sun B, Hao L, Liu C, Zhang D, Tang H, Li C, Li Y, Shi Y, Xie X, Song Y, Wang T, Li Y (2019) Genome-wide identification and comparative analysis of drought-related microRNAs in two maize inbred lines with contrasting drought tolerance by deep sequencing. PLoS One 14:e0219176

Liu X, Tan C, Cheng X, Zhao X, Li T, Jiang J (2020) miR168 targets Argonaute1A-mediated miRNA regulation pathways in response to potassium deficiency stress in tomato. BMC Plant Biol 20:1–17

Liu P, Zhang X, Zhang F, Xu M, Ye Z, Wang K et al (2021) A virus-derived siRNA activates plant immunity by interfering with ROS scavenging. Mol Plant 14(7):1088–1103

Liu R, Wen SS, Sun TT, Wang R, Zuo WT, Yang T, Wang C, Hu JJ, Lu MZ, Wang LQ (2022a) PagWOX11/12a positively regulates the PagSAUR36 gene that enhances adventitious root development in poplar. J Exp Bot 73:7298–7311. https://doi.org/10.1093/jxb/erac345

Liu G, Liu F, Wang Y, Liu X (2022b) A novel long noncoding RNA CIL1 enhances cold stress tolerance in *Arabidopsis*. Plant Sci 323:111370. https://doi.org/10.1016/j.plantsci.2022.111370

Liu N, Xu Y, Li Q, Cao Y, Yang D, Liu S et al (2022c) A lncRNA fine-tunes salicylic acid biosynthesis to balance plant immunity and growth. Cell Host Microbe 30(8):1124–1138

Liu R, Ma Y, Guo T, Li G (2023) Identification, biogenesis, function, and mechanism of action of circular RNAs in plants. Plant Commun 4(1):100430

Lu M (2020) Circular RNA: functions, applications and prospects. ExRNA 2(1):1

Lu Q, Xu W, Wang J, Wang X (2020) LncRNA improves cold resistance of winter wheat by interacting with miR398. Funct Plant Biol 47(6):544–557. https://doi.org/10.1071/FP19290

Ma Z, Hu X, Cai W, Huang W, Zhou X, Luo Q, Huang J (2014) Arabidopsis miR171-targeted scarecrow-like proteins bind to GT cis-elements and mediate gibberellin-regulated chlorophyll biosynthesis under light conditions. PLoS Genet 10(8):e1004519

Mach J (2017) The long-noncoding RNA ELENA1 functions in plant immunity. Plant Cell 29(5):1024–1025

Makkar H, Arora S, Khuman AK, Chaudhary B (2022) Target-mimicry-based miR167 diminution confers salt-stress tolerance during in vitro organogenesis of tobacco (*Nicotiana tabacum* L.). J Plant Growth Regul 41(4):1462–1480

Mantri N, Patade V, Penna S, Ford R, Pang E (2012) Abiotic stress responses in plants: present and future. In: Abiotic stress responses in plants: metabolism, productivity and sustainability. Springer, pp 1–19

Martín-Rodríguez JÁ, Leija A, Formey D, Hernández G (2018) The MicroRNA319d/TCP10 node regulates the common bean–rhizobia nitrogen-fixing symbiosis. Plant Signal Behav 13(7):e1482176

Mattick JS, Amaral P (2022) RNA, the epicenter of genetic information: A new understanding of molecular biology. CRC Press

Matzke M, Mosher RA (2014) RNA-directed DNA methylation: an epigenetic pathway of increasing complexity. Nat Rev Genet 15:394–408

Matzke MA, Kanno T, Matzke AJ (2015) RNA-directed DNA methylation: the evolution of a complex epigenetic pathway in flowering plants. Annu Rev Plant Biol 66(1):243–267

Mello CC (2007) A conversation with Craig C. Mello on the discovery of RNAi. Cell Death Differ 14(12):1981–1984

Merelo P, Ram H, Caggiano MP, Ohno C, Ott F, Straub D, Heisler MG (2016) Regulation of MIR165/166 by class II and class III homeodomain leucine zipper proteins establishes leaf polarity. Proc Natl Acad Sci 113(42):11973–11978

Millar AA, Lohe A, Wong G (2019) Biology and function of miR159 in plants. Plants 8(8):255

Mizuno T, Chou MY, Inouye M (1984) A unique mechanism regulating gene expression: translational inhibition by a complementary RNA transcript (micRNA). Proc Natl Acad Sci 81(7):1966–1970

Moghadam A, Taghizadeh MS, Haghi R, Tahmasebi A, Niazi A, Ebrahimie E (2024) Exploring novel insights: methyl jasmonate treatment reveals novel lncRNA-mediated regulation of secondary metabolite biosynthesis pathways in *Echinacea purpurea*. Food Biosci 57:103457

Mohamed A, Jin Z, Osman T, Shi N, Tör M, Jackson S, Hong Y (2022) Hotspot siRNA confers plant resistance against viral infection. Biology 11(5):714

Moison M, Pacheco JM, Lucero L, Fonouni-Farde C, Rodríguez-Melo J, Mansilla N et al (2021) The lncRNA APOLO interacts with the transcription factor WRKY42 to trigger root hair cell expansion in response to cold. Mol Plant 14(6):937–948. https://doi.org/10.1016/j.molp.2021.03.008

Nikovics K, Blein T, Peaucelle A, Ishida T, Morin H, Aida M, Laufs P (2006) The balance between the MIR164A and CUC2 genes controls leaf margin serration in *Arabidopsis*. Plant Cell 18:2929–2945

Nodine MD, Bartel DP (2010) MicroRNAs prevent precocious gene expression and enable pattern formation during plant embryogenesis. Genes Dev 24(23):2678–2692

Nogueira FTS, Timmermans MC (2007) An interplay between small regulatory RNAs patterns leaves. Plant Signal Behav 2(6):519–521

Onodera Y, Haag JR, Ream T, Costa Nunes P, Pontes O, Pikaard CS (2005) Plant nuclear RNA polymerase IV mediates siRNA and DNA methylation-dependent heterochromatin formation. Cell 120:613–622

Orek C, Mulupi K, Koima I (2023) A review of the roles of non-coding RNAs associated with drought stress response in cassava (*Manihot esculenta* Crantz). J Adv Biol Biotechnol 26(7):1–10

Pashkovskiy P, Ryazansky S, Kartashov A, Voloshin R, Khudyakova A, Kosobryukhov A, Kreslavski V, Kuznetsov VV, Allakhverdiev S (2021) Effect of red light on photosynthetic acclimation and the gene expression of certain light signalling components involved in the microRNA biogenesis in the extremophile *Eutrema salsugineum*. J Biotechnol 325:35–42

Patel P, Yadav K, Srivastava AK, Suprasanna P, Ganapathi TR (2019) Overexpression of native Musa-miR397 enhances plant biomass without compromising abiotic stress tolerance in banana. Sci Rep 9:1–15

Pikaard CS, Haag JR, Pontes OM, Blevins T, Cocklin R (2012) A transcription fork model for Pol IV and Pol V-dependent RNA-directed DNA methylation. Cold Spring Harb Symp Quant Biol 77:205–212

Pulido A, Laufs P (2010) Co-ordination of developmental processes by small RNAs during leaf development. J Exp Bot 61:1277–1291

Raza A, Charagh S, Karikari B, Sharif R, Yadav V, Mubarik MS et al (2023) miRNAs for crop improvement. Plant Physiol Biochem 201:107857

Rigo R, Bazin J, Romero-Barrios N, Moison M, Lucero L, Christ A et al (2020) The *Arabidopsis* lncRNA ASCO modulates the transcriptome through interaction with splicing factors. EMBO Rep 21(5):e48977

Rohrig H, Schmidt J, Miklashevichs E, Schell J, John M (2002) Soybean ENOD40 encodes two peptides that bind to sucrose synthase. Proc Natl Acad Sci 99(4):1915–1920

Roule T, Crespi M, Blein T (2022) Regulatory long non-coding RNAs in root growth and development. Biochem Soc Trans 50(1):403–412

Salvador-Guirao R, Baldrich P, Weigel D, Rubio-Somoza I, San Segundo B (2018) The microRNA miR773 is involved in the Arabidopsis immune response to fungal pathogens. Mol Plant-Microbe Interact 31:249–259

Sanan Mishra N, Abdul Kader Jailani A, Mandal B, Mukherjee SK (2021) Secondary siRNAs in plants: biosynthesis, various functions, and applications in virology. Front Plant Sci 12:610283

Sarkar R, Mishra GP, Premakumar, Singh A, Roy J, Shivaprasad KM et al (2025) Identification of heat tolerant lentil genotypes through stress tolerance indices. Sci Rep 15(1):3716

Saze H, Tsugane K, Kanno T, Nishimura T (2012) DNA methylation in plants: relationship to small RNAs and histone modifications, and functions in transposon inactivation. Plant Cell Physiol 53:766–784

Seo JS, Sun HX, Park BS, Huang CH, Yeh SD, Jung C, Chua NH (2017) ELF18-INDUCED LONG-NONCODING RNA associates with mediator to enhance expression of innate immune response genes in *Arabidopsis*. Plant Cell 29(5):1024–1038

Shan T, Fu R, Xie Y, Chen Q, Wang Y, Li Z, Song X, Li P, Wang B (2020) Regulatory mechanism of maize (*Zea mays* L.) miR164 in salt stress response. Russ J Genet 56:835–842

Shen E, Zhao T, Zhu QH (2024) Are miRNAs applicable for balancing crop growth and defense trade-off? New Phytol 243(5):1670–1680

Shin JH, Chekanova JA (2014) Arabidopsis RRP6L1 and RRP6L2 function in FLOWERING LOCUS C silencing via regulation of antisense RNA synthesis. PLoS Genet 10(9):e1004612

Siomi MC, Sato K, Pezic D, Aravin AA (2011) PIWI-interacting small RNAs: the vanguard of genome defence. Nat Rev Mol Cell Biol 12(4):246–258

Snigdha M, Prasath D (2021) Transcriptomic analysis to reveal the differentially expressed miRNA targets and their miRNAs in response to *Ralstonia solanacearum* in ginger species. BMC Plant Biol 21:355

Song L, Fang Y, Chen L, Wang J, Chen X (2021) Role of non-coding RNAs in plant immunity. Plant Commun 2(3):100180

Srinivasan A, Mroczko-Młotek E, Wojciechowska M (2025) Circular RNA formation and degradation are not directed by universal pathways. Int J Mol Sci 26(2):726

Statello L, Guo CJ, Chen LL, Huarte M (2021) Gene regulation by long non-coding RNAs and its biological functions. Nat Rev Mol Cell Biol 22(2):96–118

Sun X, Wang C, Xiang N, Li X, Yang S, Du J, Yang Y, Yang Y (2017) Activation of secondary cell wall biosynthesis by miR319-targeted TCP4 transcription factor. Plant Biotechnol J 15:1284–1294

Sundararajan S, Sivakumar HP, Rajendran V, Kumariah M, Ramalingam S (2023) Adventitious roots in rice, the model cereal: genetic factors and the influence of environmental cues—a mini review. Plant Cell Tissue Organ Cult 154:1–12

Suzuki N, Rivero RM, Shulaev V, Blumwald E, Mittler R (2014) Abiotic and biotic stress combinations. New Phytol 203(1):32–43

Swida-Barteczka A, Stepien A, Jarmolowski A (2023) MicroRNA172b-5p/trehalose-6-phosphate synthase module stimulates trehalose synthesis and microRNA172b-3p/AP2-like module accelerates flowering in barley upon drought stress. Front Plant Sci 14:1124785. https://doi.org/10.3389/fpls.2023.1124785

Teotia S, Tang G (2015) To bloom or not to bloom: role of microRNAs in plant flowering. Mol Plant 8(3):359–377

Tian XC, Nie S, Domingues D, Rossi Paschoal A, Jiang LB, Mao JF (2025) PlantLncBoost: key features for plant lncRNA identification and significant improvement in accuracy and generalization. New Phytol 247:1538

Tran RK, Zilberman D, de Bustos C, Ditt RF, Henikoff JG, Lindroth AM et al (2005) Chromatin and siRNA pathways cooperate to maintain DNA methylation of small transposable elements in Arabidopsis. Genome Biol 6:1–11

Tremblay BJ, Qüesta JI (2023) Mechanisms of epigenetic regulation of transcription by lncRNAs in plants. IUBMB Life 75(5):427–439

Varshney RK, Sinha P, Singh VK, Kumar A, Zhang Q, Bennetzen JL (2020) 5Gs for crop genetic improvement. Curr Opin Plant Biol 56:190–196

Vaucheret H, Voinnet O (2024) The plant siRNA landscape. Plant Cell 36(2):246–275

Vazquez F, Hohn T (2013) Biogenesis and biological activity of secondary siRNAs in plants. Scientifica 2013(1):783253

Wang JW, Schwab R, Czech B, Mica E, Weigel D (2008) Dual effects of miR156-targeted SPL genes and CYP78A5/KLUH on plastochron length and organ size in Arabidopsis thaliana. Plant Cell 20(5):1231–1243

Wang KC, Chang HY (2011) Molecular mechanisms of long noncoding RNAs. Mol Cell 43(6):904–914

Wang HLV, Chekanova JA (2016) Small RNAs: essential regulators of gene expression and defenses against environmental stresses in plants. Wiley Interdiscip Rev RNA 7(3):356–381

Wang ST, Sun CH, Guo SQ (2014) MicroRNA319 positively regulates cold tolerance by targeting OsPCF6 and OsTCP21 in rice (*Oryza sativa* L.). PLoS One 9(3):e91357. https://doi.org/10.1371/journal.pone.0091357

Wang Y, Luo X, Sun F, Hu J, Zha X, Su W, Yang J (2018a) Overexpressing lncRNA LAIR increases grain yield and regulates neighbouring gene cluster expression in rice. Nat Commun 9:3516

Wang K, Su X, Cui X, Du Y, Zhang S, Gao J (2018b) Identification and characterization of microRNA during *Bemisia tabaci* infestations in *Solanum lycopersicum* and *Solanum habrochaites*. Hortic Plant J 4:62–72

Wang J, Mei J, Ren G (2019a) Plant microRNAs: biogenesis, homeostasis, and degradation. Front Plant Sci 10:360

Wang W, Liu D, Chen D, Cheng Y, Zhang X, Song L, Hu M, Dong J, Shen F (2019b) MicroRNA414c affects salt tolerance of cotton by regulating reactive oxygen species metabolism under salinity stress. RNA Biol 16:362–375

Wang X, Chang X, Jing Y, Zhao J, Fang Q, Sun M et al (2020) Identification and functional prediction of soybean circRNAs involved in low-temperature responses. J Plant Physiol 250:153188

Wang R, Yang X, Guo S, Wang Z, Zhang Z, Fang Z (2021) MiR319-targeted OsTCP21 and OsGAmyb regulate tillering and grain yield in rice. J Integr Plant Biol 63:1260–1272

Wang Y, Zuo Y, Zhang Y, Gao S (2023) The long non-coding RNA PILNCR2 increases low phosphate tolerance in maize by interfering with miRNA399-guided cleavage of ZmPHT1s. Mol Plant 16(7):1146–1159. https://doi.org/10.1016/j.molp.2023.04.003

Wang Y, Tang C, Wang Y, Yu H (2024) PHD17 acts as a target of miR1320 to negatively control cold tolerance via JA-activated signaling in rice. Crop J. https://doi.org/10.1016/j.cj.2024.07.012

Wassenegger M, Heimes S, Riedel L, Sänger HL (1994) RNA-directed de novo methylation of genomic sequences in plants. Cell 76(3):567–576

Wei L, Lu W, Yang S et al (2022) Functional analysis of long non-coding RNAs involved in alkaline stress responses in wheat. J Exp Bot 73(16):5698–5714. https://doi.org/10.1093/jxb/erac259

Wielogórska M, Rucińska A, Kloc Y, Boczkowska M (2025) Long non-coding RNAs in the cold-stress response of horticultural plants: molecular mechanisms and potential applications. Int J Mol Sci 26(21):10464

Wierzbicki AT (2012) The role of long non-coding RNAs in transcriptional gene silencing. Curr Opin Plant Biol 15:517–522

Wierzbicki AT, Haag JR, Pikaard CS (2008) Noncoding transcription by RNA polymerase Pol IVb/Pol V mediates transcriptional silencing of overlapping and adjacent genes. Cell 135(4):635–648

Wierzbicki AT, Blevins T, Swiezewski S (2021) Long noncoding RNAs in plants. Annu Rev Plant Biol 72(1):245–271

Williams L, Grigg SP, Xie M, Christensen S, Fletcher JC (2005) Regulation of Arabidopsis shoot apical meristem and lateral organ formation by microRNA miR166g and its AtHD-ZIP target genes. Development 132(16):3657–3668

Wu G, Park MY, Conway SR, Wang JW, Weigel D, Poethig RS (2009) The sequential action of miR156 and miR172 regulates developmental timing in Arabidopsis. Cell 138(4):750–759

Wu Y, Lv W, Hu L, Rao W, Zeng Y, Zhu L, He Y, He G (2017) Identification and analysis of brown planthopper-responsive microRNAs in resistant and susceptible rice plants. Sci Rep 7:8712

Xie Q, Frugis G, Colgan D, Chua NH (2000) Arabidopsis NAC1 transduces auxin signal downstream of TIR1 to promote lateral root development. Genes Dev 14:3024–3036

Xu J, Xu S (2013) microRNAs: macro regulators in biological networks. In: The analysis of regulatory DNA: current developments, knowledge and applications uncovering gene regulation. Bentham Science Publishers, pp 86–100

Xu M, Hu T, Zhao J, Park MY, Earley KW, Wu G et al (2016a) Developmental functions of miR156-regulated SQUAMOSA PROMOTER BINDING PROTEIN-LIKE (SPL) genes in Arabidopsis thaliana. PLoS Genet 12(8):e1006263

Xu M, Hu T, Smith MR, Poethig RS (2016b) Epigenetic regulation of vegetative phase change in Arabidopsis. Plant Cell 28:28–41

Xu X, Wang K, Pan J, Chen X (2019) Small RNA sequencing identifies cucumber miRNA roles in waterlogging-triggered adventitious root primordia formation. Mol Biol Rep 46:6381–6389

Xu J, Xian Q, Zhang N, Wang K, Zhou X, Li Y, Dong J, Chen X (2021) Identification of miRNA-target gene pairs responsive to Fusarium wilt of cucumber via an integrated analysis of miRNA and transcriptome profiles. Biomolecules 11:1620

Yadav A, Singh A, Sharma R (2024) Overexpression of miR166 in response to root rhizobacteria enhances drought adaptive efficacy by targeting HD-ZIP III family genes in chickpea. J Soil Sci Plant Nutr 24(3):6024–6039. https://doi.org/10.1007/s42729-024-01585-z

Yan J, Zhao C, Zhou J, Yang Y, Wang P, Zhu X, Zhu JK (2016) The miR165/166 mediated regulatory module plays critical roles in ABA homeostasis and response in Arabidopsis thaliana. PLoS Genet 12(11):e1006416

Ye Z, Zeng J, Long L, Ye L, Zhang G (2021) Identification of microRNAs in response to low potassium stress in the shoots of Tibetan wild barley and cultivated. Curr Plant Biol 25:100193

Yu Y, Zhang Y, Chen X, Chen Y (2019) Plant noncoding RNAs: hidden players in development and stress responses. Annu Rev Cell Dev Biol 35(1):407–431.

Yuan H, Xu L, Jin Y et al (2024) miR396b/GRF6 module contributes to salt tolerance in rice. Plant Biotechnol J. https://doi.org/10.1111/pbi.14326

Yue E, Cao H, Liu B (2020) OsmiR535, a potential genetic editing target for drought and salinity stress tolerance in *Oryza sativa*. Plants 9(10):1337. https://doi.org/10.3390/plants9101337

Zamecnik P (2005) From protein synthesis to genetic insertion. Annu Rev Biochem 74(1):1–28

Zandalinas SI, Balfagón D, Arbona V, Gómez-Cadenas A, Inupakutika MA, Mittler R (2016) ABA is required for the accumulation of APX1 and MBF1c during a combination of water deficit and heat stress. J Exp Bot 67(18):5381–5390

Zhang P, Dai M (2022) CircRNA: a rising star in plant biology. J Genet Genomics 49(12):1081–1092

Zhang Z, Zhang X (2012) Argonautes compete for miR165/166 to regulate shoot apical meristem development. Curr Opin Plant Biol 15:652–658

Zhang H, Zhu JK (2011) RNA-directed DNA methylation. Curr Opin Plant Biol 14:142–147

Zhang YC, Yu Y, Wang CY, Li ZY, Liu Q (2013) Overexpression of microRNA OsmiR397 improves rice yield by increasing grain size and promoting panicle branching. Nat Biotechnol 31:848–852. https://doi.org/10.1038/nbt.2646

Zhang C, Ding Z, Wu K, Yang L, Li Y, Yang Z, Shi S, Liu X, Zhao S, Yang Z et al (2016) Suppression of jasmonic acid-mediated defense by viral-inducible microRNA319 facilitates virus infection in rice. Mol Plant 9:1302–1314

Zhang Q, Li Y, Zhang Y, Wu C, Wang S, Hao L, Wang S, Li T (2017) MdmiR156ab and Md-miR395 target WRKY transcription factors to influence apple resistance to leaf spot disease. Front Plant Sci 8:526

Zhang X, Dong J, Deng F, Wang W, Cheng Y, Song L et al (2019) The long non-coding RNA lncRNA973 is involved in cotton response to salt stress. BMC Plant Biol 19(1):459

Zhang P, Li S, Chen M (2020) Characterization and function of circular RNAs in plants. Front Mol Biosci 7:91

Zhang LL, Huang YY, Zheng YP, Liu XX, Zhou SX, Yang XM et al (2022) Osa-miR535 targets SQUAMOSA promoter binding protein-like 4 to regulate blast disease resistance in rice. Plant J 110(1):166–178

Zhang L, Lin T, Zhu G, Wu B, Zhang C, Zhu H (2023) LncRNAs exert indispensable roles in orchestrating the interaction among diverse noncoding RNAs and enrich the regulatory network of plant growth and its adaptive environmental stress response. Hortic Res 10(12):uhad234

Zhang D, Ma Y, Naz M, Ahmed N, Zhang L, Zhou JJ et al (2024) Advances in circRNAs in the past decade: review of circRNAs biogenesis, regulatory mechanisms, and functions in plants. Genes 15(7):958

Zhang M, Ju B, Wang X, Zhou S, Zhao H, Wang Z, Li X (2025) Global knowledge mapping and emerging research trends in non-coding RNAs related to animal and plant male sterility: a visual analysis of CiteSpace maps. Medicine 104(23):e42612

Zhao YF, Peng T, Sun HZ, Teotia S, Wen HL, Du YX et al (2019) miR1432-OsACOT (Acyl-CoA thioesterase) module determines grain yield via enhancing grain filling rate in rice. Plant Biotechnol J 17(4):712–723

Zhao Y, Xie J, Wang S, Xu W, Chen S, Song X, Lu M, El-Kassaby YA, Zhang D (2021) Synonymous mutation in growth regulating factor 15 of miR396a target sites enhances photosynthetic efficiency and heat tolerance in poplar. J Exp Bot 72:4502–4519

Zhao C, Yang X, Zhao J (2024) Drought-triggered repression of miR166 promotes drought tolerance in soybean. Crop J 12(1):154–163. https://doi.org/10.1016/j.cj.2023.06.010

Zheng Z, Xing Y, He XJ, Li W, Hu Y, Yadav SK et al (2010) An SGS3-like protein functions in RNA-directed DNA methylation and transcriptional gene silencing in Arabidopsis. Plant J 62(1):92–99

Zhong X, Du J, Hale CJ, Gallego-Bartolome J, Feng S, Vashisht AA et al (2014) Molecular mechanism of action of plant DRM de novo DNA methyltransferases. Cell 157(5):1050–1060

Zhou M, Tang W (2019) MicroRNA156 amplifies transcription factor-associated cold stress tolerance in plant cells. Mol Gen Genomics 294:379–393. https://doi.org/10.1007/s00438-019-01528-2

Zhou GK, Kubo M, Zhong R, Demura T, Ye ZH (2007) Overexpression of miR165 affects apical meristem formation, organ polarity establishment and vascular development in Arabidopsis. Plant Cell Physiol 48:391–404

Zhou Y, Honda M, Zhu H, Zhang Z, Guo X, Li T, Li Z, Peng X, Nakajima K, Duan L et al (2015) Spatiotemporal sequestration of miR165/166 by Arabidopsis Argonaute10 promotes shoot apical meristem maintenance. Cell Rep 10:1819–1827

Zhou M, Law JA, Jacobsen SE (2018) Establishing, maintaining and modifying DNA methylation patterns in plants and animals. Nat Rev Genet 19(8):489–506

Zhou Y, Liu Y, Feng Y (2020) Integration of sRNA, degradome, transcriptome analysis and functional investigation reveals gma-miR398c negatively regulates drought tolerance via GmCSDs and GmCCS in transgenic Arabidopsis and soybean. BMC Plant Biol 20:1–19. https://doi.org/10.1186/s12870-020-02339-8

Zhu QH, Helliwell CA (2011) Regulation of flowering time and floral patterning by miR172. J Exp Bot 62(2):487–495

Zhu QH, Upadhyaya NM, Gubler F, Helliwell CA (2009) Overexpression of miR172 causes loss of spikelet determinacy and floral organ abnormalities in rice (*Oryza sativa*). BMC Plant Biol 9:149

Zhu L, Ow DW, Dong Z (2018) Transfer RNA-derived small RNAs in plants. Sci China Life Sci 61(2):155–161

Zhu H, Zhang Y, Tang R, Qu H, Duan X, Jiang Y (2019) Banana sRNAome and degradome identify microRNAs functioning in differential responses to temperature stress. BMC Genomics 20:1–16

Zilberman D, Cao X, Johansen LK, Xie Z, Carrington JC, Jacobsen SE (2004) Role of Arabidopsis ARGONAUTE4 in RNA-directed DNA methylation triggered by inverted repeats. Curr Biol 14(13):1214–1220

Zorb C, Geilfus CM, Dietz KJ (2019) Salinity and crop yield. Plant Biol 21:31–38

The Intricate Roles of Epigenetic Modifications in Shaping the Plant Transcriptome

Nibedita Chakraborty, Arnab Mukherjee, Suman Kumar Saha, Debashis Oraon, Shreyashi Nandi, and Sujit Roy

Abstract The dynamic area of plant transcriptome–epigenetic regulatory interaction explores how gene expression is modulated beyond the DNA sequence itself, influencing cellular differentiation and various responses to environmental cues. The complex interplay between transcriptional networks and epigenetic modifications is a key regulator of gene expression in plants. Epigenetic mechanisms, including dynamic changes in chromatin structure, DNA methylation, histone modifications, and RNA-associated silencing, influence gene activity by recruiting transcription factors, RNA polymerase, and other regulatory elements. Recent advances in high-throughput sequencing technologies, such as RNA–seq and ChIP–seq, have elucidated significant insights into this interplay, enabling researchers to map epigenetic modifications and their impact on gene expression at both transcriptional and post-transcriptional levels. The complexity of this interaction remains an active area of research, with ongoing efforts to fully elucidate the range of epigenetic modifications and their effects on gene regulation. Plants can finely tune how they respond to environmental cues and developmental signals by modulating gene activity through various epigenetic modifiers. Understanding the variation in the plant transcriptome and epigenetic regulation not only provides insights into fundamental plant biology but also offers potential strategies for improving crop resilience and productivity through targeted epigenetic interventions. This knowledge has implications for development, environmental adaptation, and crop improvement, making it a vital research area in modern plant biology.

Keywords Transcriptomics · NGS technology · RNA–seq · Epigenetic modifications · DNA methylation · Histone modifications · Crop improvement

---

Nibedita Chakraborty and Arnab Mukherjee contributed equally with all other contributors.

N. Chakraborty · A. Mukherjee · S. K. Saha · D. Oraon · S. Nandi · S. Roy (✉)
Department of Botany, UGC Centre for Advanced Studies, The University of Burdwan, Golapbag Campus, Burdwan, West Bengal, India

L. M. Vaschetto (ed.), *Omics Technologies in Plant Science*, Interdisciplinary Biotechnological Advances, https://doi.org/10.1007/978-981-92-0533-2_4

Abbreviations

ABA	Abscisic acid
ABI3	ABSCISIC ACID INSENSITIVE3
ABI3	ABSCISIC ACID INSENTITIVE 3
AG	AGAMOS
AMV	Alfalfa Mosaic Virus
AS1	ASYMMETRIC LEAVES1
BABA	β-Aminobutyric acid
BAF60	BRG1/BRM-associated factor 60
BRM	BRAMHA
CBF	C REPEAT BINDING FACTOR
ChIP–seq	Chromatin Immunoprecipitation Sequencing
CLF	CURLY LEAF
CLV	CLAVATA
CMT	CHROMOMETHYLASE
COOLAIR	Cold Induced Long Antisense Intragenic RNA
COR	COLD RESPONSIVE
CTL	Chitinase-Like
CUL4	CULLIN 4
DME	DEMETER
DMR	Differentially Methylated Region
DOG1	DELAY OF GERMINATION 1
DRM2	DNA methyltransferase
EC	Evening complex
ECT12	EVOLUTIONARILY CONSERVED C-TERMINAL REGION 12
ETC1	ENHANCER OF TRY AND CPC 1
FIS2	FERTILIZATION-INDEPENDENT SEED2
FLC	FLOWERING LOCUS C
FLS2	Flagellin Sensing 2
FT	FLOWERING LOCUS T
FUS3	FUSCA 3
FUS3	FUSCA3
GA	Gibberellin
GA3ox1	GIBBERELLIN 3-OXIDASE 1
GCN5	GENERAL CONTROL NON-REPRESSED 5 PROTEIN
GL2	GLABRA 2
GOLS3	GALACTINOL SYNTHASE3
HAD	Histone Deacetylases
HD2D	HISTONE DEACETYLASE 2D
HDA6	HISTONE DEACETYLASE 6
HDACs	Histone Deacetylases
HIRA	HISTONE REGULATORY HOMOLOG 1
HKT	HIGH-AFFINITY POTASSIUM TRANSPORTER

HOS15	HIGH EXPRESSION OF OSMOTICALLY RESPONSIVE GENE 15
HSFA	Heat Stress Responsive Transcription Factors
ICE1	INDUCER OF CBF EXPRESSION
JMJ17JU	MONJI DOMAIN-CONTAINING PROTEIN17
KNAT1	KNOTTED-LIKE FROM ARABIDOPSIS THALIANA 1
KNOX	KNOTTED-LIKE HOMOBOX
KNU	KNUCKLE
KYP	KRYPTONITE
LDL1	LYSINE-SPECIFIC DEMETHYLASE 1-LIKE1
LEC1/2	LEAFY COTYLEDON1/2
LEC1	LEAFY COTYLEDON 1
LHP1	LIKE HETEROCHROMATIN PROTEIN 1
LOB	LATERAL ORGAN BOUNDARIES
LUX	LUX ARRHYTHMO
MBD	Methyl domain binding
MEA	MEDEA
MET1	METHYLTRANSFERASE 1
miRNAs	microRNAs
NCED3	NINE CIS-EPOXYCAROTENOID DIOXYGENASE 3
NGS	Next-generation sequencing
NPR1	Non-expressor of PR genes
PcG	Polycomb Group
PGX3	POLYGALACTURONASE INVOLVED IN EXPANSION-3
PHD-PRC2	Plant HomeoDomain-Polycomb Repressive Complex 2
PHE 1	PHERES 1
PHYB	PHYTOCHROME B
PIF1	PHYTOCHROME INTERACTING FACTOR1
PIL5	Phytochrome Interacting Factor 3-Like 5
PKL	PICKLE
PLT	PLETHORAs
PR1	PATHOGENESIS-RELATED GENE 1
PRC2	POLYCOMB-REPRESSIVE COMPLEX 2
PRR	Plant Pattern Recognition Receptor
PTM	Posttranslational modification
PYL	PYRABACTIN RESISTANCE 1-LIKE 6
QC	Quiescent centre
qRT–PCR	Quantitative real-time PCR
RdDM	RNA-directed DNA methylation
RISC	RNA-induced silencing complex
RNA–seq	RNA sequencing
ROS1	REPRESSOR OF SILENCING 1
SAM	Shoot Apical Meristem
SCN	Soybean Cyst Nematode
SCN	Stem Cell Niche
SDC	SUPPRESSOR OF DRM1 DRM2 CMT3

siRNAs	small interfering RNAs
SNI1	Suppressor of NPR1 Inducible 1
SNPs	Single-nucleotide polymorphisms
SOC1	SUPPRESSOR OF OVEREXPRESSION OF CONSTANS 1
SOS	Salt Overly Sensitive
ssRNA	Single-Stranded RNA
SUP	SUPERMAN
SWI/SNF	SWItch/Sucrose Non-Fermentable
VAL1	VP1/ABI3-LIKE1
VIL1	VIN3-LIKE 1
VIN3	VERNALIZATION INSENSITIVE 3
WOX5	WUSCHEL-RELATED HOMEOBOX 5
WUS	WUSCHEL
YUC	YUCCA

1 Introduction

Transcriptome encompasses all forms of RNA molecules, including mRNA, rRNA, tRNA, and various non-coding RNAs, transcribed from the genome of an organism in a cell or tissue (Raza et al. 2021; Yang et al. 2021), making transcriptome analysis essential for understanding the functional segments of the genome (Zhuang et al. 2014). Methods of transcriptomics have been used extensively to identify genes involved in plant growth and development, and genes that responded differently under various biotic and abiotic stress conditions (Agarwal et al. 2014; Rastogi et al. 2019; Tyagi et al. 2022). The advancement of transcriptomics has evolved significantly over the years from primitive Northern blotting to modern RNA sequencing (RNA–seq). The key methods driving the progress include microarray analysis and quantitative real-time PCR (qRT–PCR), with RNA–seq emerging as a leading approach. RNA–seq, a key next-generation sequencing (NGS) technology, enables comprehensive analysis of transcriptomes from both model and non-model plants, with the data deposited in open resources such as GenBank, NCBI, etc. Integrating this data with small RNA assessments, molecular markers, and epigenome analyses can provide insights into the mechanisms involved in plant development and stress responses (Velculescu et al. 1995; Lu et al. 2013). Furthermore, NGS has been used to analyze plant transcriptomes lacking a reference genome, establishing its potential in transcriptomics. This technology has also enabled the detection of low-abundance transcripts, non-coding RNAs, and analysis of mutations like single-nucleotide polymorphisms (SNPs) and insertions/deletions (Gao et al. 2013; Agarwal et al. 2007). Therefore, the progression in transcriptomics has revolutionized plant transcriptomics, enabling a detailed analysis of transcriptomes and epigenomes, and assure for accelerating crop improvement and enhancing resistance to environmental stresses through the development of climate-resilient, high-yielding, and nutrient-rich crops. The insights gained from plant transcriptomics and epigenomics will eventually enrich our understanding of the intricate interplay between DNA sequence, epigenetic modifications, and transcriptional regulation.

At the molecular level, gene transcription is influenced by both the DNA sequence and the chromatin architecture (Agarwal et al. 2020). In plants and other eukaryotic organisms, compactness within the chromatin creates a physical barrier that prevents gene transcription, making modification in chromatin structure crucial for gene control. These modifications can be passed down through meiotic or mitotic cell division, and play a crucial role in regulating plant growth, development, and stress responses (Lauria and Rossi 2011). The epigenetic changes can influence the chromatin state, controlling the accessibility of transcription factors to specific genes and ultimately regulating their expression (Abdulraheem et al. 2024; Duarte-Aké et al. 2023). In plants, there are various types of epigenetic modifications, namely, DNA methylation, post-translational modification (PTM) of histones, and small RNA-mediated gene silencing, which respond to environmental cues and internal signals (Lauria and Rossi 2011; Saraswat et al. 2017). Among these, DNA methylation, often coupled with DNA replication, can be stably inherited through generations. It has also been linked to the suppression of gene transcription because it can attract chromatin remodeling complexes that create a more compact chromatin structure by interacting with methyl domain binding (MBD) proteins (Jogam et al. 2022; Lämke and Bäurle 2017). A wide range of histone PTMs primarily occur in the amino-terminal tail, including acetylation, methylation, phosphorylation, ubiquitination, sumoylation, and ADP ribosylation, ultimately altering chromatin structure (Li et al. 2007; Nunez-Vazquez et al. 2022). Except for methylation, these modifications can alter the net charge of nucleosomes, affecting the electrostatic or intra-nucleosomal interactions. Histone PTMs can directly influence the formation of a transcriptionally active state by promoting the recruitment of transcription factors (TFs), or their deposition can be directed by transcription itself, such as through histone exchange during elongation (Lauria and Rossi 2011). Similarly, small RNAs, such as microRNAs (miRNAs) and small interfering RNAs (siRNAs), guide the RNA-induced silencing complex (RISC) to target specific mRNAs for degradation or translational repression, regulating plant stress responses (Park et al. 2017; Begum 2022). In plants, small RNAs are often derived from stress-responsive genes or transposable elements that become activated under stress conditions. This mechanism helps plants fine-tune their gene expression in response to various biotic and abiotic stresses, for example, pathogen resistance or drought tolerance, enhancing their adaptability and survival. Additionally, small RNAs are involved in chromatin remodeling by directing DNA methylation and histone modifications to specific genomic regions. These epigenetic regulations can either activate or silence genes, contributing to the stress response and overall stress resilience in plants (Jagadhesan et al. 2022; Zhan and Meyers 2023). This section will explore the current understanding of plant transcriptomics, epigenetics, and chromatin regulation, with a focus on their roles in plant development, stress responses, and crop improvement.

2 Functional Interplay Between Epigenetic Modification and Transcriptomics in Plant Development

The interrelation between epigenetic modification and transcriptomics is essential to plant development, influencing gene expression and cellular differentiation. This intricate relationship directs various aspects of plant growth and development, from seed germination to flowering and senescence. Numerous studies have detailed the functional characterization of target genes and their chromatin modifiers in plant growth and developmental stages, which are summarized in Table 1. This table highlights the complex interplay between epigenetic modifications and transcriptomics, which will be discussed in further detail below.

2.1 *Seed Dormancy and Germination*

Seed dormancy is a survival approach in plants that inhibits premature germination under unfavorable conditions. During the process of embryogenesis, the embryo stores the reserves for subsequent germination, followed by a desiccation phase where dehydration occurs and the seed remains dormant (Nonogaki 2014). Alterations in temperature around the seed bank cause a progressive shift in seed dormancy during post-maturation. During minimal dormancy, changes in light and temperature remove the germination barriers, and adequate water availability induces the protrusion of the radical; thus, germination occurs (Batlla and Benech-Arnold 2015). Several studies have shown the role of abscisic acid (ABA) and gibberellin (GA) signaling in regulating gene expression by acting as regulators between molecular signals and environmental stimuli. The essential factors in regulating seed dormancy and germination are the ratio of ABA to GA concentration and sensitivity (Tognacca and Botto 2021). *DELAY OF GERMINATION 1* (*DOG1*) acts as a main regulator during seed dormancy, as inactivation of *DOG1* results in inhibition of dormancy. Several factors regulate seed dormancy by modulating the expression of *DOG1*. For example, during the process of seed maturation, lower temperature enhances the tenure of seed dormancy, which is positively correlated with increased *DOG1* expression (Nakabayashi et al. 2012). Moreover, *DOG1* expression is tightly regulated by several epigenetic modifiers. For instance, during post-maturation, the expression of *DOG1* is suppressed by histone demethylases [LYSINE-SPECIFIC DEMETHYLASE 1-LIKE1 (LDL1) and LDL2] and histone methyltransferases (KRYPTONITE (KYP)/SUVH4/SUVH5) by reducing the level of H3K4 methylation at *DOG1* chromatin (Zheng et al. 2012; Zhao et al. 2015). In addition, losing function mutations in plant-specific histone deacetylases HD2A and HD2B results in prolonged seed dormancy, leading to hyperacetylation at the *DOG1* locus followed by enhancement in the *DOG1* expression during seed maturation. This suggests that H2DA and HD2B negatively regulate *DOG1* expression by affecting seed dormancy and promoting seed germination during seed imbibition

Table 1 List of target genes and their chromatin modifiers functionally characterized in different plant developmental stages

Developmental pathways	Target gene	Epigenetic modifiers	Function	Ref.
Seed dormancy	*DOG1*	1. Histone demethylases LDL1 and LDL2 2. Histone deacetylase HD2A, HD2B	It acts as a genetic barrier during seed germination	Zhao et al. (2015); Han et al. (2023); Zheng et al. (2012); Liu et al. (2007)
Seed germination	*PHYA* and *PHYB*	Histone demethylase JMJ20 and JMJ22	*PHYA* and *PHYB* play a significant role in seed germination by regulating the content of ABA and the GA ratio under light signal	Cho et al. (2012)
	DOG1	H3K9 methyltransferase SUVH5	*DOG1* expression is inhibited by the deposition of H3K9me2	Gu et al. (2019)
Seedling establishment	*PROTO-CHLOROPHYLLIDE OXIDOREDUCTASE C* (*PORC*)	H3K4me3 demethylase JMJ17	JMJ17 significantly regulates the expression of the chlorophyll biosynthesis gene *PORC* by demethylation of H3K4 at the promoter region	Islam et al. (2021)
Leaf development	*KNOX* (class I)	PRC2 mediated H3K27me3 deposition	Essential for vegetative development in plants	Lodha et al. (2013)
Shoot meristem development	*WUS*	Histone deacetylases (HDACs)	Plays a crucial role by regulating the stem cell population within the SAM	Ma et al. (2019)
	WUSCHEL-related homeobox 5 (*WOX5*)	Histone deacetylases 19 (HDA19)	It acts as a negative regulator in shoot meristem development	Lee et al. (2022)

(continued)

Table 1 (continued)

Developmental pathways	Target gene	Epigenetic modifiers	Function	Ref.
Root meristem development	*ETC1* and *GL2*	Histone deacetylase 6 (HDA6)	The epigenetic modification of *ETC1* and *GL2* is crucial for root meristem development	Li et al. (2015)
	FUS3, *LEC1*/*2* and *ABI3*	PRC1 RING-finger proteins AtRING1a/b and AtBMI1a/b/c	PRC1 RING-finger proteins recruit repressive H2Aub1 mark, which in turn results in primary root development	Bratzel et al. (2010); Yang et al. (2013)
Flowering transition	*FLC*	PHD-PRC2 complex	*FLC* is inhibited by the deposition of repressive H3K27me3 marks by PHD-PRC2 complex	De Lucia et al. (2008)
	FT and *SOC1*	BRAHMA, chromatin remodeler	Regulation of photoperiodic flowering	Farrona et al. (2004)
Floral meristem development	*AG*	H3K27 methyltransferase	*AG* acts as a positive regulator of Floral meristem development	Sun et al. (2014)
	SUPERMAN	H3K27 methyltransferase	Floral meristem development	Xu et al. (2018)
Seed development and maturation	*FIS2*	H3K27 methyltransferase	FIS forms a complex with PRC2 and recruits the repressive histone marks to regulate seed development	Luo et al. (2000)
	MEA	DNA glycosylase DEMETER (DME) activity	Regulation of seed development	Köhler and Makarevich (2006)
Senescence	*WRKY53*	H3K4 methyltransferase	WRKY53 induces senescence in plants	Ay et al. (2009)
	EIN2 and *ORE1*	H3K27me3 demethylation	Regulators of leaf senescence	Lu et al. (2011)

(Han et al. 2023). The regulation of *DOG1* is also governed by the recruitment of the components of polycomb group proteins, such as CURLY LEAF (CLF) and LIKE HETEROCHROMATIN PROTEIN 1 (LHP1), to the *DOG1* promoter. These components deposit repressive H3K27me3 marks, causing a subsequent reduction in *DOG1* expression (Chen et al. 2020). PICKLE (PKL), an ATP-dependent chromatin-remodeling factor, is also known to regulate *DOG1* expression by recruiting the H3K27me3 mark at *DOG1* chromatin (Zhang et al. 2008, 2012). Moreover, PKL physically interacts with LUX ARRHYTHMO (LUX), a component of the evening complex (EC), and deposits repressive histone marks, reducing *DOG1* expression. Studies on *lux* and *pkl* mutants indicated a reduction in H3K27me3 at *DOG1* and enhanced dormancy compared with wild-type plants, suggesting that PKL is a negative regulator of *DOG1* (Zha et al. 2019).

Seed germination is a crucial physiological response that mediates the transition from the silent state of seeds to the active status of seedlings (Ding et al. 2022). During germination, less compaction occurs in the condensed chromatin state, creating access for transcription factors to activate gene expression and metabolic processes that induce seed germination (Van Zanten et al. 2013). Moreover, seeds can manipulate their germination strategies in response to external environmental stimuli, including light signals (Ding et al. 2022). PHYTOCHROME B (PHYB)-PHYTOCHROME INTERACTING FACTOR1 (PIF1)/PIL5 pathway plays an essential role in seed germination under light conditions. In Arabidopsis, PIF1 acts as a negative regulator of seed germination. For instance, when red light activates PHYB, it destabilizes PIF1 to regulate the genes involved in the ABA and GA pathways (Ding et al. 2022). The PHYB-dependent seed germination pathway is extensively regulated by several epigenetic modifiers. During imbibition, SUVH5 (H3K9 methyltransferase) positively regulates Arabidopsis seed germination by suppressing ABA biosynthesis and recruiting H3K9me2 at the *DOG1* locus (Gu et al. 2019). In contrast, the recruitment of HDA15 by PIF1 results in the prevention of light-inducible seed germination by reducing the level of H3 acetylation at corresponding genes involved in light signaling pathways (Gu et al. 2017). Cho et al. (2012) reported the function of two histone demethylases, JMJ20 and JMJ22, which act as a positive regulator of PHYB-dependent seed germination by removing repressive histone arginine methylation at *GIBBERELLIN 3-OXIDASE 1* (*GA3ox1*) and *GA3ox2* to induce seed germination. This suggests that the PHYB-PIF1-JMJ20/22 module may partially regulate the light-mediated seed germination process in Arabidopsis. Furthermore, the existence of a dynamic genome-wide reprogramming of DNA methylation has been reported during seed dormancy to seed germination processes in Arabidopsis (Kawakatsu et al. 2017). In addition, enhanced hypermethylation of mCHH around transposable elements was found during seed development and continued throughout stratification. This hypermethylation is mediated by the RdDM and CHROMOMETHYLASE (CMT2) pathway. On the other hand, hypermethylation in desiccated seeds was clearly evident to be reprogrammed by REPRESSOR OF SILENCING 1 (ROS1)-independent passive CHH demethylation (Kawakatsu et al. 2017) during the seed germination stage of Arabidopsis.

2.2 Seedling Establishment

In Arabidopsis, etiolated seedlings undergo skotomorphogenesis with closed cotyledons, apical hook, elongated hypocotyl, and pale plastids or etioplasts under dark conditions. Upon exposure to light, de-etiolated seedlings experience photomorphogenesis with open cotyledons, a shorter hypocotyl, and green chloroplasts (Von Arnim and Deng 1996). In the plant life cycle, cotyledon greening serves as a major phenotypic observation of the seedling establishment stage (Islam et al. 2021). Interestingly, *Arabidopsis* JUMONJI DOMAIN-CONTAINING PROTEIN17 (JMJ17), a positive histone mark H3K4me3 demethylase, is involved in cotyledon greening during the seedling establishment stage. Loss-of-function mutation of *JMJ17* displayed reduced cotyledon greening and decreased chlorophyll compared to wild-type de-etiolated seedlings. In contrast, etiolated seedlings of *jmj17* were reported to accumulate a much higher concentration of the precursor of chlorophyll biosynthesis. Moreover, a ChIP–qPCR study revealed that JMJ17 binds to the promoter region of genes related to chlorophyll synthesis in the dark and dissociates from the promoter upon light irradiation. This suggests that JMJ17 demethylates positive histone marks at the promoter regions of chlorophyll biosynthesis genes in the dark, whereas the H3K4me3 demethylation activity of JMJ17 is diminished at the promoters of several chlorophyll biosynthesis genes upon light exposure (Islam et al. 2021).

2.3 Vegetative Growth

The vegetative development in a plant comprises the growth and differentiation of non-reproductive organs such as leaves, shoots, and roots. This stage is crucial in the plant life cycle as it establishes the plant vigor before entering the reproductive stage.

2.3.1 Leaf Development

The development of a leaf starts with the formation of leaf primordia, which occurs at the peripheral zone of the apical meristem (Hung et al. 2023). The downregulation of the class I *KNOTTED-LIKE HOMOBOX* (*KNOX*) gene mainly facilitates this process (Jackson et al. 1994; Vollbrecht et al. 2000; Scofield and Murray 2006). Two transcription factors, ASYMMETRIC LEAVES1 (AS1) and the AS2/LATERAL ORGAN BOUNDARIES (AS2/LOB) domain protein AS2, form a transcription complex and are found to recruit H3K27me3 mediated by POLYCOMB-REPRESSIVE COMPLEX 2 (PRC2) to downregulate the expression of the *KNOX* gene for the promotion of leaf development (Lodha et al. 2013). Moreover, AS1/2 is also known to deposit HISTONE REGULATORY HOMOLOG 1 (HIRA) to regulate target transcripts during organogenesis (Guo et al. 2008). In recent years, Hung et al. (2023) proposed the involvement of the AS1/2 transcription complex in leaf development by interacting with Arabidopsis H3K9 methyltransferases KYP/SUVH5/6 that repress the homeobox gene *KNOTTED-LIKE FROM ARABIDOPSIS THALIANA 1* (*KNAT1*) and *KNAT2* by modifying the level of histone H3 acetylation and H3K9 dimethylation at the corresponding gene loci.

2.3.2 Shoot Meristem Development

Shoot development in plants is a vital, highly regulated process that involves a complex interplay between epigenetic modifications and changes in gene expression. Epigenetic modifications play an essential role in regulating the expression of genes involved in shoot apical meristem (SAM) maintenance, differentiation, and initiation of organogenesis. SAM, which harbors stem cells, is a major initiation center during post-embryonic development in plants (Nguyen and Gutzat 2022). The WUSCHEL (WUS)/CLAVATA3 (CLV3) feedback loop, a key regulatory mechanism in SAM, involves WUS specifying stem cell fate and activating stem cell integrating factor, CLAVATA3 (CLV3) (Clark et al. 1995), which represses WUS expression through the heterodimer CLV1/CLV2 receptor complex, thereby maintaining stem cell pool size (Song et al. 2021). This genetic circuit is further modulated by epigenetic modifications, which contribute to the precise regulation of stem cell maintenance and differentiation. For instance, the WUS transcription factor regulates auxin signaling pathway genes by interacting with histone deacetylases (HDACs), highlighting the epigenetic role of auxin in stem cell maintenance (Ma et al. 2019). Furthermore, transposon activation represents a supplementary epigenetic layer, contributing to SAM stem cell plasticity and meristem development upon developmental and environmental cues (Nguyen and Gutzat 2022).

2.3.3 Root Meristem Development

A fundamental stage in a plant's life cycle is the development of the root, which governs physical support, water absorption, nutrient uptake, and response to various environmental signals. In the Arabidopsis embryo, the pattern of cell division produces many initial cells inside the root promeristem, establishing the primary root meristem organization during embryogenesis (Scheres et al. 1994). The initial cells, surrounded by the quiescent center (QC), make up the stem cell niche (SCN) that generates different root cells, including stele, epidermis, cortex, and endodermis, lateral root cap, and columella (Dolan et al. 1993). Several histone modifiers are involved in regulating gene expression associated with root development. For instance, histone acetylation is essential for maintenance and homeostasis of SCN, where expression of *PLETHORAs* (*PLT*) and *WUSCHEL-RELATED HOMEOBOX 5* (*WOX5*) is modulated to control columella stem cell fate (Aida et al. 2004; Burkart et al. 2022). Likewise, HISTONE DEACETYLASE 6 (HDA6) is evidenced to be involved in regulating two core transcription factors, namely, *ENHANCER OF TRY AND CPC 1* (*ETC1*) and *GLABRA 2* (*GL2*), by binding to the promoter region of these corresponding genes for the determination of epidermal cell fate in root development (Li et al. 2015). In addition, plant-specific HISTONE DEACETYLASE 2D (HD2D) serves functions in elongated lateral root formation (Han et al. 2016). Apart from histone acetylation, H3K4 methyltransferase is also recognized as a regulator of SCN maintenance for the partitioning of root cells (Napsucialy-Mendivil et al. 2014). PRC1 RING-finger proteins AtRING1a/b and AtBMI1a/b/c play an important role in root development by recruiting the repressive histone monoubiquitination (H2Aub1) mark to seed maturation genes, for example *FUSCA3* (*FUS3*), *LEAFY COTYLEDON1/2* (*LEC1/2*), and *ABSCISIC ACID INSENSITIVE3* (*ABI3*). As a result, the conversion from embryonic to post-germinative growth is started in the primary root developmental stage (Bratzel et al. 2010; Yang et al. 2013). Interestingly, PKL is also reported to maintain root meristem activity as the *pkl* mutant shows reduced root growth and decreased meristem development (Aichinger et al. 2011). Notably, CLF and PKL control the root meristem activity antagonistically, with loss of CLF enhancing root stem cell activity (Aichinger et al. 2011). The SWItch/Sucrose Non-Fermentable (SWI/SNF) subunit BRAMHA (BRM) and BRG1/BRM-associated factor 60 (BAF60) are also evidenced to regulate root growth through ABA and cytokinin signaling pathways, respectively (Han et al. 2012; Jégu et al. 2015).

2.4 Transition to Flowering

The transition from vegetative to reproductive stage is crucial for successful reproduction. *FLOWERING LOCUS C* (*FLC*) encodes a MADS box transcription factor that plays a significant role in flowering transition in winter-annual Arabidopsis (Michaels and Amasino 1999). The upregulated expression of *FLC* and *FLC*-like genes prevents precocious flowering by inhibiting the expression of flowering integrator genes, including *FLOWERING LOCUS T* (*FT*) and *SUPPRESSOR OF OVEREXPRESSION OF CONSTANS 1* (*SOC1*) (Sheldon et al. 2000; Michaels et al. 2005). However, a prolonged period of cold temperature (called vernalization) causes stable repression of *FLC*, enabling early flowering at favorable day length during the spring (Sheldon et al. 2008). The vernalization-mediated *FLC* repression is mitotically stable and involves epigenetic regulation (Sung and Amasino 2004; Sheldon et al. 2008). Before vernalization, the *FLC* locus is enriched with active histone marks such as H3K4me3, H3K36me3, and histone acetylation, recruited by *FRIGIDA* and its complexes (Kim et al. 2005); however, at the early stage of vernalization, COOLAIR (cold-induced long antisense intragenic RNA), the antisense non-coding RNA of *FLC*, reduces the level of H3K36me3 by physically interacting with *FLC* chromatin (Csorba et al. 2014). Moreover, *VP1/ABI3-LIKE1* (*VAL1*) is likely to recruit HDA19 to remove acetylation at the *FLC* chromatin (Qüesta et al. 2016). At the late stage of vernalization, the PHD finger domain protein

VERNALIZATION INSENSITIVE 3 (*VIN3*) forms a heterodimer with *VIN3-LIKE 1* (*VIL1*) to recruit the PHD-PRC2 complex at the first intron region of *FLC*, which in turn increases the level of repressive H3K27me3 marks at the *FLC* locus and inhibits its expression (De Lucia et al. 2008). Interestingly, a low level of *FLC* expression causes the activation of *FT* and *SOC1* required for the transition to the reproductive stage that ultimately results in flowering induction.

2.5 Floral Meristem (FM) Development and Determinacy

Upon receiving signals to flower, the vegetative meristem transforms into a reproductive meristem, which develops floral meristems and activates flowering genes. Unlike SAM, which indefinitely produces leaves and axillary buds, the floral meristem is determinate in its function and ends after flower development. The MADS box transcription factor *AGAMOUS* (*AG*) is known as a crucial floral identity gene that functions in floral primordia development, reproductive organ differentiation, and FM determination (Bowman et al. 1989). In order to generate mature stamens, the initiation of *AG* expression from stages 5–6 to stage 10 is essential (Guo et al. 2015). During floral stem cell maintenance, AG inhibits the expression of *WUS* by hindering suppressive PRC2 deposition at the promoter region of *KNUCKLE* (*KNU*), the repressor of *WUS*. In stage 6, AG directly binds to the promoter of *KNU*, causing cell-division-mediated dilution of the repressive histone mark H3K27me3, collectively resulting in increased *KNU* expression and direct suppression of *WUS* transcription at the floral meristem (Sun et al. 2014). Besides the formation of floral organ identity, including stamen and carpel, *AG* is also found to be involved in FM termination. During late developmental stage 3, AG induces *SUPERMAN* (*SUP*) expression that epigenetically suppresses auxin biosynthesis genes *YUC1/4* in floral whorl 3/4 periphery regions (Xu et al. 2018). Another direct *AG* target, *CRC*, is found to end FM by modulating auxin homeostasis and promoting gynoecium formation in the later flower developmental stage (Yamaguchi et al. 2018).

2.6 Seed Development and Maturation

Endosperm and embryo are two distinct structures that form angiosperm seeds. The endosperm acts as a reservoir that provides all the nutrients to the developing embryo until it can use photosynthesis to make its own food. Eventually, the embryo develops into a new plant. Plant seed development signifies a complex interacting network between transcription factors and hormone signaling, such as ABA, GA, and auxin, regulated by several epigenetic modifications (Luján-Soto and Dinkova 2021). In Arabidopsis and other crops, several main regulators, including *LEAFY COTYLEDON 1* (*LEC1*), *LEC2*, *FUSCA 3* (*FUS3*), and *ABSCISIC ACID INSENSITIVE 3* (*ABI3*), are known to govern seed development, nutrient accumulation, and seed maturity (Sreenivasulu and Wobus 2013; Sun et al. 2010). Moreover, *MEDEA* (*MEA*) and *FERTILIZATION-INDEPENDENT SEED2* (*FIS2*) exhibit a key role in genomic imprinting during endosperm development (Köhler and Makarevich 2006). PKL is involved in seed development by repressing the expression of *LEC1* and *FUS3*, as altered expression was noted in the *pkl* mutant seeds during imbibition (Ogas et al. 1999). DNA methylation plays a critical role in transcription silencing, transposon suppression, and gene imprinting during embryo and endosperm development in plants (Gehring et al. 2009; Gehring 2013). Interestingly, distinct methylation patterns occur during embryo and endosperm development, likely associated with the differential hypomethylation state of the maternal allele in the endosperm (Ibarra et al. 2012). The *FIS* gene complex encodes polycomb group (PcG) proteins that direct histone methyltransferase activity by recruiting H3K27me3 to target loci (Köhler and Makarevich 2006). Interestingly, a mutation study on *MEA* and *FIS2* genes revealed that these two genes are regulated by genomic imprinting (Luo et al. 2000). For instance, in the female gametophyte, the maternally imprinted *PHERES 1* (*PHE 1*) gene is suppressed by the FIS PcG complex protein, which recruits methylation on the H3K27 residue, while the paternal *PHE 1* allele remains active with the removal of the repressive H3K27me3 mark. Likewise, the paternally imprinted gene *MEA* is methylated by DNA METHYLTRANSFERASE 1 (MET1) and remains suppressed by the FIS complex. The central cell of the female gametophyte contains DNA glycosylase DEMETER (DME) activity, which removes DNA methylation from the maternal *MEA* allele. Thus, the interaction between the FIS complex and DNA methylation plays a significant role in genomic imprinting during endosperm development (Köhler and Makarevich 2006).

2.7 Senescence and Programmed Cell Death

Senescence is the last developmental stage in the plant life cycle. The genes upregulated in this stage are known as senescence-activated genes (SAGs), while the reduction in genes involved especially in photosynthesis and chloroplast developmental pathways is called senescence downregulated genes (SDGs) (Miryeganeh 2022). It is well accepted that aging in plants is associated with DNA demethylation. The methylation-sensitive DNA fragmentation (MS–AFLP) assay in Arabidopsis shoot revealed the reduction in cytosine methylation of DNA in the aging regions, and consequently, the downregulation of *CMT3* and *MET1* methyltransferase genes was observed along with the upregulated expression of demethylase genes, including *ROS1*, *DME*, *DML2*, and *DML3* (Ogneva et al. 2016). Ay et al. (2009) reported that overexpression of SUPPRESSION (VAR) 3–9 homolog2 (SUVH2) histone methyltransferase results in RdDM-mediated repression of the senescence process by repressing the main TF *WRKY53*, which promotes leaf senescence. The increased level of H3K4me2/3 at the coding region of *WRKY53* and several SAGs was found to be associated with developmental leaf senescence (Ay et al. 2009). Moreover, RELATIVE OF EARLY FLOWERING 6 (REF6) was involved in the leaf senescence process by recruiting H3K27me3 demethylation at the promoter and coding region of senescence regulators, including *ETHYLENE INSENSITIVE 2* (*EIN2*), *ORESARA 1* (*ORE1*), and *NAP* (NAC-like, activated by AP3/P1), which promoted chloroplast degradation by activating these genes (Lu et al. 2011). In addition, REF6 also assists in the recruitment of BRM-containing SWI/SNF complexes that regulate cellular senescence by modulating the chromatin structure of the target genes (Ostrowska-Mazurek et al. 2020).

3 Functional Interrelation Between Epigenetic Modifications and Transcriptional Alterations upon Different Environmental Cues

Plants are sessile organisms and therefore face exposure to many environmental stresses, which can cause alterations in epigenetic modifications and modify the expression of stress-responsive genes to facilitate plant adaptation to stressful conditions (Ma et al. 2024). Environmental stress is generally considered as two types: abiotic stress and biotic stress. Abiotic stresses include extreme temperatures, water deficiency, high salinity, nutrient deficits, and heavy metal toxicity, while biotic stress is caused by viruses, microbes, insect pests, diseases, and competition with other plants cause biotic stress. Plants develop their tolerance to these harsh environmental conditions through alterations in epigenetic modifications (Ashapkin et al. 2020).

3.1 Abiotic Stress

3.1.1 Extreme Temperatures

Temperature is a crucial environmental signal that influences plant growth and development. Plants respond differently to temperature variations, which can be categorized into two main types: sub-optimal conditions and excessive temperatures (Sung et al. 2003; Penfield 2008). Recent studies showed that epigenetic modifications such as dynamic regulation of histone modifications (Pajoro et al. 2017; Tasset et al. 2018; Shen et al. 2019; Cui et al. 2021), chromatin remodeling (Zha et al. 2017), and histone variant exchanges (Xue et al. 2021; Zhao et al. 2023) are involved in thermomorphogenesis.

3.1.2 Cold Stress

One of the most well-characterized mechanisms responsible for the cold stress response in plants is the C REPEAT BINDING FACTOR (CBF)-COLD RESPONSIVE (COR) pathway (Chang et al. 2020). *INDUCER OF CBF EXPRESSION* (*ICE1*) and *ICE2* are essential for the induction of CBFs (Fig. 1a). In Arabidopsis, freezing tolerance and the expression level of genes involved in the CBF–COR pathway are inversely correlated with DNA methylation levels at the *ICE1* locus (Xie et al. 2018). In addition, cold tolerance can increase either through the suppression of DNA methylation or loss-of-function mutation in DNA methyltransferase (*DRM2*), indicating that DNA demethylation is necessary to induce cold tolerance in Arabidopsis (Xie et al. 2018). Similarly, Kwon et al. (2009) demonstrated that cold treatment mediated the gradual removal of the repressive H3K27me3 mark at the *COR* genes *COR15A* and *GALACTINOL SYNTHASE3* (*ATGOLS3*). Notably, low temperature triggers degradation of the histone deacetylase HD2C, which binds to the promoter of *COR* (COLD-REGULATED) genes. This enhances H3 acetylation and activates *COR* gene expression (Park et al. 2018). Moreover, PKL, a subunit of the Mi-2/CHD3 subfamily of ATP-dependent chromatin remodelers, likely modulates the functional activity of CBF3 to alter cold adaptation (Yang et al. 2019). This PKL is involved in RdDM and H3K27me3 deposition, signifying that H3K27me3 and DNA methylation may act as memory tags for cold-induced freeze tolerance (Yang et al. 2017).

3.1.3 Heat Stress

Cellular homeostasis can be strongly affected by extremely high temperatures, which can severely hinder plant growth and development or even cause plant death (Kotak et al. 2007). Under heat stress, the compacted heterochromatin is decondensed in *Arabidopsis* (Pecinka et al. 2010; Tittel-Elmer et al. 2010). In tomatoes, *HSFA1a*, a heat stress transcription factor, responds to heat stress by transiently forming enhancer-promoter chromatin loops, which increases the expression of heat stress-responsive genes (Fig. 1b) (Huang et al. 2023). This heat stress transcription factor is regulated by phosphorylation, dephosphorylation, SUMOylation, and protein-protein interactions (Ohama et al. 2017). Although both heat stress-responsive transcription factors *HSFA1* and *HSFA2* are necessary for ONSEN activation, ONSEN overactivation and transposition are also inhibited by the RdDM pathway (Cavrak et al. 2014; Niu et al. 2022). According to Popova et al. (2013), the RdDM pathway and the Rpd3-type histone deacetylase HDA6 play diverse, important functions in basal heat tolerance. Moreover, histone deacetylation is also responsible for regulating heat stress-responsive gene expression. Heat-responsive genes such as *HSFA3* and *HSP101* become induced with increased H4K16 acetylation and inactivation of histone deacetylase HD2C. Thus, HD2C may be detrimental by preventing uncontrolled heat-induced gene activation in plants (Buszewicz et al. 2016).

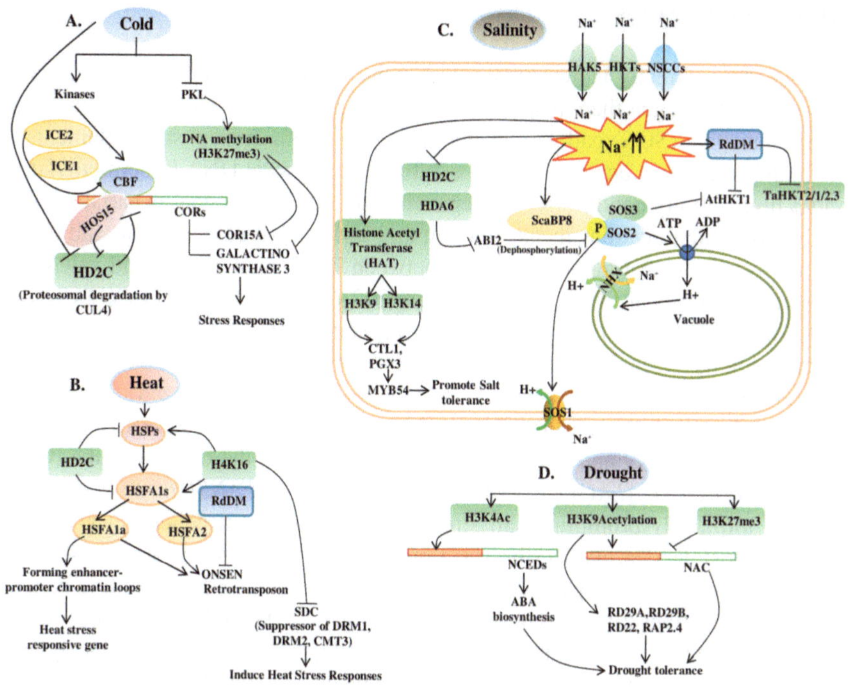

Fig. 1 Integration of epigenetic mechanisms in transcriptional regulation of abiotic stress response. (**a**) In presence of cold stress, HIGH EXPRESSION OF OSMOTICALLY RESPONSIVE GENE 15 (HOS15) recruits CULLIN 4 (CUL4) for ubiquitination and subsequent degradation of HD2C, leading to increased H3 acetylation on COR promoters. However, HOS15 recruits C REPEAT BINDING FACTOR (CBF) transcription factors (TFs) to the COR promoters to induce the expression of COR genes. Additionally, the chromatin state of COR genes and other cold-responsive genes may be regulated by the chromatin regulator PICKLE (PKL) through H3K27me3-dependent silencing. (**b**) Heat stress-induced proteins are known as heat shock proteins (HSPs). HSFA2 is one of the direct targets of HSFA1s (HEAT SHOCK TRANSCRIPTION FACTORs), which are key transcription factors in heat stress. In response to heat stress, H3K9Ac and H3K4me3 are elevated in certain HSP genes. Moreover, SUPPRESSOR OF DRM1 DRM2 CMT3 (SDC) is also found to be regulated by the RNA-directed DNA methylation (RdDM) pathway. The transcription of the ONSEN retrotransposon is also activated by heat stress, and the RdDM pathway is necessary to prevent heat stress-induced ONSEN transgenerational retrotransposition. (**c**) Under salt stress, the salt overly sensitive (SOS) pathway is activated in the presence of high concentrations of Na^+ ion, resulting in the phosphorylation of SCaBP8 by SOS2. Salt stress causes RNA-directed DNA methylation (RdDM) in wheat, which suppresses TaHKT2.1/2.3 expression in a tissue- and genotype-specific manner. Salt stress inhibits HISTONE DEACETYLASE 2C (HD2C), which interacts with HDA6. HDA6 deacetylates histones, suppressing the expression of ABA-responsive genes *ABI2* during normal circumstances. *ABI2* may inhibit SOS2 or dephosphorylate the downstream target ion transporters of SOS2. Salt stress also enhances H3K9 and H3K14 acetylation that activates the chitinase-like (CTL) gene *CTL1*, *POLYGALACTURONASE INVOLVED IN EXPANSION-3* (*PGX3*) and *MYB54*, boosting salt stress tolerance. (**d**) Drought stress stimulates H3K4me3 and H3K9Ac levels in the promoter regions of ABA biosynthesis genes including *NINE CIS-EPOXYCAROTENOID DIOXYGENASE 3* (*NCED3*) and drought-responsive genes *RD29A*, *RD29B*, *RD22*, and *RAP2.4*. However, H3K27me3 deposition on AtNAC transcription factors (TFs) represses the expression of drought tolerance genes, whereas the histone acetylation promotes H3K9Ac deposition on the NAC genes to enhance the expression of drought-resistant genes

3.1.4 Salinity Stress

Salinity stress directly induces two major types of stress: osmotic stress and ionic stress, which trigger their respective signaling pathways, such as ABA and salt overly sensitive (SOS) pathway (Yang and Guo 2018). Transposable Elements in the mangrove *Bruguiera gymnorhiza* exhibit genome-wide DNA hypermethylation under high-salinity conditions (Miryeganeh et al. 2022). Salt causes variable DNA methylation modifications in salt-tolerant and salt-sensitive rice, wheat, and chickpea cultivars, suggesting a connection between DNA methylation and salt stress resistance (Wang et al. 2015; Gupta and Garg 2023). *HIGH-AFFINITY POTASSIUM TRANSPORTER* (*HKT*) genes are crucial regulators of salt resistance, and DNA methylation has been shown to regulate their expression (Kumar et al. 2017; Wang et al. 2020). In Arabidopsis, salt stress promotes cell wall integrity and salt tolerance by inducing GENERAL CONTROL NON-REPRESSED 5 PROTEIN (GCN5), which deposits H3K9 and H3K14 acetylation to activate the Chitinase-like (CTL) gene *CTL1*, *POLYGALACTURONASE INVOLVED IN EXPANSION-3* (*PGX3*), and *MYB54*, ultimately promoting salt stress tolerance (Fig. 1c) (Zheng et al. 2019). Similarly, histone deacetylases have various functions in plant salt stress tolerance. The class I histone deacetylase HDA19 reduces salt tolerance by decreasing histone acetylation and transcription at the ABA receptor genes *PYRABACTIN RESISTANCE 1-LIKE 6* (*PYL6*), *PYL5*, and *PYL4*. Meanwhile, class II histone deacetylases HDA18/15/14/5 positively regulate salt resistance in Arabidopsis (Mehdi et al. 2015; Ueda et al. 2017). Additionally, plant salt stress tolerance is regulated by dynamic RNA modifications like m^6A methylation, which stabilizes the mRNA of salt stress-responsive genes, possibly involving the m^6A reader EVOLUTIONARILY CONSERVED C-TERMINAL REGION 12 (ECT12); notably, m6A methyltransferase FIONA1 has been shown to enhance salt resistance in *Arabidopsis* (Anderson et al. 2018; Amara et al. 2024; Cai et al. 2024).

3.1.5 Drought Stress

One of the severe environmental pressures affecting plant development and productivity is drought. Many studies have shown drought-induced changes in DNA methylation dynamics in multiple crop species, including rice, wheat, maize, barley, mungbean, and cotton (Chwialkowska et al. 2016; Lu et al. 2017; Rajkumar et al. 2020; Naderi et al. 2024). Similarly, in maize roots, overall DNA methylation levels rise under drought stress, while the methylome of drought-tolerant inbred lines exhibits significantly greater stability compared to that of drought-sensitive inbred lines (Wang et al. 2021). HDA6 and HDA9 negatively modulate the drought stress response in *Arabidopsis* by inhibiting H3K9 acetylation and the transcription of several genes responsive to water shortage stress (Fig. 1d) (Zheng et al. 2016). Transcriptome sequencing revealed a strong association between gene expression and methylation levels, with excessive methylation leading to gene silencing. In drought-stressed plants, about 7400 genes show increased methylation levels, and

10,300 genes exhibit decreased methylation levels compared to control plants. During drought treatment, 1156 genes encoding transcription factors (TFs), including those in the MYB, AP2, WRKY, NAC, and bHLH families, showed decreased DNA methylation and increased gene expression (Ashapkin et al. 2020). Moreover, at the RNA modification level, the Arabidopsis m6A reader protein ECT8 is found to inhibit drought tolerance by sequestering the m6A-modified transcripts of the ABA receptor gene *PYL7* in stress granules, which reduces *PYL7* mRNA availability for translation in the cytoplasm (Wu et al. 2024). Interestingly, the absence of putative RNA demethylases in tomato or Arabidopsis has been shown to modulate drought tolerance (Huong et al. 2020; Shen et al. 2024).

3.2 Biotic Stress

3.2.1 Viruses

In contrast to animals, plants lack adaptive immune systems. Instead, they have developed a complex RNA-based system for regulating genes and defending against foreign nucleic acids. Most of the plant viruses have a single-stranded RNA (ssRNA) genome. The plant immunological response elicited by RNA viruses often leads to degradation of their genomes into siRNAs (Ashapkin et al. 2020). Recent findings indicate that m^6A-specific methylation of the RNA genome in the alfalfa mosaic virus (AMV) regulates infection in *Arabidopsis* (Martínez-Pérez et al. 2017). The *Arabidopsis* protein ALKBH9B (At2g17970) exhibits demethylase activity that eliminates m6A from single-stranded RNA molecules of alfalfa mosaic virus (AMV) in vitro and localizes to cytoplasmic siRNA bodies. This indicates that ALKBH9B functions as an m6A demethylase implicated in mRNA silencing and degradation processes.

3.2.2 Bacteria and Fungi

Infection by the bacterial pathogen *Pseudomonas syringae pv. tomato (Pst)* in *Arabidopsis* elicits a fundamental defense response that is inhibited by bacterial virulence factors (Dowen et al. 2012). However, DNA methylation of pathogen-responsive genes may serve as a crucial strategy for plant defense (Fig. 2). For instance, CG- and CHG-specific methylation is evidenced to be changed in plants subjected to salicylic acid or avirulent or virulent *Pseudomonas syringae* strains (Ashapkin et al. 2020). Flagellin Sensing 2 (FLS2) is an extensively studied plant pattern recognition receptor (PRR) that detects the bacterial flagellin-derived peptide flg22 and modulates the expression of multiple genes, including the DNA demethylase gene *REPRESSOR OF SILENCING 1* (*ROS1*). The functional mutation in *ROS1* causes moderate bacterial proliferation in response to the *Pst* strain DC3000, corroborating the function of active DNA demethylation in antibacterial

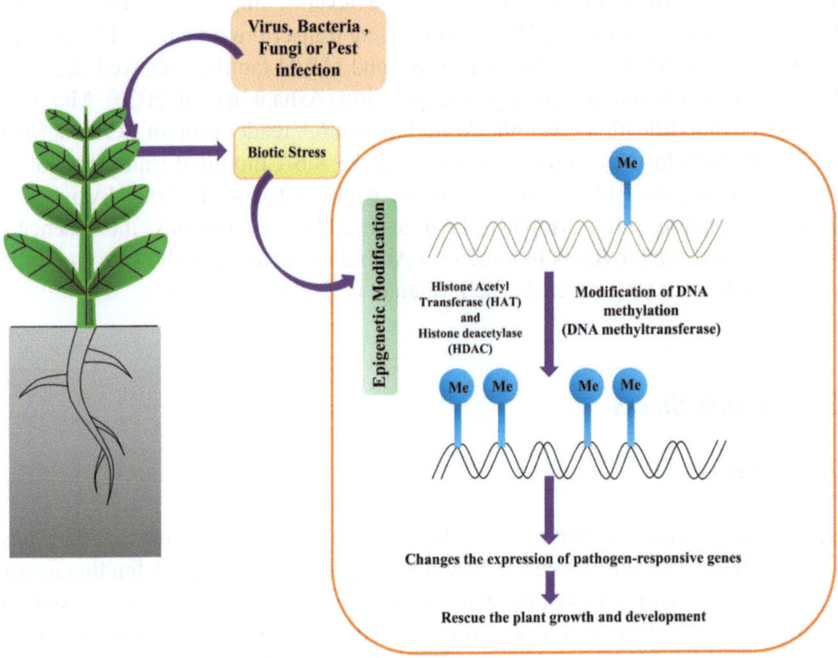

Fig. 2 Epigenetic changes shaping transcriptional responses under biotic stress. Biotic stress signals initiate signaling cascades that regulate the activity of DNA methyltransferase, histone acetyltransferases (HATs), and histone deacetylases (HDACs). Acetyl groups are generally attached to histone lysine residues by histone acetyltransferases (HATs) to activate target genes, while histone deacetylases (HDACs) inhibit genes by eliminating the acetyl groups (not shown), and DNA-methyltransferase attaches a methyl group at multiple CG and CHG sites and regulates the pathogen-responsive gene expression

resistance. Moreover, *ros1* mutant plants show reduced SA-dependent defensive response, as shown by flg22-induced lower activation of *PATHOGENESIS-RELATED GENE 1* (*PR1*). This indicates that ROS1-mediated DNA demethylation is imperative for the flg22-triggered defense response in plants (Yu et al. 2013; Ashapkin et al. 2020).

In *Arabidopsis*, DNA methylation mutants infected with the obligatory biotrophic downy mildew oomycete *Hyaloperonospora arabidopsidis* (Hpa) were used to investigate the potential role of DNA methylation in resistance against biotrophic infections (Sánchez et al. 2016). The selected *Hpa*-resistant lines exhibited no growth defects and showed higher PR1 expression in the presence of Hpa infection, compared with wild-type plants, indicating a priming of SA-inducible responses (Huang and Jin 2022). Furthermore, reduced resistance to the necrotrophic fungus *Plectosphaerella cucumerina* was detected in the mutants of the RdDM pathway, which were linked to suppressed responses of JA-inducible defense genes (Sánchez et al. 2016). Additionally, β-aminobutyric acid (BABA) treatment primes potato defense and enhances resistance against the oomycete *Phytophthora infestans* through the epigenetic regulation of H3K4me2 and H3K27me3 dynamics. H3K4me2 was transiently enhanced in *SNI1* (*Suppressor of NPR1 Inducible 1*) and *NPR1* (*Non-expressor of PR genes*), resulting in the induction of SA-responsive genes and the upregulation of the defense genes *WRKY1*, *PR1*, and *PR2* in plants (Meller et al. 2018).

3.2.3 Pests

Plant pests and diseases can diminish the quality and productivity of crops and timber. Epigenetic regulation may play a crucial role in controlling pest or pathogen infection. For instance, the soybean cyst nematode (SCN; *Heterodera glycines*) invades the soybean roots to elicit the development of a multinucleated feeding structure, the syncytium. Surprisingly, both the susceptible (S) and resistant (R) varieties of SCN share a common genetic background, yet they differ in their DNA methylomes. Furthermore, SCN infection has been shown to promote differential methylation, suggesting its involvement in regulating diverse gene expression patterns. Similar epigenetic mechanisms have been reported to contribute to long-term defense priming and immune memory in plant-nematode interactions, and environmental cues can modulate these epigenetic marks, affecting stress memory and adaptive responses that may even be transmitted to subsequent generations (Banihashemian and Mirmajlessi 2025). A detailed study on SCN-infected plants showed the association of 6903 differentially expressed genes, among which the involvement of hypoDMR (Differentially Methylated Region) genes was predominant over hyperDMR (Ashapkin et al. 2020). Moreover, lower DNA methylation levels in S plants and higher levels in R plants clearly indicate the significant role of DNA methylation during the host-pathogen interaction in soybean (Rambani et al. 2020; Ashapkin et al. 2020).

4 Conclusion

The interrelationship between transcriptome profiling and epigenetic modifications in plants is a complex and dynamic process that directs plant developmental transitions, adaptation, and responses to environmental signals. Epigenetic modifications, including histone modifications, DNA methylation, and non-coding RNAs, act as regulatory mechanisms that govern transcriptional changes by influencing chromatin structure, enabling plants to adapt to changing environments and optimize growth and phenotypic flexibility.

Environmental cues, such as biotic and abiotic constraints, induce epigenetic modifiers that can influence both stable and transient changes in transcriptional patterns. The resulting epigenetic memory allows the plant to adjust to the changing environment and, in some cases, pass advantageous alterations from one generation to the next. The synergistic relationship between transcriptional changes and epigenetic modifications strengthens the remarkable capacity of plants to regulate gene expression in response to developmental and environmental stresses.

The advancement in omics technologies has greatly enhanced our knowledge regarding the role of epigenetic modifications in regulating the transcriptional dynamics that have governed intricate networks linking the external environmental cues with the internal genetic process. The role of DNA methylation in crop improvement is a promising mechanism for studying gene function and modifying plants to generate superior varieties capable of surviving under stress conditions. Further exploration of the crosstalk between transcriptome and epigenome in plants will help researchers to develop improved crops with enhanced resilience to biotic and abiotic stresses, increased yields, and greater sustainability.

Acknowledgments The authors gratefully acknowledge Council of Scientific and Industrial Research, Govt. of India, (Ref. No. 38(1587)/16/EMR-II, dated: 17/05/2016; Ref. No. 38(1526)/23/EMR-II, dated 05/07/2023 to SR), UGC, Govt. of India Start-Up research grant No.F.30-158/2015 (BSR) and SERB, DST, Govt of India (Ref. No. ECR/2016/000539; Ref No. CRG/2021/000989-G to SR) for providing financial support related to the topic discussed in this review. *NC* is thankful to DST, WOS-A, Govt. of India (DST/WOS-A/LS-207/2021) for a post-doctoral fellowship. *AM* is thankful to UGC (Ref. No. 221610047029), Govt. of India, for the research fellowship. *SN* (DST/INSPIRE Fellowship/2021/IF200219) is thankful for the DST-INSPIRE fellowship. *DO* is thankful to CSIR (38(1526)/23/EMR-II), Govt. of India, for the research fellowship. *SKS* is thankful to UGC (231620253469), Govt. of India, for the research fellowship. We apologize to those authors whose work could not be cited due to space limitations.

Conflict of Interest The authors have no conflicts of interest to declare.

References

Abdulraheem MI, Xiong Y, Moshood AY, Cadenas-Pliego G, Zhang H, Hu J (2024) Mechanisms of plant epigenetic regulation in response to plant stress: recent discoveries and implications. Plants 13(2):163. https://doi.org/10.3390/plants13020163

Agarwal P, Arora R, Ray S, Singh AK, Singh VP, Takatsuji H, Kapoor S, Tyagi AK (2007) Genome-wide identification of C2H2 zinc-finger gene family in rice and their phylogeny and expression analysis. Plant Mol Biol 65(4):467–485. https://doi.org/10.1007/s11103-007-9199-y

Agarwal P, Parida SK, Mahto A, Das S, Mathew IE, Malik N, Tyagi AK (2014) Expanding frontiers in plant transcriptomics in aid of functional genomics and molecular breeding. Biotechnol J 9(12):1480–1492. https://doi.org/10.1002/biot.201400063

Agarwal G, Kudapa H, Ramalingam A, Choudhary D, Sinha P, Garg V, Singh VK, Patil GB, Pandey MK, Nguyen HT, Guo B, Sunkar R, Niederhuth CE, Varshney RK (2020) Epigenetics and epigenomics: underlying mechanisms, relevance, and implications in crop improvement. Funct Integr Genomics 20(6):739–761. https://doi.org/10.1007/s10142-020-00756-7

Aichinger E, Villar CBR, Di Mambro R, Sabatini S, Köhler C (2011) The CHD3 chromatin remodeler PICKLE and polycomb group proteins antagonistically regulate meristem activity in the Arabidopsis root. Plant Cell 23(3):1047–1060. https://doi.org/10.1105/tpc.111.083352

Aida M, Beis D, Heidstra R, Willemsen V, Blilou I, Galinha C, Nussaume L, Noh YS, Amasino R, Scheres B (2004) The PLETHORA genes mediate patterning of the Arabidopsis root stem cell niche. Cell 119(1):109–120. https://doi.org/10.1016/j.cell.2004.09.018

Amara U, Hu J, Park SJ, Kang H (2024) ECT12, an YTH-domain protein, is a potential mRNA m6A reader that affects abiotic stress responses by modulating mRNA stability in Arabidopsis. Plant Physiol Biochem 206:108255. https://doi.org/10.1016/j.plaphy.2023.108255

Anderson SJ, Kramer MC, Gosai SJ, Yu X, Vandivier LE, Nelson ADL, Anderson ZD, Beilstein MA, Fray RG, Lyons E, Gregory BD (2018) N6-methyladenosine inhibits local ribonucleolytic cleavage to stabilize mRNAs in Arabidopsis. Cell Rep 25(5):1146–1157.e3. https://doi.org/10.1016/j.celrep.2018.10.020

Ashapkin VV, Kutueva LI, Aleksandrushkina NI, Vanyushin BF (2020) Epigenetic mechanisms of plant adaptation to biotic and abiotic stresses. Int J Mol Sci 21(20):7457. https://doi.org/10.3390/ijms21207457

Ay N, Irmler K, Fischer A, Uhlemann R, Reuter G, Humbeck K (2009) Epigenetic programming via histone methylation at WRKY53 controls leaf senescence in Arabidopsis thaliana. Plant J 58(2):333–346. https://doi.org/10.1111/j.0960-7412.2009.03782.x

Banihashemian SN, Mirmajlessi SM (2025) Epigenetic modifications, immune control processes, and plant responses to nematodes. Agriculture 15(7):742. https://doi.org/10.3390/agriculture15070742

Batlla D, Benech-Arnold RL (2015) A framework for the interpretation of temperature effects on dormancy and germination in seed populations showing dormancy. Seed Sci Res 25(2):147–158. https://doi.org/10.1017/s0960258514000452

Begum Y (2022) Regulatory role of microRNAs (miRNAs) in the recent development of abiotic stress tolerance of plants. Gene 821:146283. https://doi.org/10.1016/j.gene.2022.146283

Bowman JL, Smyth DR, Meyerowitz EM (1989) Genes directing flower development in Arabidopsis. Plant Cell 1(1):37–52. https://doi.org/10.1105/tpc.1.1.37

Bratzel F, López-Torrejón G, Koch M, Del Pozo JC, Calonje M (2010) Keeping cell identity in Arabidopsis requires PRC1 RING-finger homologs that catalyze H2A monoubiquitination. Curr Biol 20(20):1853–1859. https://doi.org/10.1016/j.cub.2010.09.046

Burkart RC, Strotmann VI, Kirschner GK, Akinci A, Czempik L, Dolata A, Maizel A, Weidtkamp-Peters S, Stahl Y (2022) PLETHORA-WOX5 interaction and subnuclear localization control Arabidopsis root stem cell maintenance. EMBO Rep 23(6):e54105. https://doi.org/10.15252/embr.202154105

Buszewicz D, Archacki R, Palusiński A, Kotliński M, Fogtman A, Iwanicka-Nowicka R, Sosnowska K, Kuciński J, Pupel P, Olędzki J, Dadlez M, Misicka A, Jerzmanowski A, Koblowska MK (2016) HD2C histone deacetylase and a SWI/SNF chromatin remodelling complex interact and both are involved in mediating the heat stress response in Arabidopsis. Plant Cell Environ 39(10):2108–2122. https://doi.org/10.1111/pce.12756

Cai J, Hu J, Xu T, Kang H (2024) FIONA1-mediated mRNA m6A methylation regulates the response of Arabidopsis to salt stress. Plant Cell Environ 47(3):900–912. https://doi.org/10.1111/pce.14807

Cavrak VV, Lettner N, Jamge S, Kosarewicz A, Bayer LM, Scheid OM (2014) How a retrotransposon exploits the plant's heat stress response for its activation. PLoS Genet 10(1):e1004115. https://doi.org/10.1371/journal.pgen.1004115

Chang Y, Zhu C, Jiang J, Zhang H, Zhu J, Duan C (2020) Epigenetic regulation in plant abiotic stress responses. J Integr Plant Biol 62(5):563–580. https://doi.org/10.1111/jipb.12901

Chen N, Wang H, Abdelmageed H, Veerappan V, Tadege M, Allen RD (2020) HSI2/VAL1 and HSL1/VAL2 function redundantly to repress DOG1 expression in Arabidopsis seeds and seedlings. New Phytol 227(3):840–856. https://doi.org/10.1111/nph.16559

Cho JN, Ryu JY, Jeong YM, Park J, Song JJ, Amasino RM, Noh B, Noh YS (2012) Control of seed germination by light-induced histone arginine demethylation activity. Dev Cell 22(4):736–748. https://doi.org/10.1016/j.devcel.2012.01.024

Chwialkowska K, Nowakowska U, Mroziewicz A, Szarejko I, Kwasniewski M (2016) Water-deficiency conditions differently modulate the methylome of roots and leaves in barley (Hordeum vulgare L.). J Exp Bot 67(4):1109–1121. https://doi.org/10.1093/jxb/erv552

Clark SE, Running MP, Meyerowitz EM (1995) CLAVATA3 is a specific regulator of shoot and floral meristem development affecting the same processes as CLAVATA1. Development 121(7):2057–2067. https://doi.org/10.1242/dev.121.7.2057

Csorba T, Questa JI, Sun Q, Dean C (2014) Antisense COOLAIR mediates the coordinated switching of chromatin states at FLC during vernalization. Proc Natl Acad Sci 111(45):16160–16165. https://doi.org/10.1073/pnas.1419030111

Cui X, Zheng Y, Lu Y, Issakidis-Bourguet E, Zhou DX (2021) Metabolic control of histone demethylase activity involved in plant response to high temperature. Plant Physiol 185(4):1813–1828. https://doi.org/10.1093/plphys/kiab020

De Lucia F, Crevillen P, Jones AME, Greb T, Dean C (2008) A PHD-polycomb repressive complex 2 triggers the epigenetic silencing of FLC during vernalization. Proc Natl Acad Sci 105(44):16831–16836. https://doi.org/10.1073/pnas.0808687105

Ding X, Jia X, Xiang Y, Jiang W (2022) Histone modification and chromatin remodeling during the seed life cycle. Front Plant Sci 13:865361. https://doi.org/10.3389/fpls.2022.865361

Dolan L, Janmaat K, Willemsen V, Linstead P, Poethig S, Roberts K, Scheres B (1993) Cellular organisation of the Arabidopsis thaliana root. Development 119(1):71–84. https://doi.org/10.1242/dev.119.1.71

Dowen RH, Pelizzola M, Schmitz RJ, Lister R, Dowen JM, Nery JR, Dixon JE, Ecker JR (2012) Widespread dynamic DNA methylation in response to biotic stress. Proc Natl Acad Sci 109(32):E2183–E2191. https://doi.org/10.1073/pnas.1209329109

Duarte-Aké F, Us-Camas R, De-La-Peña C (2023) Epigenetic regulation in heterosis and environmental stress: the challenge of producing hybrid epigenomes to face climate change. Epigenomes 7(3):14. https://doi.org/10.3390/epigenomes7030014

Farrona S, Hurtado L, Bowman JL, Reyes JC (2004) The Arabidopsis thaliana SNF2 homolog AtBRM controls shoot development and flowering. Development 131(20):4965–4975. https://doi.org/10.1242/dev.01363

Gao Y, Xu H, Shen Y, Wang J (2013) Transcriptomic analysis of rice (Oryza sativa) endosperm using the RNA-Seq technique. Plant Mol Biol 81(4–5):363–378. https://doi.org/10.1007/s11103-013-0009-4

Gehring M (2013) Genomic imprinting: insights from plants. Annu Rev Genet 47(1):187–208. https://doi.org/10.1146/annurev-genet-110711-155527

Gehring M, Bubb KL, Henikoff S (2009) Extensive demethylation of repetitive elements during seed development underlies gene imprinting. Science 324(5933):1447–1451. https://doi.org/10.1126/science.1171609

Gu D, Chen C-Y, Zhao M, Zhao L, Duan X, Duan J, Wu K, Liu X (2017) Identification of HDA15-PIF1 as a key repression module directing the transcriptional network of seed germination in the dark. Nucleic Acids Res 45(12):7137–7150. https://doi.org/10.1093/nar/gkx283

Gu D, Ji R, He C, Peng T, Zhang M, Duan J, Xiong C, Liu X (2019) Arabidopsis histone methyltransferase SUVH5 is a positive regulator of light-mediated seed germination. Front Plant Sci 10:841. https://doi.org/10.3389/fpls.2019.00841

Guo M, Thomas J, Collins G, Timmermans MCP (2008) Direct repression of KNOXLoci by the ASYMMETRIC LEAVES1 complex of Arabidopsis. Plant Cell 20(1):48–58. https://doi.org/10.1105/tpc.107.056127

Guo S, Sun B, Looi L-S, Xu Y, Gan E-S, Huang J, Ito T (2015) Co-ordination of flower development through epigenetic regulation in two model species: rice and Arabidopsis. Plant Cell Physiol 56(5):830–842. https://doi.org/10.1093/pcp/pcv037

Gupta K, Garg R (2023) Unravelling differential DNA methylation patterns in genotype dependent manner under salinity stress response in chickpea. Int J Mol Sci 24(3):1863. https://doi.org/10.3390/ijms24031863

Han SK, Sang Y, Rodrigues A, Wu MF, Rodriguez PL, Wagner D (2012) The SWI2/SNF2 chromatin remodeling ATPase BRAHMA represses abscisic acid responses in the absence of the stress stimulus in Arabidopsis. Plant Cell 24(12):4892–4906. https://doi.org/10.1105/tpc.112.105114

Han Z, Yu H, Zhao Z, Hunter D, Luo X, Duan J, Tian L (2016) AtHD2D gene plays a role in plant growth, development, and response to abiotic stresses in Arabidopsis thaliana. Front Plant Sci 7:310. https://doi.org/10.3389/fpls.2016.00310

Han Y, Georgii E, Priego-Cubero S, Wurm CJ, Hüther P, Huber G, Koller R, Becker C, Durner J, Lindermayr C (2023) Arabidopsis histone deacetylase HD2A and HD2B regulate seed dormancy by repressing DELAY OF GERMINATION 1. Front Plant Sci 14:1124899. https://doi.org/10.3389/fpls.2023.1124899

Huang CY, Jin H (2022) Coordinated epigenetic regulation in plants: a potent managerial tool to conquer biotic stress. Front Plant Sci 12:795274. https://doi.org/10.3389/fpls.2021.795274

Huang Y, An J, Sircar S, Bergis C, Lopes CD, He X, Da Costa B, Tan FQ, Bazin J, Antunez-Sanchez J, Mammarella MF, Devani R-S, Brik-Chaouche R, Bendahmane A, Frugier F, Xia C, Rothan C, Probst AV, Mohamed Z, Bergounioux C, Delarue M, Zhang Y, Zheng S, Crespi M, Fragkostefanakis S, Mahfouz MM, Ariel F, Gutierrez-Marcos J, Raynaud C, Latrasse D, Benhamed M (2023) HSFA1a modulates plant heat stress responses and alters the 3D chromatin organization of enhancer-promoter interactions. Nat Commun 14(1):469. https://doi.org/10.1038/s41467-023-36227-3

Hung FY, Feng YR, Hsin KT, Shih YH, Chang CH, Zhong W, Lai YC, Xu Y, Yang S, Sugimoto K, Cheng YS, Wu K (2023) Arabidopsis histone H3 lysine 9 methyltransferases KYP/SUVH5/6 are involved in leaf development by interacting with AS1-AS2 to repress KNAT1 and KNAT2. Commun Biol 6(1):219. https://doi.org/10.1038/s42003-023-04607-6

Huong TT, Ngoc LNT, Kang H (2020) Functional characterization of a putative RNA demethylase ALKBH6 in Arabidopsis growth and abiotic stress responses. Int J Mol Sci 21(18):6707. https://doi.org/10.3390/ijms21186707

Ibarra CA, Feng X, Schoft VK, Hsieh T-F, Uzawa R, Rodrigues JA, Zemach A, Chumak N, Machlicova A, Nishimura T, Rojas D, Fischer RL, Tamaru H, Zilberman D (2012) Active DNA demethylation in plant companion cells reinforces transposon methylation in gametes. Science 337(6100):1360–1364. https://doi.org/10.1126/science.1224839

Islam MT, Wang L, Chen I, Lo K, Lo W (2021) Arabidopsis JMJ17 promotes cotyledon greening during de-etiolation by repressing genes involved in tetrapyrrole biosynthesis in etiolated seedlings. New Phytol 231(3):1023–1039. https://doi.org/10.1111/nph.17327

Jackson D, Veit B, Hake S (1994) Expression of maize KNOTTED1 related homeobox genes in the shoot apical meristem predicts patterns of morphogenesis in the vegetative shoot. Development 120(2):405–413. https://doi.org/10.1242/dev.120.2.405

Jagadhesan B, Das S, Singh D, Jha SK, Durgesh K, Sathee L (2022) Micro RNA mediated regulation of nutrient response in plants: the case of nitrogen. Plant Physiol Rep 27(3):345–357. https://doi.org/10.1007/s40502-022-00653-0

Jégu T, Domenichini S, Blein T, Ariel F, Christ A, Kim SK, Crespi M, Boutet-Mercey S, Mouille G, Bourge M, Hirt H, Bergounioux C, Raynaud C, Benhamed M (2015) A SWI/SNF chromatin remodelling protein controls cytokinin production through the regulation of chromatin architecture. PLoS One 10(10):e0138276. https://doi.org/10.1371/journal.pone.0138276

Jogam P, Sandhya D, Alok A, Peddaboina V, Allini VR, Zhang B (2022) A review on CRISPR/Cas-based epigenetic regulation in plants. Int J Biol Macromol 219:1261–1271. https://doi.org/10.1016/j.ijbiomac.2022.08.182

Kawakatsu T, Nery JR, Castanon R, Ecker JR (2017) Dynamic DNA methylation reconfiguration during seed development and germination. Genome Biol 18(1):171. https://doi.org/10.1186/s13059-017-1251-x

Kim SY, He Y, Jacob Y, Noh Y-S, Michaels S, Amasino R (2005) Establishment of the vernalization-responsive, winter-annual habit in Arabidopsis requires a putative histone H3 methyl transferase [W]. Plant Cell 17(12):3301–3310. https://doi.org/10.1105/tpc.105.034645

Köhler C, Makarevich G (2006) Epigenetic mechanisms governing seed development in plants. EMBO Rep 7(12):1223–1227. https://doi.org/10.1038/sj.embor.7400854

Kotak S, Larkindale J, Lee U, Von Koskull-Döring P, Vierling E, Scharf KD (2007) Complexity of the heat stress response in plants. Curr Opin Plant Biol 10(3):310–316. https://doi.org/10.1016/j.pbi.2007.04.011

Kumar S, Beena AS, Awana M, Singh A (2017) Salt-induced tissue-specific cytosine methylation downregulates expression of HKT genes in contrasting wheat (Triticum aestivum L.) genotypes. DNA Cell Biol 36(4):283–294. https://doi.org/10.1089/dna.2016.3505

Kwon CS, Lee D, Choi G, Chung W (2009) Histone occupancy-dependent and -independent removal of H3K27 trimethylation at cold-responsive genes in Arabidopsis. Plant J 60(1):112–121. https://doi.org/10.1111/j.1365-313x.2009.03938.x

Lämke J, Bäurle I (2017) Epigenetic and chromatin-based mechanisms in environmental stress adaptation and stress memory in plants. Genome Biol 18(1):124. https://doi.org/10.1186/s13059-017-1263-6

Lauria M, Rossi V (2011) Epigenetic control of gene regulation in plants. Biochim Biophys Acta Gene Regul Mech 1809(8):369–378. https://doi.org/10.1016/j.bbagrm.2011.03.002

Lee K, Won JH, Seo PJ (2022) Overexpression of the *WOX5* gene inhibits shoot development. Plant Signal Behav 17(1):2050095. https://doi.org/10.1080/15592324.2022.2050095

Li B, Carey M, Workman JL (2007) The role of chromatin during transcription. Cell 128(4):707–719. https://doi.org/10.1016/j.cell.2007.01.015

Li DX, Chen WQ, Xu ZH, Bai SN (2015) HISTONE DEACETYLASE6-defective mutants show increased expression and acetylation of ENHANCER OF TRIPTYCHON AND CAPRICE1 and GLABRA2 with small but significant effects on root epidermis cellular pattern. Plant Physiol 168(4):1448–1458. https://doi.org/10.1104/pp.15.00821

Liu P, Montgomery TA, Fahlgren N, Kasschau KD, Nonogaki H, Carrington JC (2007) Repression of AUXIN RESPONSE FACTOR10 by microRNA160 is critical for seed germination and post-germination stages. Plant J 52(1):133–146. https://doi.org/10.1111/j.1365-313x.2007.03218.x

Lodha M, Marco CF, Timmermans MCP (2013) The ASYMMETRIC LEAVES complex maintains repression of KNOX homeobox genes via direct recruitment of Polycomb-repressive complex 2. Genes Dev 27(6):596–601. https://doi.org/10.1101/gad.211425.112

Lu F, Cui X, Zhang S, Jenuwein T, Cao X (2011) Arabidopsis REF6 is a histone H3 lysine 27 demethylase. Nat Genet 43(7):715–719. https://doi.org/10.1038/ng.854

Lu X, Chen D, Shu D, Zhang Z, Wang W, Klukas C, Chen LL, Fan Y, Chen M, Zhang C (2013) The differential transcription network between embryo and endosperm in the early developing maize seed. Plant Physiol 162(1):440–455. https://doi.org/10.1104/pp.113.214874

Lu X, Wang X, Chen X, Shu N, Wang J, Wang D, Wang S, Fan W, Guo L, Guo X, Ye W (2017) Single-base resolution methylomes of upland cotton (Gossypium hirsutum L.) reveal epigenome modifications in response to drought stress. BMC Genomics 18(1):297. https://doi.org/10.1186/s12864-017-3681-y

Luján-Soto E, Dinkova TD (2021) Time to wake up: epigenetic and small-RNA-mediated regulation during seed germination. Plants 10(2):236. https://doi.org/10.3390/plants10020236

Luo M, Bilodeau P, Dennis ES, Peacock WJ, Chaudhury A (2000) Expression and parent-of-origin effects for FIS2, MEA, and FIE in the endosperm and embryo of developing Arabidopsis seeds. Proc Natl Acad Sci 97(19):10637–10642. https://doi.org/10.1073/pnas.170292997

Ma Y, Miotk A, Šutiković Z, Ermakova O, Wenzl C, Medzihradszky A, Gaillochet C, Forner J, Utan G, Brackmann K, Galván-Ampudia CS, Vernoux T, Greb T, Lohmann JU (2019) WUSCHEL acts as an auxin response rheostat to maintain apical stem cells in Arabidopsis. Nat Commun 10(1):5093. https://doi.org/10.1038/s41467-019-13074-9

Ma L, Xing L, Li Z, Jiang D (2024) Epigenetic control of plant abiotic stress responses. J Genet Genomics 52:129–144. https://doi.org/10.1016/j.jgg.2024.09.008

Martínez-Pérez M, Aparicio F, López-Gresa MP, Bellés JM, Sánchez-Navarro JA, Pallás V (2017) Arabidopsis m6A demethylase activity modulates viral infection of a plant virus and the m6A abundance in its genomic RNAs. Proc Natl Acad Sci 114(40):10755–10760. https://doi.org/10.1073/pnas.1703139114

Mehdi S, Derkacheva M, Ramström M, Kralemann L, Bergquist J, Hennig L (2015) The WD40 domain protein MSI1 functions in a histone deacetylase complex to fine-tune abscisic acid signaling. Plant Cell 28(1):42–54. https://doi.org/10.1105/tpc.15.00763

Meller B, Kuźnicki D, Arasimowicz-Jelonek M, Deckert J, Floryszak-Wieczorek J (2018) BABA-primed histone modifications in potato for intergenerational resistance to Phytophthora infestans. Front Plant Sci 9:1228. https://doi.org/10.3389/fpls.2018.01228

Michaels SD, Amasino RM (1999) FLOWERING LOCUS C encodes a novel MADS domain protein that acts as a repressor of flowering. Plant Cell 11(5):949–956. https://doi.org/10.1105/tpc.11.5.949

Michaels SD, Himelblau E, Kim SY, Schomburg FM, Amasino RM (2005) Integration of flowering signals in winter-annual Arabidopsis. Plant Physiol 137(1):149–156. https://doi.org/10.1104/pp.104.052811

Miryeganeh M (2022) Epigenetic mechanisms of senescence in plants. Cells 11(2):251. https://doi.org/10.3390/cells11020251

Miryeganeh M, Marlétaz F, Gavriouchkina D, Saze H (2022) De novo genome assembly and in natura epigenomics reveal salinity-induced DNA methylation in the mangrove tree Bruguiera gymnorhiza. New Phytol 233(5):2094–2110. https://doi.org/10.1111/nph.17738

Naderi S, Maali-Amiri R, Sadeghi L, Hamidi A (2024) Physio-biochemical and DNA methylation analysis of the defense response network of wheat to drought stress. Plant Physiol Biochem 209:108516. https://doi.org/10.1016/j.plaphy.2024.108516

Nakabayashi K, Bartsch M, Xiang Y, Miatton E, Pellengahr S, Yano R, Seo M, Soppe WJJ (2012) The time required for dormancy release in Arabidopsis is determined by DELAY OF GERMINATION1 protein levels in freshly harvested seeds. Plant Cell 24(7):2826–2838. https://doi.org/10.1105/tpc.112.100214

Napsucialy-Mendivil S, Alvarez-Venegas R, Shishkova S, Dubrovsky JG (2014) ARABIDOPSIS HOMOLOG of TRITHORAX1 (ATX1) is required for cell production, patterning, and morphogenesis in root development. J Exp Bot 65(22):6373–6384. https://doi.org/10.1093/jxb/eru355

Nguyen V, Gutzat R (2022) Epigenetic regulation in the shoot apical meristem. Curr Opin Plant Biol 69:102267. https://doi.org/10.1016/j.pbi.2022.102267

Niu X, Chen L, Kato A, Ito H (2022) Regulatory mechanism of a heat-activated retrotransposon by DDR complex in Arabidopsis thaliana. Front Plant Sci 13:1048957. https://doi.org/10.3389/fpls.2022.1048957

Nonogaki H (2014) Seed dormancy and germination—emerging mechanisms and new hypotheses. Front Plant Sci 5:233. https://doi.org/10.3389/fpls.2014.00233

Nunez-Vazquez R, Desvoyes B, Gutierrez C (2022) Histone variants and modifications during abiotic stress response. Front Plant Sci 13:984702. https://doi.org/10.3389/fpls.2022.984702

Ogas J, Kaufmann S, Henderson J, Somerville C (1999) PICKLE is a CHD3 chromatin-remodeling factor that regulates the transition from embryonic to vegetative development in Arabidopsis. Proc Natl Acad Sci 96(24):13839–13844. https://doi.org/10.1073/pnas.96.24.13839

Ogneva ZV, Dubrovina AS, Kiselev KV (2016) Age-associated alterations in DNA methylation and expression of methyltransferase and demethylase genes in Arabidopsis thaliana. Biol Plant 60(4):628–634. https://doi.org/10.1007/s10535-016-0638-y

Ohama N, Sato H, Shinozaki K, Yamaguchi-Shinozaki K (2017) Transcriptional regulatory network of plant heat stress response. Trends Plant Sci 22(1):53–65. https://doi.org/10.1016/j.tplants.2016.08.015

Ostrowska-Mazurek A, Kasprzak P, Kubala S, Zaborowska M, Sobieszczuk-Nowicka E (2020) Epigenetic landmarks of leaf senescence and crop improvement. Int J Mol Sci 21(14):5125. https://doi.org/10.3390/ijms21145125

Pajoro A, Severing E, Angenent GC, Immink RGH (2017) Histone H3 lysine 36 methylation affects temperature-induced alternative splicing and flowering in plants. Genome Biol 18(1):102. https://doi.org/10.1186/s13059-017-1235-x

Park JJ, Dempewolf E, Zhang W, Wang ZY (2017) RNA-guided transcriptional activation via CRISPR/dCas9 mimics overexpression phenotypes in Arabidopsis. PLoS One 12(6):e0179410. https://doi.org/10.1371/journal.pone.0179410

Park J, Lim CJ, Shen M, Park HJ, Cha JY, Iniesto E, Rubio V, Mengiste T, Zhu J-K, Bressan RA, Lee SY, Lee B-H, Jin JB, Pardo JM, Kim W-Y, Yun D-J (2018) Epigenetic switch from repressive to permissive chromatin in response to cold stress. Proc Natl Acad Sci 115(23):E5400–E5409. https://doi.org/10.1073/pnas.1721241115

Pecinka A, Dinh HQ, Baubec T, Rosa M, Lettner N, Scheid OM (2010) Epigenetic regulation of repetitive elements is attenuated by prolonged heat stress in Arabidopsis. Plant Cell 22(9):3118–3129. https://doi.org/10.1105/tpc.110.078493

Penfield S (2008) Temperature perception and signal transduction in plants. New Phytol 179(3):615–628. https://doi.org/10.1111/j.1469-8137.2008.02478.x

Popova OV, Dinh HQ, Aufsatz W, Jonak C (2013) The RDDM pathway is required for basal heat tolerance in Arabidopsis. Mol Plant 6(2):396–410. https://doi.org/10.1093/mp/sst023

Qüesta JI, Song J, Geraldo N, An H, Dean C (2016) Arabidopsis transcriptional repressor VAL1 triggers Polycomb silencing at FLC during vernalization. Science 353(6298):485–488. https://doi.org/10.1126/science.aaf7354

Rajkumar MS, Shankar R, Garg R, Jain M (2020) Bisulphite sequencing reveals dynamic DNA methylation under desiccation and salinity stresses in rice cultivars. Genomics 112(5):3537–3548. https://doi.org/10.1016/j.ygeno.2020.04.005

Rambani A, Pantalone V, Yang S, Rice JH, Song Q, Mazarei M, Arelli PR, Meksem K, Stewart CN, Hewezi T (2020) Identification of introduced and stably inherited DNA methylation variants in soybean associated with soybean cyst nematode parasitism. New Phytol 227(1):168–184. https://doi.org/10.1111/nph.16511

Rastogi S, Shah S, Kumar R, Vashisth D, Akhtar MQ, Kumar A, Dwivedi UN, Shasany AK (2019) Ocimum metabolomics in response to abiotic stresses: cold, flood, drought and salinity. PLoS One 14(2):e0210903. https://doi.org/10.1371/journal.pone.0210903

Raza A, Tabassum J, Kudapa H, Varshney RK (2021) Can omics deliver temperature resilient ready-to-grow crops? Crit Rev Biotechnol 41(8):1209–1232. https://doi.org/10.1080/07388551.2021.1898332

Sánchez AL, Stassen JHM, Furci L, Smith LM, Ton J (2016) The role of DNA (de)methylation in immune responsiveness of Arabidopsis. Plant J 88(3):361–374. https://doi.org/10.1111/tpj.13252

Saraswat S, Yadav AK, Sirohi P, Singh NK (2017) Role of epigenetics in crop improvement: water and heat stress. J Plant Biol 60(3):231–240. https://doi.org/10.1007/s12374-017-0053-8

Scheres B, Wolkenfelt H, Willemsen V, Terlouw M, Lawson E, Dean C, Weisbeek P (1994) Embryonic origin of the Arabidopsis primary root and root meristem initials. Development 120(9):2475–2487. https://doi.org/10.1242/dev.120.9.2475

Scofield S, Murray JAH (2006) KNOX gene function in plant stem cell niches. Plant Mol Biol 60(6):929–946. https://doi.org/10.1007/s11103-005-4478-y

Sheldon CC, Rouse DT, Finnegan EJ, Peacock WJ, Dennis ES (2000) The molecular basis of vernalization: the central role of FLOWERING LOCUS C (FLC). Proc Natl Acad Sci 97(7):3753–3758. https://doi.org/10.1073/pnas.97.7.3753

Sheldon CC, Hills MJ, Lister C, Dean C, Dennis ES, Peacock WJ (2008) Resetting of FLOWERING LOCUS C expression after epigenetic repression by vernalization. Proc Natl Acad Sci 105(6):2214–2219. https://doi.org/10.1073/pnas.0711453105

Shen Y, Lei T, Cui X, Liu X, Zhou S, Zheng Y, Guérard F, Issakidis-Bourguet E, Zhou D (2019) Arabidopsis histone deacetylase HDA15 directly represses plant response to elevated ambient temperature. Plant J 100(5):991–1006. https://doi.org/10.1111/tpj.14492

Shen H, Zhou Y, Liao C, Xie Q, Chen G, Hu Z, Wu T (2024) The AlkB homolog SlALKBH10B negatively affects drought and salt tolerance in Solanum lycopersicum. Int J Mol Sci 25(1):173. https://doi.org/10.3390/ijms25010173

Song X-F, Hou X-L, Liu C-M (2021) CLE peptides: critical regulators for stem cell maintenance in plants. Planta 255(1):5. https://doi.org/10.1007/s00425-021-03791-1
Sreenivasulu N, Wobus U (2013) Seed-development programs: a systems biology–based comparison between dicots and monocots. Annu Rev Plant Biol 64(1):189–217. https://doi.org/10.1146/annurev-arplant-050312-120215
Sun X, Shantharaj D, Kang X, Ni M (2010) Transcriptional and hormonal signaling control of Arabidopsis seed development. Curr Opin Plant Biol 13(5):611–620. https://doi.org/10.1016/j.pbi.2010.08.009
Sun B, Looi L-S, Guo S, He Z, Gan E-S, Huang J, Xu Y, Wee W-Y, Ito T (2014) Timing mechanism dependent on cell division is invoked by polycomb eviction in plant stem cells. Science 343(6170):1248559. https://doi.org/10.1126/science.1248559
Sung S, Amasino RM (2004) Vernalization in Arabidopsis thaliana is mediated by the PHD finger protein VIN3. Nature 427(6970):159–164. https://doi.org/10.1038/nature02195
Sung DY, Kaplan F, Lee KJ, Guy CL (2003) Acquired tolerance to temperature extremes. Trends Plant Sci 8(4):179–187. https://doi.org/10.1016/s1360-1385(03)00047-5
Tasset C, Yadav AS, Sureshkumar S, Singh R, Van Der Woude L, Nekrasov M, Tremethick D, Van Zanten M, Balasubramanian S (2018) POWERDRESS-mediated histone deacetylation is essential for thermomorphogenesis in Arabidopsis thaliana. PLoS Genet 14(3):e1007280. https://doi.org/10.1371/journal.pgen.1007280
Tittel-Elmer M, Bucher E, Broger L, Mathieu O, Paszkowski J, Vaillant I (2010) Stress-induced activation of heterochromatic transcription. PLoS Genet 6(10):e1001175. https://doi.org/10.1371/journal.pgen.1001175
Tognacca RS, Botto JF (2021) Post-transcriptional regulation of seed dormancy and germination: current understanding and future directions. Plant Commun 2(4):100169. https://doi.org/10.1016/j.xplc.2021.100169
Tyagi P, Singh D, Mathur S, Singh A, Ranjan R (2022) Upcoming progress of transcriptomics studies on plants: an overview. Front Plant Sci 13:1030890. https://doi.org/10.3389/fpls.2022.1030890
Ueda M, Matsui A, Tanaka M, Nakamura T, Abe T, Sako K, Sasaki T, Kim J-M, Ito A, Nishino N, Shimada H, Yoshida M, Seki M (2017) The distinct roles of class I and II RPD3-like histone deacetylases in salinity stress response. Plant Physiol 175(4):1760–1773. https://doi.org/10.1104/pp.17.01332
Van Zanten M, Liu Y, Soppe WJJ (2013) Epigenetic signalling during the life of seeds. In: Signaling and communication in plants. Springer, pp 127–153
Velculescu VE, Zhang L, Vogelstein B, Kinzler KW (1995) Serial analysis of gene expression. Science 270(5235):484–487. https://doi.org/10.1126/science.270.5235.484
Vollbrecht E, Reiser L, Hake S (2000) Shoot meristem size is dependent on inbred background and presence of the maize homeobox gene, knotted1. Development 127(14):3161–3172. https://doi.org/10.1242/dev.127.14.3161
Von Arnim A, Deng X-W (1996) Light control of seedling development. Annu Rev Plant Physiol Plant Mol Biol 47(1):215–243. https://doi.org/10.1146/annurev.arplant.47.1.215
Wang W, Huang F, Qin Q, Zhao X, Li Z, Fu B (2015) Comparative analysis of DNA methylation changes in two rice genotypes under salt stress and subsequent recovery. Biochem Biophys Res Commun 465(4):790–796. https://doi.org/10.1016/j.bbrc.2015.08.089
Wang J, Nan N, Li N, Liu Y, Wang TJ, Hwang I, Liu B, Xu ZY (2020) A DNA methylation reader–chaperone regulator–transcription factor complex activates OsHKT1;5 expression during salinity stress. Plant Cell 32(11):3535–3558. https://doi.org/10.1105/tpc.20.00301
Wang Q, Xu J, Pu X, Lv H, Liu Y, Ma H, Wu F, Wang Q, Feng X, Liu T, Tang Q, Liu Y, Lu Y (2021) Maize DNA methylation in response to drought stress is involved in target gene expression and alternative splicing. Int J Mol Sci 22(15):8285. https://doi.org/10.3390/ijms22158285
Wu X, Su T, Zhang S, Zhang Y, Wong CE, Ma J, Shao Y, Hua C, Shen L, Yu H (2024) N6-methyladenosine-mediated feedback regulation of abscisic acid perception via

phase-separated ECT8 condensates in Arabidopsis. Nat Plants 10(3):469–482. https://doi.org/10.1038/s41477-024-01638-7

Xie H, Sun Y, Cheng B, Xue S, Cheng D, Liu L, Meng L, Qiang S (2018) Variation in ICE1 methylation primarily determines phenotypic variation in freezing tolerance in Arabidopsis thaliana. Plant Cell Physiol 60(1):152–165. https://doi.org/10.1093/pcp/pcy197

Xu Y, Prunet N, Gan E, Wang Y, Stewart D, Wellmer F, Huang J, Yamaguchi N, Tatsumi Y, Kojima M, Kiba T, Sakakibara H, Jack TP, Meyerowitz EM, Ito T (2018) SUPERMAN regulates floral whorl boundaries through control of auxin biosynthesis. EMBO J 37(11):e97499. https://doi.org/10.15252/embj.201797499

Xue M, Zhang H, Zhao F, Zhao T, Li H, Jiang D (2021) The INO80 chromatin remodeling complex promotes thermomorphogenesis by connecting H2A.Z eviction and active transcription in Arabidopsis. Mol Plant 14(11):1799–1813. https://doi.org/10.1016/j.molp.2021.07.001

Yamaguchi N, Huang J, Tatsumi Y, Abe M, Sugano SS, Kojima M, Takebayashi Y, Kiba T, Yokoyama R, Nishitani K, Sakakibara H, Ito T (2018) Chromatin-mediated feed-forward auxin biosynthesis in floral meristem determinacy. Nat Commun 9(1):5290. https://doi.org/10.1038/s41467-018-07763-0

Yang Y, Guo Y (2018) Elucidating the molecular mechanisms mediating plant salt-stress responses. New Phytol 217(2):523–539. https://doi.org/10.1111/nph.14920

Yang C, Bratzel F, Hohmann N, Koch M, Turck F, Calonje M (2013) VAL- and AtBMI1-mediated H2Aub initiate the switch from embryonic to postgerminative growth in Arabidopsis. Curr Biol 23(14):1324–1329. https://doi.org/10.1016/j.cub.2013.05.050

Yang R, Zheng Z, Chen Q, Yang L, Huang H, Miki D, Wu W, Zeng L, Liu J, Zhou J-X, Ogas J, Zhu J-K, He X-J, Zhang H (2017) The developmental regulator PKL is required to maintain correct DNA methylation patterns at RNA-directed DNA methylation loci. Genome Biol 18(1):103. https://doi.org/10.1186/s13059-017-1226-y

Yang R, Hong Y, Ren Z, Tang K, Zhang H, Zhu J-K, Zhao C (2019) A role for PICKLE in the regulation of cold and salt stress tolerance in Arabidopsis. Front Plant Sci 10:900. https://doi.org/10.3389/fpls.2019.00900

Yang Y, Saand MA, Huang L, Abdelaal WB, Zhang J, Wu Y, Li J, Sirohi MH, Wang F (2021) Applications of multi-omics technologies for crop improvement. Front Plant Sci 12:563953. https://doi.org/10.3389/fpls.2021.563953

Yu A, Lepère G, Jay F, Wang J, Bapaume L, Wang Y, Abraham A-L, Penterman J, Fischer RL, Voinnet O, Navarro L (2013) Dynamics and biological relevance of DNA demethylation in Arabidopsis antibacterial defense. Proc Natl Acad Sci 110(6):2389–2394. https://doi.org/10.1073/pnas.1211757110

Zha P, Jing Y, Xu G, Lin R (2017) PICKLE chromatin-remodeling factor controls thermosensory hypocotyl growth of Arabidopsis. Plant Cell Environ 40(10):2426–2436. https://doi.org/10.1111/pce.13049

Zha P, Liu S, Li Y, Ma T, Yang L, Jing Y, Lin R (2019) The evening complex and the chromatin-remodeling factor PICKLE coordinately control seed dormancy by directly repressing DOG1 in Arabidopsis. Plant Commun 1(2):100011. https://doi.org/10.1016/j.xplc.2019.100011

Zhan J, Meyers BC (2023) Plant small RNAs: their biogenesis, regulatory roles, and functions. Annu Rev Plant Biol 74(1):21–51. https://doi.org/10.1146/annurev-arplant-070122-035226

Zhang H, Rider SD, Henderson JT, Fountain M, Chuang K, Kandachar V, Simons A, Edenberg HJ, Romero-Severson J, Muir WM, Ogas J (2008) The CHD3 remodeler PICKLE promotes trimethylation of histone H3 lysine 27. J Biol Chem 283(33):22637–22648. https://doi.org/10.1074/jbc.m802129200

Zhang H, Bishop B, Ringenberg W, Muir WM, Ogas J (2012) The CHD3 remodeler PICKLE associates with genes enriched for trimethylation of histone H3 lysine 27. Plant Physiol 159(1):418–432. https://doi.org/10.1104/pp.112.194878

Zhao M, Yang S, Liu X, Wu K (2015) Arabidopsis histone demethylases LDL1 and LDL2 control primary seed dormancy by regulating DELAY OF GERMINATION 1 and ABA signaling-related genes. Front Plant Sci 6:159. https://doi.org/10.3389/fpls.2015.00159

Zhao F, Xue M, Zhang H, Li H, Zhao T, Jiang D (2023) Coordinated histone variant H2A.Z eviction and H3.3 deposition control plant thermomorphogenesis. New Phytol 238(2):750–764. https://doi.org/10.1111/nph.18738

Zheng J, Chen F, Wang Z, Cao H, Li X, Deng X, Soppe WJJ, Li Y, Liu Y (2012) A novel role for histone methyltransferase KYP/SUVH4 in the control of Arabidopsis primary seed dormancy. New Phytol 193(3):605–616. https://doi.org/10.1111/j.1469-8137.2011.03969.x

Zheng Y, Ding Y, Sun X, Xie S, Wang D, Liu X, Su L, Wei W, Pan L, Zhou D-X (2016) Histone deacetylase HDA9 negatively regulates salt and drought stress responsiveness in Arabidopsis. J Exp Bot 67(6):1703–1713. https://doi.org/10.1093/jxb/erv562

Zheng M, Liu X, Lin J, Liu X, Wang Z, Xin M, Yao Y, Peng H, Zhou D, Ni Z, Sun Q, Hu Z (2019) Histone acetyltransferase GCN5 contributes to cell wall integrity and salt stress tolerance by altering the expression of cellulose synthesis genes. Plant J 97(3):587–602. https://doi.org/10.1111/tpj.14144

Zhuang J, Zhang J, Hou X-L, Wang F, Xiong A-S (2014) Transcriptomic, proteomic, metabolomic and functional genomic approaches for the study of abiotic stress in vegetable crops. Crit Rev Plant Sci 33(2–3):225–237. https://doi.org/10.1080/07352689.2014.870420

Advances in Plant Metabolomics: Techniques and Key Applications

Krystle Angelique Aguda Santiago and Adeline Su Yien Ting

Abstract Metabolomics is a powerful and rapidly evolving tool to unravel the complex biochemical nature of plants. Through this, the plant metabolome is continuously explored to understand plant physiology, stress responses, and the production of valuable secondary metabolites. This chapter provides a comprehensive overview of plant metabolomics, encompassing sample preparation strategies, extending to widely used analytical platforms such as capillary electrophoresis–mass spectrometry (CE–MS), gas chromatography–mass spectrometry (GC–MS), liquid chromatography–mass spectrometry (LC–MS), and nuclear magnetic resonance (NMR), as well as the typical workflow from data acquisition to bioinformatics-driven data analysis. This chapter also highlights the diverse applications of metabolomics in plant research, including studies relevant to plant development, stress response, crop improvement, and the discovery of bioactive compounds. Key challenges in plant metabolomics research, such as metabolite annotation, lack of standardized protocols, limited spectral libraries, and the complexity of the plant metabolome, are also discussed, along with emerging trends and future directions that continue to shape metabolomics as an indispensable tool in plant science research.

Keywords Analytical techniques · Metabolite profiling · Metabolomics workflow · Plant metabolomics · Secondary metabolites

---

K. A. A. Santiago · A. S. Y. Ting (✉)
School of Science, Monash University Malaysia, Bandar Sunway, Selangor, Malaysia
e-mail: adeline.ting@monash.edu

L. M. Vaschetto (ed.), *Omics Technologies in Plant Science*, Interdisciplinary Biotechnological Advances, https://doi.org/10.1007/978-981-92-0533-2_5

Abbreviations

APCI	Atmospheric pressure chemical ionization
APPI	Atmospheric pressure photoionization
CE–MS	Capillary electrophoresis-mass spectrometry
CI	Chemical ionization
DART	Direct analysis in real time
DESI	Desorption electrospray ionization
EI	Electron ionization
ESI	Electrospray ionization
FT–ICR	Fourier transform ion cyclotron resonance
FT–MS	Fourier transform mass spectrometry
GC × GC	Two dimensional gas chromatography
GC–ITMS	Gas chromatography–ion trap mass spectrometry
GC–MS	Gas chromatography–mass spectrometry
GC–QMS	Gas chromatography–quadrupole mass spectrometry
GC–SIM–MS	Gas chromatography–selected ion monitoring–mass spectrometry
GC–TOF–MS	Gas chromatography time-of-flight mass spectrometry
HCA	Hierarchical cluster analysis
HILIC	Hydrophilic interaction liquid chromatography
LC–MS	liquid chromatography-mass spectrometry
MALDI	Matrix-assisted laser desorption ionization
MS	Mass spectrometry
MS/MS	Tandem mass spectrometry
MSI	Metabolomics Standards Initiative
NMR	Nuclear magnetic resonance
NP–LC	Normal-phase liquid chromatography
OPLS–DA	Orthogonal partial least squares-discriminant analysis
PCA	Principal component analysis
Q-TOF	Quadruple time-of-flight
RCBD	Randomized complete block design
RF	Random forest
RP–LC	Reversed-phase liquid chromatography
S/N	Signal-to-noise ratio
TOF	Time-of-flight
TQ	Triple quadruples
VSN	Variance stabilization normalization

1 Introduction

Metabolomics is a powerful tool for studying the complete set of small molecules in biological samples (Sun and Xia 2024). It is broadly classified into targeted and non-targeted approaches. Targeted metabolomics focuses on the precise quantification of known metabolites, while non-targeted metabolomics aims to detect as many metabolites as possible in a single analysis, enabling the discovery of significant and/or novel compounds and their metabolic changes (Ribbenstedt et al. 2018). As a rapidly advancing field in plant research, metabolomics reveals essential information about the dynamic chemical changes in plants driven by diverse biochemical reactions, molecular mechanisms, and biological pathways (Yan et al. 2022). Plant metabolites serve as key indicators of physiological status and act as a bridge between the genome and the observable phenotype, providing insights into underlying biological mechanisms (Zhao et al. 2024). Plants are particularly vulnerable to environmental threats. To cope with these challenges, they rely on their metabolism as a defense mechanism against biotic and abiotic stresses (Shen et al. 2023). Given its importance, the integration of metabolomics into OMICS technologies (alongside genomics, transcriptomics, and proteomics) (Veenstra 2021) significantly enhances our understanding of plant metabolism and its role in adaptation, development, and stress responses.

Plant metabolomics has revolutionized various fields, from fundamental biology to applied disciplines such as agricultural sciences (Emam et al. 2025), medicinal plant sciences (Waris et al. 2022), and plant ecology (Rivas-Ubach et al. 2025). One of its key applications is detecting changes in plant metabolic pathways and their metabolite profiles in response to environmental stress factors. For example, metabolomics analysis of root tissues of *Pistia stratiotes* (an aquatic floating herb) exposed to cadmium (Cd) revealed significant alterations in eight metabolic pathways, including those related to unsaturated fatty acids, amino acids, nucleotides, sulfur compounds, and flavonoids (Wei et al. 2023). Plant metabolomics also plays a crucial role in the discovery of bioactive compounds with medicinal potential. A metabolomics study identified bioactive compounds, such as alkaloids, isoflavonoids, saponins, fatty acyls, flavonoids, and prenol lipids, in the "underused" tissues (i.e., roots, leaves, branches, and pods) of the soy plant (Bragagnolo et al. 2021), highlighting its potential for pharmaceutical and nutraceutical applications. Furthermore, metabolomics is also applied in agriculture to improve crop yield and quality amidst climate change. For example, metabolic markers associated with heat tolerance in chickpea (*Cicer arietinum* L.) were identified through targeted metabolomics analysis, providing valuable insights for crop improvement strategies (Jha et al. 2025). Metabolomics is also instrumental in studying plant–pathogen interactions, offering a comprehensive understanding of host–pathogen dynamics at the metabolic level. Ren et al. (2021) analyzed the metabolic interactions between wheat (*Triticum aestivum* L.) and the pathogen *Tilletia controversa* Kühn, the causative agent of dwarf bunt, revealing a decrease in plant metabolites and an increase in fungal metabolites following infection. Table 1 shows recent studies in 2024 and

Table 1 Recent studies (2024–2025) in plant metabolomics and their significance in advancing the field

Plant species	Metabolomics platform[a]	Significance	References
Broccoli (family Brassicaceae)	LC–MS	Functional food and medicine	Ibrahim et al. (2024)
White tea	LC–MS and GC–MS	Metabolite profiling; integration of omics technologies	Liu et al. (2024)
Flax (*Linum usitatissimum*)	NMR and LC–MS	Biocontrol agent; integration of omics technologies	Benamar et al. (2024)
Witchweed (*Striga asiatica* (L.) O. Kuntze)	LC–MS	Artificial cultivation of *S. asiatica*	Tang et al. (2024)
Wheat (*Triticum aestivum*)	NMR and LC–MS	Bio stimulants for crop growth and yield improvement; integration of omics technologies	Hamade et al. (2024)
Shiny-leaf prickly-ash (*Zanthoxylum nitidum* (Roxb.) DC)	LC–MS	Plant origin authentication	Xiao et al. (2025)
Ginseng varieties (*Panax japonicus* and *P. japonicus* var. *major*	LC–MS	Species identification	Jiang et al. (2025)
Cathedral bells (*Kalanchoe pinnata* (Lam.) Pers.)	LC–MS	Metabolite profiling for environmental response	Pinheiro et al. (2025)
Fennel (*Foeniculum vulgare* Mill.)	NMR and LC–MS	Metabolome characterization	Crescenzi et al. (2025)
Whip grass (*Hemarthria compressa*)	LC–MS	Effects of heavy metal stress	Jing et al. (2025)
Salt marsh iris (*Iris halophila*)	LC–MS	Effects of salinity	Liu et al. (2025)

[a]Metabolomics Platform: *GC–MS* gas chromatography–mass spectrometry, *LC–MS* liquid chromatography–mass spectrometry, *NMR* nuclear magnetic resonance

2025 focusing on metabolite profiling, species authentication, crop protection, and multi-omics integration, illustrating the increasing importance and popularity of plant metabolomics. Overall, plant metabolomics serves as a powerful tool for unraveling plant physiology, elucidating adaptive responses to environmental challenges, discovering bioactive compounds, and advancing agricultural sustainability.

The design of plant sample preparation and metabolite extraction methods is a critical step in plant metabolomics research, as incomplete extraction of metabolites can significantly impact the overall quality and reliability of the study (Bijttebier et al. 2016). The selection of plant tissues for metabolite extraction, along with the choice of the solvent for extraction, plays a crucial role in optimizing metabolite yield and diversity. Additionally, other important parameters, such as the methods for cell disruption, the number of replicates, and tissue-to-solvent ratio, are to be carefully determined before the study begins to ensure consistency and reliability

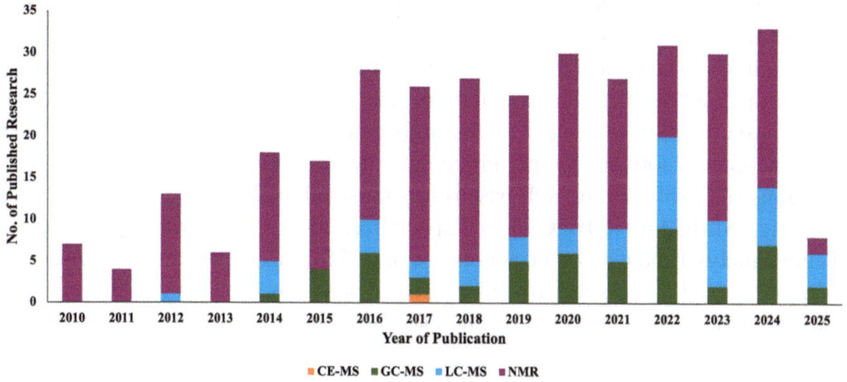

Fig. 1 The number of published research articles on plant metabolomics using capillary electrophoresis-mass spectrometry (CE–MS), gas chromatography-mass spectrometry (GC–MS), liquid chromatography-mass spectrometry (LC–MS), and nuclear magnetic resonance (NMR). The bibliometric data from Google Scholar, retrieved from 2010 to 2025, were retrieved using Publish or Perish software (Harzing 2024). The search was performed using the keywords "plant metabolomics," "gas chromatography," "capillary electrophoresis-mass spectrometry," "mass spectrometry," and "nuclear magnetic resonance." Technical reports or protocol papers and review papers were excluded in the search

(Wu et al. 2014). Such methodological considerations are essential for accurately characterizing the biochemical composition of plant samples, ultimately enabling meaningful biological interpretations.

The success of plant metabolomics is also highly driven by advancements in analytical techniques, particularly nuclear magnetic resonance (NMR) and mass spectrometry (MS)-based platforms, such as gas chromatography–mass spectrometry (GC–MS), liquid chromatography–mass spectrometry (LC–MS), and capillary electrophoresis–mass spectrometry (CE–MS). In the last 15 years, there has been a notable increase in plant metabolomics studies, reflecting the increased use of these analytical techniques (Fig. 1). Among these techniques, NMR has been the preferred analytical method in plant metabolomics (Fig. 1) due to its high reproducibility, non-destructive nature, minimal sample preparation and the relatively rapid identification of metabolites (Munjal et al. 2022). Following NMR is LC–MS (Fig. 1), which offers rapid measurement with high sensitivity for ion detection and the ability to separate and identify ionized analytes based on their mass-to-charge ratios (m/z) (Heiles 2021). The third most widely used technique is GC–MS (Fig. 1). Known for its robustness, separation capability, selectivity, sensitivity, and reproducibility, GC–MS is limited to the analysis of low-molecular-weight (i.e., 50–600 Da) and volatile compounds (Beale et al. 2018). CE–MS, on the other hand, is the least commonly employed technique in plant metabolomics (Fig. 1). While CE–MS holds potential for profiling primary and secondary metabolites, this technique has stability and sensitivity concerns compared to other analytical platforms, hindering its widespread adoption (Harada and Fukusaki 2009). As such, selecting a compatible and reliable analytical technique is essential for the successful identification and quantification of plant metabolites.

In addition to analytical techniques, the software tools for processing, analyzing, and annotating "big data" are essential. As analytical techniques advance, data analytics must co-evolve to manage increasingly complex datasets (Misra 2021). These tools encompass software for pre-processing raw data (e.g., MZMine, MS–DIAL), platforms (i.e., software or web-based) for statistical analyses and visualization of data (e.g., SIMCA-P, ROBPCA), and specialized tools for pathway analysis (e.g., MetaboAnalyst) and metabolite annotation (e.g., MetFrag). There are also tools that offer end-to-end processing, from raw data pre-processing to data visualization and interpretation. For example, the web-based MetaboAnalyst (Pang et al. 2024) is widely used in plant metabolomics and supports raw data processing, multivariate statistical analysis, annotation, pathway mapping, and network analysis. Similarly, the web-based platform XCMS Online (Tautenhahn et al. 2012) and the software MetaboScape (Bruker) provide integrated environments for raw data processing, statistical analysis, pathway mapping, and data interpretation. Additionally, commonly used metabolite databases, such as METLIN, PubChem, and KNapSAck (Nakamura et al. 2013), play a crucial role in ensuring accurate and reliable metabolite identification. These tools enhance the precision, reproducibility, and reliability of data generated from plant metabolomics research, enabling deeper insights into metabolic profiles and pathways and their biological significance.

Despite remarkable innovations in plant research, metabolomics still faces several challenges. The inherent complexity of the plant metabolome, the lack of standardized protocols for sample preparation, and the limited availability of databases for metabolite annotation and identification continue to hinder data reproducibility and accuracy. To overcome these limitations, plant metabolomics adopts integrative and collaborative approaches, most notably through the integration of the multi-omics framework. By combining metabolomics with other OMICS technologies, such as genomics, transcriptomics, and/or proteomics, researchers can gain a more comprehensive understanding of plant biology. The continuous improvement of bioinformatics tools will also play a critical role in enhancing data integration, interpretation, and the overall impact of plant metabolomics.

This chapter provides a comprehensive overview of the key techniques used in plant metabolomics, including sample preparation, metabolite extraction methods, and analytical platforms. The essential techniques used for data mining, such as pre-processing, pre-treatment, statistical analysis, and metabolite annotation, are also outlined. The diverse applications of metabolomics in plant research are highlighted, demonstrating its significance in understanding plant physiology, stress responses, and biochemical diversity. Furthermore, key challenges in plant metabolomics research, such as data standardization, metabolite identification, and computational limitations, are addressed. Finally, emerging trends and future directions in the field are explored, highlighting the evolving role of metabolomics in advancing plant research.

2 Metabolomics for Plant Studies

The first application of metabolomics in plants was reported in the year 2000 (Fiehn et al. 2000). Using gas chromatography–mass spectrometry (GC–MS), 326 distinct compounds were quantified from the leaf extracts of *Arabidopsis thaliana*. This pioneering work demonstrated the potential of metabolite profiling as a powerful tool for comparative analysis of gene function, providing comprehensive insight into complex regulatory networks and their influence on phenotypic traits. The study showed that plant samples could be grouped not only by genotype but also by their unique metabolic profiles. Building on this foundational work, current metabolomics studies place strong emphasis on methodological rigor, encompassing pre-analytical (i.e., sample preparation, metabolite extraction), analytical, and post-analytical (i.e., data mining) techniques to ensure reliable and reproducible results.

When conducting plant metabolomic studies, the collection and preparation of biological samples is carefully planned, as these initial steps are crucial for ensuring reliable downstream analyses. Ideally, the metabolomics workflow would follow the Metabolomics Standards Initiative (MSI) (Sumner et al. 2007) and the minimum reporting standards for plant metabolomics (Fiehn et al. 2007) (Fig. 2). The general workflow encompasses sample preparation, data acquisition, data mining, visualization and interpretation (Fig. 2). Certain steps, such as data mining, can be modified depending on the selected analytical technique. For example, the pre-processing of raw data generated from an MS-based analytical technique differs from that of data generated by NMR. Additionally, data mining approaches vary between targeted and non-targeted metabolomics.

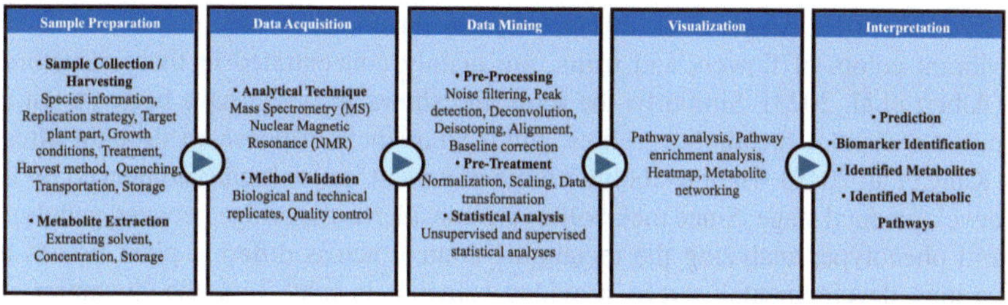

Fig. 2 The general workflow for a plant metabolomics study. Some steps (e.g., data pre-processing and pre-treatment) may be modified based on the analytical technique selected for the study

2.1 Pre-analytical Technique: Plant Sample Preparation

2.1.1 Plant Species Information and Targeted Plant Tissue

It is essential to obtain complete details about the plant, including accurate species identification, when conducting plant metabolomic studies. Proper taxonomic classification enables meaningful comparisons with previous studies and ensures reproducible results. This is especially critical when studying herbal or medicinal plants, where closely related species may exhibit vastly different metabolite profiles. Different countries often have unique local names for the same plant, which can lead to misidentification and inconsistencies in metabolomic data. Using the taxonomic name ensures clarity, consistency, and accurate identification across regions and scientific studies. For example, a metabolomics study authenticated Uncariae Rammulus Cum Uncis (Gou-Teng), a well-known traditional Chinese herbal medicine for cardiovascular diseases and neurological disorders (Pan et al. 2020). The study revealed significant chemical variations among *Uncaria* species commonly sold under the same name, highlighting the need for plant authentication. Without proper plant species information, metabolomics interpretation may lead to erroneous conclusions, which may significantly affect downstream applications in plant science. Species identification also provides insights into expected metabolite profiles, as plant metabolism is shaped by both genetic variation and environmental factors (Waris et al. 2022).

The selection of plant tissues to be sampled (e.g., stem, leaves, roots, flowers, fruits) is another critical step in sample preparation, as different tissues exhibit distinct metabolite profiles. In medicinal science, different plant tissues are often thought to harbor different bioactive compounds throughout the plant (Xiao et al. 2022). For example, alkaloids are predominantly found in roots and leaves of plants (Li et al. 2015; Singh et al. 2021), whereas flavonoids, which contribute to the vibrant colors of flowers and fruits, are mainly concentrated in these structures (Albert et al. 2023). Similarly, the roots and shoots of wheat have been shown to possess distinct metabolic profiles, influencing their response to drought stress (Kang et al. 2019). The metabolite profile of a plant is also highly influenced by its developmental stage. Since metabolites are considered the bridge between genotype and phenotype, analyzing the metabolite content across different plant tissues at various developmental stages provides valuable insights into the dynamics of metabolite production and accumulation (Shen et al. 2023). For example, a metabolomics analysis of 8- and 25-year-old *Camellia sinensis* (tea plant) revealed clear metabolic differentiation between young and mature leaves (Mozumder et al. 2020). Certain metabolites, such as theanine, glutamine, catechin, and gallocatechin, have been found to be age-dependent, highlighting the importance of plant developmental stage in metabolomics studies.

2.1.2 Plant Growth Conditions, Randomization, Replication Strategy, and Treatment Application

In plant metabolomics, carefully designed experimental set-ups are essential to ensure that the metabolite profiles reflect true biological responses rather than random environmental fluctuations. Most plant metabolomics studies employ either controlled plant-chamber conditions or field-grown (*in planta*) settings to evaluate plant metabolic responses, although both approaches are used in combination to generate more accurate and reliable data. Regardless of the setting, it is essential to standardize key growth parameters such as temperature, water availability, humidity, and illumination or light intensity (Fiehn et al. 2007; Rodrigues et al. 2019). For example, a metabolomics study was conducted to determine how *Arabidopsis thaliana* guard cells respond to red light and how these metabolic changes relate to stomatal behavior (Zhu et al. 2019). This study used a growth chamber set at precise environmental conditions, including fixed light cycles (e.g., 8 h light/16 h night) and temperature (e.g., 22 °C/20 °C) to ensure experimental consistency, reliability, and reproducibility. Their findings showed that plant metabolites are dependent on light intensities, thus emphasizing the need for strict control of growth parameters.

In field-based metabolomics studies, where environmental conditions are naturally fluctuating, the design of plant randomization and replication is essential. Randomization helps minimize the influence of uncontrolled variables and potential disturbances, reducing experimental bias (Fig. 3) (Jacyna et al. 2019). A widely adopted strategy is the randomized complete block design (RCBD), which integrates randomization with blocking to account for environmental variability (Fig. 3). In RCBD, each treatment is replicated within a block, and plots are arranged randomly along a known environmental gradient (e.g., light intensity), ensuring all treatments experience similar environmental fluctuations (Hartung et al. 2019). For example, an RCBD was employed in the metabolomics analysis of barley (*Hordeum vulgare* L.) to evaluate its physiological and metabolic responses to drought stress (Hein et al. 2016), highlighting the importance of well-structured field experiments in capturing meaningful metabolic responses while minimizing confounding environmental effects.

The incorporation of replication (i.e., biological and technical replicates) in plant metabolomic studies is also crucial for assessing the impact of random errors and biological variability (Jacyna et al. 2019; Rodrigues et al. 2019). Technical replicates enhance the reliability of the experimental protocol and instrumentation, ensuring consistency in measurements, but do not improve the statistical power of the results. In contrast, biological replicates represent independent samples from the population under study, directly influencing the biological interpretation and generalizability of the findings (Rodrigues et al. 2019). For example, a metabolomics study was conducted to determine the biological variability within a single accession of *Arabidopsis thaliana* (Mönchgesang et al. 2016). The findings emphasized the importance of accounting for individual biological variability when designing metabolomics experiments. A minimum number of six biological replicates is generally considered sufficient, although increasing the number of replicates is

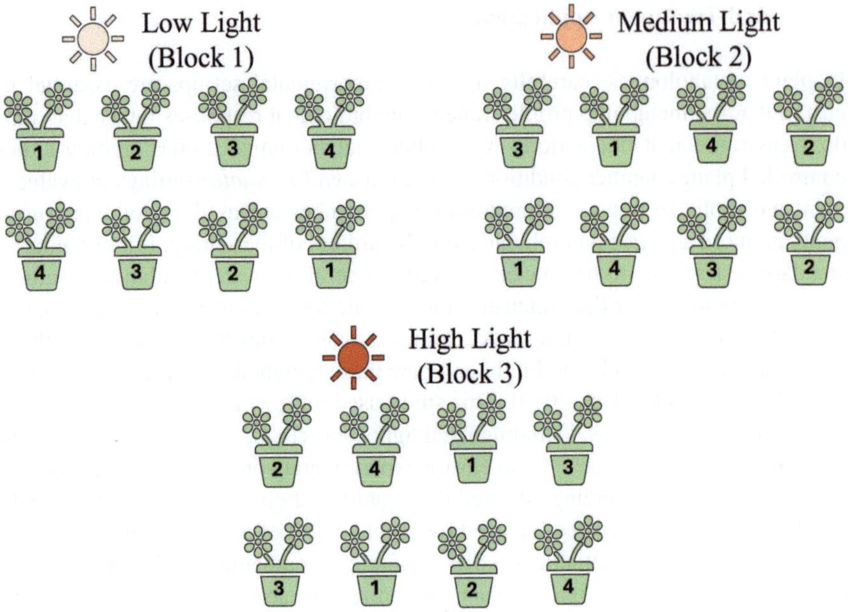

Fig. 3 Sample set-up of randomized complete block design (RCBD), where an experiment is divided into blocks arranged along a known environmental gradient (e.g., light intensity). Numbers represent different treatments

recommended whenever possible to enhance statistical robustness and reliability (Lisec et al. 2006). The number of technical replicates, on the other hand, should be determined based on the specific objectives of the study and the expected sources of variability. Of the two, biological replication is significantly more important than technical replication, as it captures natural variability and ensures the robustness of biological conclusions (Fernie et al. 2011).

Careful planning and standardization of treatment applications are as important as controlling growth conditions. Treatments are commonly applied to assess plant responses to abiotic or biotic stresses. For example, in wild soybean (*Glycine soja*), amino acid, carbohydrate, and polyol metabolisms were enhanced under neutral-salt stress, whereas alkali-salt stress primarily influenced the metabolism of organic acid, amino acid, and tricarboxylic acid (Jiao et al. 2018). The study employed a standardized protocol in which three soybean seeds were sown per pot, salt stress was applied uniformly, and eight biological replicates were maintained under controlled incubation conditions. Similarly, an LC–MS-based metabolomics study of two tomato cultivars and their pathogens, *Meloidogyne incognita* (Southern root-knot nematode) and *Fusarium oxysporum* (Fusarium wilt fungus), utilized standardized germination, transplanting, incubation, and controlled temperature, light, and humidity (Sikandar et al. 2024).

2.1.3 Harvesting, Quenching, Relocation, and Storage of Plant Samples

The harvesting of whole plants or specific plant tissues requires careful planning and precise execution, as it marks the critical point at which metabolism is halted (Fiehn et al. 2007). Proper timing, handling, and storage of harvested specimens are essential to prevent metabolic changes that could compromise the accuracy of downstream analyses. Harvesting should be conducted as quickly as possible under controlled conditions to minimize diurnal variations (i.e., fluctuations or changes that occur within a 24-h cycle) and prevent metabolite loss (Lisec et al. 2006; Jorge et al. 2015). Throughout a 24-h period, environmental factors such as light intensity and quality, temperature, and humidity fluctuate, significantly influencing nutrient intake and photosynthesis. These changes directly affect the availability and distribution of carbon and nitrogen, which are key components in metabolite biosynthesis and processing (Liebelt et al. 2019). Plants subjected to treatments and high replication should be harvested at the same time, under similar environmental conditions. The harvesting of leaf tissue, for example, is suggested to be done at the photoperiod midpoint, as most detectable and quantifiable leaf metabolites fluctuate significantly throughout the day due to diurnal rhythms (Lisec et al. 2006). By harvesting at a specific time, a more stable and representative metabolic profile can be captured while minimizing variations caused by natural daily fluctuations. It is also a recommended practice when harvesting plants for host–pathogen interaction studies to harvest healthy plants first before handling infected ones, minimizing the risk of pathogen spread. Important information, such as harvest time and date, plant growth stage, metabolism quenching method, harvest method, and sample storage, is suggested to be recorded (Fiehn et al. 2007). Whenever possible, provide detailed environmental conditions during harvest (Rodrigues et al. 2019).

After harvest, plant samples are recommended to undergo immediate quenching to halt metabolic activities (e.g., enzymatic and chemical activities) and prevent further changes (Jorge et al. 2016; Rodrigues et al. 2019; Alseek et al. 2021). This critical step preserves metabolite stability by preventing degradation and fluctuations in concentration, chemical composition, or physical properties (Rodrigues et al. 2019). This is typically achieved by rapidly freezing the samples at extremely low temperatures using liquid nitrogen (shock freezing), followed by storage at −80 °C until metabolite extraction (Jorge et al. 2016; Rodrigues et al. 2019). It is important to ensure that debris (e.g., soil from root samples) is removed before quenching, as in the practice when conducting oil palm research (Nurazah et al. 2021). However, if liquid nitrogen or dry ice is not available (e.g., when the sampling site is far from the laboratory), harvested samples can also be transported in a cooler with ice packs until they can be properly stored and frozen (Uthe et al. 2021).

2.2 Pre-analytical Technique: Plant Metabolite Extraction

The extraction of plant metabolites is a critical step in plant metabolomics, as it significantly influences both the range of metabolites detected and the yield of each metabolite recovered from plant tissues (Rodrigues et al. 2019). Currently, organic solvents are the primary choice for this purpose (Table 2). An ideal extraction solvent should be non-toxic, possess high solubilizing power, and exhibit broad extraction capability to maximize metabolite recovery (Mushtaq et al. 2014; Rodrigues et al. 2019). However, solvent selection must also align with the specific objectives of the study and the analytical platform to be employed (i.e., NMR, GC–MS, LC–MS). Polar metabolites are typically extracted using polar solvents, such as methanol or methanol–water mixtures, whereas non-polar solvents are better suited for the extraction of non-polar metabolites (Table 2) (Jorge et al. 2016). Among these, methanol–water mixtures are widely preferred in metabolomics studies due to their demonstrated ability to extract a broad range of metabolites, including alkaloids, amino acids, organic acids, phenolic compounds, and sugars (Table 2) (Mushtaq et al. 2014). This mixture has a balanced polarity as both methanol and water have different polarities. Furthermore, hydrophobic compounds such as terpenoids, fatty acids, and lipids have been extracted using chloroform, ethyl acetate, and *n*-hexane (Mushtaq et al. 2014). As such, no single solvent extract can provide a complete spectrum of metabolites (Jorge et al. 2016; Martin et al. 2014).

Following metabolite extraction, solvents are typically evaporated to concentrate the metabolite. This is commonly performed using a vacuum concentrator, such as a rotary evaporator, and often at room temperature to prevent thermal degradation. It is suggested to avoid high temperatures (generally above 40 °C), as these can lead to metabolite degradation. Additionally, when working with extracts containing metabolites sensitive to oxidative changes, a nitrogen flow evaporator is recommended (Salem et al. 2020). Once dried, the crude extracts are resuspended in a solvent compatible with the intended analytical platform (e.g., LC–MS grade methanol). In the case of GC–MS, dried samples often require chemical

Table 2 The commonly used solvents for metabolite extraction in plant metabolomics research

Solvent	Plant species (plant tissues)	Properties evaluated	References
Ethyl acetate	Malvaceae plants (leaves and stems); *Toona sinensis* (shoots), and *Firmiana simplex* (barks)	Antibacterial, antioxidant, anti-infective	Abdelhafez et al. (2019), Yang et al. (2022)
70% methanol	*Tithonia diversifolia* (leaves, stems, roots, and inflorescences)	Influence of abiotic factors	Sampaio et al. (2016)
Methanol	*Spinacia oleracea* (spinach, roots and leaves); *Jacobaea vulgaris* (leaf); *Valeriana officinalis, Melissa officinalis, Hypericum perforatum*, and *Passiflora incarnata* (leaves and roots); tobacco (leaves)	Phytotoxicity; plant–soil interactions; neurobiological; plant tolerance to micronutrient supplementation/toxicity; environmental influence	Zhang et al. (2019), Huberty et al. (2022), Gonulalan et al. (2020), Zhao et al. (2015)

derivatization, as many naturally occurring metabolites are non-volatile and cannot be directly analyzed by GC–MS (Moros et al. 2016). If immediate analysis of the crude extracts is not required, it is generally not advisable to store them in liquid form (whether containing aqueous or organic solvents) even at low temperatures (−20 °C), as this may compromise metabolite stability (Fernie et al. 2011). The preferred approach is to store the extracts in a dried form. However, the optimal storage method ultimately depends on the chemical stability of the specific metabolites present in the extract (Fernie et al. 2011).

2.3 Analytical Techniques

The successful implementation of plant metabolomics heavily depends on the careful selection of an appropriate analytical technique, as plant extracts contain a chemically diverse array of metabolites with varying abundance, polarity, and chemical properties. No single technique can capture the entire plant metabolome, as each possesses its own unique strengths and limitations (Table 3). Among the techniques used in plant metabolomics are capillary electrophoresis-mass spectrometry (CE–MS), gas chromatography–mass spectrometry (GC–MS), liquid chromatography–mass spectrometry (LC–MS), and nuclear magnetic resonance (NMR) (Table 3). These analytical techniques differ in their detection principles, sample requirements, sensitivity, resolution, and suitability for specific metabolite classes (Table 3).

Table 3 Overview of analytical techniques used in plant metabolomics

Analytical technique	Detection principle	Sample requirement	Sensitivity and resolution	Target metabolites
Capillary electrophoresis–mass spectrometry (CE–MS)	Differential mobility of charged species in an electrical field	Very small sample volumes (in nanoliter)	High	Polar and charged metabolites
Gas chromatography–mass spectrometry (GC–MS)	Separation of volatile compounds via the gas phase	Requires chemical derivatization for non-volatile compounds	High	Volatile organic compounds; for primary metabolites
Liquid chromatography–mass spectrometry (LC–MS)	Mass-to-charge (*m/z*) ratio via ionization	Requires extraction with suitable solvent	High	Various metabolites detected based on the column used
Nuclear magnetic resonance (NMR)	Spins of atomic nuclei	Larger sample volume; does not require chemical derivatization	Moderate	Polar and non-polar metabolites

CE–MS is particularly suitable for analyzing polar and charged metabolites, offering high sensitivity and resolution and minimal sample preparation (Table 3). GC–MS, on the other hand, is highly effective for detecting volatile organic compounds, such as primary metabolites including sugars and amino acids (Table 3). Its high resolution and sensitivity make it a valuable tool in plant metabolomics, though it typically requires chemical derivatization of samples before the analysis. LC–MS excels at analyzing various metabolites, depending on the type of column used, without the need for chemical derivatization (Table 3). NMR spectroscopy, while less sensitive than LC–MS, provides highly reproducible quantitative data and structural information without the need for chemical derivatization (Table 3). The selection of an analytical technique for a plant metabolomic study is guided not only by the chemical nature of the target metabolites but also by key factors such as analytical sensitivity, resolution, and sample preparation requirements.

2.3.1 Capillary Electrophoresis–Mass Spectrometry (CE–MS)

Capillary electrophoresis (CE) separates metabolites based on their charge-to-size ratio under the influence of an electric field. It is effective for resolving small, polar, and ionic compounds that may be challenging for other methods (Li et al. 2024). Its narrow capillaries enhance separation efficiency, enabling rapid, high-resolution analyses with minimal sample and reagent use, often without organic solvents. CE uses simple fused-silica capillaries, making it a cost-effective and environmentally friendly technique (Ramautar et al. 2008). When coupled with mass spectrometry (CE–MS), it provides highly selective and sensitive metabolite detection in complex plant matrices (Harada and Fukusaki 2009). However, CE–MS is rarely employed as a standalone platform in plant metabolomics and is usually integrated with other analytical techniques. For example, the integration of CE–MS and gas chromatography–selected ion monitoring–mass spectrometry (GC–SIM–MS) revealed differences in metabolic processes in tobacco leaves from Chinese regions (Xuchang and Dali) (Zhao et al. 2015), demonstrating the high complementarity of CE–MS to GC–MS in capturing metabolites less detectable by other techniques.

While CE–MS has its unique advantages, it is still considered technically challenging and less reproducible compared to other analytical techniques, making it an under-represented platform in metabolic profiling studies (Zhang and Ramautar 2020). A notable limitation arises from the volume requirements of commercial CE instruments, which typically need 5-10 μl (or more) of sample in the vial, despite only nanoliter-scale volumes being injected into the capillary. This discrepancy between required and injected volumes can hinder the analytical performance of CE–MS in metabolomics studies where sample quantity is highly limited (Zhang and Ramautar 2020). Nevertheless, developments were made in recent years, allowing enhanced accuracy and efficiency in separation processes (Li et al. 2024).

2.3.2 Gas Chromatography–Mass Spectrometry (GC–MS)

Gas chromatography–mass spectrometry (GC–MS) is a well-established analytical technique widely used in metabolomics due to its affordability, reproducibility, stability, and ease of data processing (Lisec et al. 2006; Putri et al. 2022). In this technique, gas chromatography (GC) separates volatile compounds, which are then identified by mass spectrometry (MS) based on their *m/z* ratios and the relative abundance of the molecular and fragment ions (Mota et al. 2021). However, GC–MS is limited to the analysis of volatile and thermally stable metabolites, and non-volatile compounds must undergo chemical derivatization to be converted into volatile forms suitable for analysis (Jorge et al. 2016).

Recent technological advances in GC have significantly enhanced its application in metabolomics. These include the development of gas chromatography time-of-flight mass spectrometry (GC–TOF–MS), gas chromatography–quadrupole mass spectrometry (GC–QMS), and gas chromatography–ion trap mass spectrometry (GC–ITMS). Among these, GC–TOF–MS has emerged as a widely adopted platform in metabolomics due to its fast acquisition speed, high mass resolution, and excellent sensitivity, making it well-suited for untargeted metabolite profiling where detection of a wide range of compounds is essential (Rohloff 2015). For example, GC–TOF–MS-based metabolomics revealed that nitrogen deficiency alters the physiological response and metabolic profiles of both tea (*Camellia sinensis* (L.)) leaves and roots (Lin et al. 2021). In addition to advances in instrumentation, two main ionization techniques are commonly used in GC–MS: electron ionization (EI) and chemical ionization (CI) (Beale et al. 2018). Of these, EI is the most widely applied in metabolomics due to the availability of comprehensive mass spectral libraries, which facilitate compound identification (Beale et al. 2018).

Interestingly, comprehensive two-dimensional gas chromatography (GC × GC) has been developed and used in studying plants, particularly the analysis of essential oils (Omar et al. 2012). This is an advanced chromatographic technique designed to significantly improve the separation of complex mixtures by combining two different separation mechanisms in a sequential and continuous manner (Mondello et al. 2025). This system uses two columns with different stationary phases, typically a non-polar column in the first dimension (1D) and a polar column in the second dimension (2D), which allows for greater peak capacity and improved resolution of co-eluting compounds. GC × GC can resolve hundreds of thousands of compounds in a single run, making it particularly valuable for profiling highly complex plant metabolomes (Wong et al. 2017).

While GC × GC offers enhanced resolution for highly complex plant samples, the conventional GC–MS remains widely used and continues to play an essential role in plant metabolomics due to its accessibility, robustness, and proven effectiveness across diverse plant species. One example includes the evaluation of cuticular wax composition in four wheat (*Triticum aestivum*) cultivars, where a total of 263 putative compounds were detected, including 58 wax-related metabolites such as alkanes and fatty acids (Lavergne et al. 2018). In another application, GC–MS was used to investigate the phytochemical composition of *Wedelia trilobata*, an invasive weed traditionally used to treat inflammation and bacterial infections (Azizan et al. 2015). As such, GC–MS serves as a versatile tool in plant metabolomics that uncovers relevant plant metabolites that contribute to fundamental and applied plant science research.

2.3.3 Liquid Chromatography–Mass Spectrometry (LC–MS)

Liquid chromatography–mass spectrometry (LC–MS) is another widely used analytical technique in plant metabolomics. It is highly valued for its excellent sensitivity, broad metabolite coverage, and capability to analyze a diverse range of polar, semi-polar, and non-volatile compounds. LC–MS does not require chemical derivatization, making it particularly advantageous for the analysis of plant metabolites (Waris et al. 2022). The liquid chromatography (LC) component separates complex mixtures based on differences in polarity or other physicochemical properties, and not volatility or boiling points as in GC–MS. This is achieved through the interaction of metabolites with a stationary phase and a liquid mobile phase. Once separated, the metabolites are introduced into the mass spectrometer (MS), where they are ionized and converted into charged particles. These ions are then separated according to their *m/z* ratios, enabling their detection, identification, and quantification (Kruve et al. 2015; Sarkar et al. 2025). Furthermore, tandem mass spectrometry (MS/MS) enhances structural elucidation by selecting precursor ions, fragmenting them into product ions, and analyzing the resulting fragments. This technique is particularly useful in plant metabolomics for characterizing individual metabolites and resolving components within mixtures (de Hoffman 2006; Albrecht et al. 2012).

In LC–MS, metabolites are ionized through ionization techniques that convert neutral molecules into charged ions. Several ionization methods are commonly used, including electrospray ionization (ESI), matrix-assisted laser desorption ionization (MALDI), atmospheric pressure chemical ionization (APCI), atmospheric pressure photo ionization (APPI), desorption electrospray ionization (DESI), and direct analysis in real time (DART) (Awad et al. 2015). The ions are then separated and analyzed by mass analyzers, such as single-quadrupole, triple-quadrupole (TQ), time-of-flight (TOF), ion traps, orbitrap, and Fourier transform ion cyclotron resonance (FT–ICR) (Lu et al. 2008; Himmelsbach 2012). Quadrupole analyzers (e.g., single-quadruple, triple quadrupole) use oscillating electrical fields to selectively filter ions based on their *m/z* ratio, allowing only selected ions to pass through the detector (Patel et al. 2012). In contrast, TOF analyzers accelerate ions through an electric field and measure the time they take to reach the detector, with lighter ions arriving faster than heavier ones (Patel et al. 2012). This enables high-speed acquisition of accurate mass data, making TOF particularly useful for untargeted metabolomics. A hybrid configuration combining a quadrupole mass filter with a TOF analyzer, known as Q-TOF, has become widely adopted in plant metabolomics due to its high resolution, sensitivity, and ability to perform tandem MS (MS/MS) experiments (Glauser et al. 2013; Mais et al. 2018; Otify et al. 2023a).

The ability to detect a wide range of metabolites is largely attributed to advances in LC column technology. Reversed-phase liquid chromatography (RP–LC), particularly using C18 columns, is the most employed separation mode and offers effective resolution of nonpolar and weakly polar compounds. For example, RP–LC–MS-based metabolomics of *Hass* avocados from Peru, Chile, and Spain revealed distinct metabolic profiles by origin (Serrano-Garcia et al. 2023), emphasizing the important role of RP–LC in distinguishing closely related plant species on their unique metabolic signatures. However, despite this advantage, this approach exhibits limited retention of highly hydrophilic, ionic, and polar metabolites (Tang et al. 2014). To address this limitation, normal-phase (NP) liquid chromatography has been explored due to its enhanced ability to retain polar compounds. However, NP–LC has not gained widespread use in plant metabolomics, primarily because many polar analytes are not readily soluble in the typical NP solvent systems used. To overcome this drawback, hydrophilic interaction liquid chromatography (HILIC) was developed as an alternative approach for enhancing the retention and separation of highly polar metabolites. HILIC utilizes a polar stationary phase and a predominantly organic mobile phase, enabling strong interactions with polar metabolites and thus improving their chromatographic resolution (Tang et al. 2014; Waris et al. 2022). This makes HILIC particularly valuable in a metabolomics workflow where comprehensive coverage of both polar and non-polar metabolites is essential (Ito et al. 2014).

Like other analytical techniques, LC–MS also has its limitations. One key challenge is the reproducibility of metabolite detection and spectral data (Gika et al. 2014). In contrast to GC–MS, where the EI produces consistent and reproducible fragmentation patterns that serve as reliable "fingerprints" for metabolite identification, LC–MS spectra can be highly variable. This variability is largely due to the unpredictable formation of adducts and molecular clusters, as well as inconsistent fragmentation patterns, even across instruments of the same type (Gika et al. 2014). Therefore, the development and implementation of standardized protocols, robust quality control procedures, and method validation are essential to improve data reliability. Additionally, integrating multiple analytical platforms can help overcome individual limitations and provide a more comprehensive metabolomics profile (Zhang et al. 2012; Gika et al. 2014).

2.3.4 Nuclear Magnetic Resonance (NMR)

The nuclear magnetic resonance (NMR) is a powerful analytical technique that can identify and simultaneously quantify a wide range of organic compounds in the micro-molar range (Zhang et al. 2012). Its metabolite detection is based on the interaction between the magnetic moments of atomic nuclei and an external magnetic field. These magnetic moments arise from a property known as nuclear spin, which is an intrinsic form of angular momentum possessed by certain nuclei. The spin quantum number defines the value of nuclear spin, and only nuclei with non-zero spin (e.g., ^{1}H, ^{13}C, ^{15}N, ^{31}P) are NMR-active. The magnetic moment associated with the nuclear spin is determined by the intrinsic properties of the nucleus, including its charge and mass, as well as its spin number (Mlynárik 2017).

While NMR has lower detection sensitivity compared to mass spectrometry (MS), it offers several advantages in terms of structural information content, reproducibility, and quantification (Schripsema 2010). One of the key strengths of NMR is that it is non-discriminatory and less biased, as the results are not influenced by ionization conditions, unlike MS-based techniques (Salem et al. 2020). Furthermore, NMR methods are highly reproducible, non-invasive, and non-destructive, allowing for repeated analyses of the same sample without compromising its chemical integrity (Gautam et al. 2025). This allows metabolomics analyses to be performed directly on intact plant tissues, preserving them for subsequent tests or applications (Gautam et al. 2025). NMR also requires minimal sample preparation, as the sample does not come into physical contact with any component of the instrument, such as a chromatographic column, which is often required in MS-based workflows (Salem et al. 2020). As such, NMR has been utilized in several plant metabolomics studies, from comprehensive profiling of tissue-specific metabolites in plants (Li et al. 2018) to identifying plant disease-related biomarkers (Pontes et al. 2016).

2.4 *Post-analytical Technique: Data Mining*

The generation of "Big Data" through analytical techniques commonly used in plant metabolomics necessitates robust and efficient data mining approaches to fully extract and interpret all information embedded in complex datasets (Tugizimana et al. 2016). Data mining includes data pre-processing, data pre-treatment, statistical analysis, and metabolite annotation, which can be achieved using available software (Table 4).

A variety of software and web-based platforms are available to support plant metabolomics research (Table 4). Each tool is typically tailored to a specific type of raw data (e.g., MS or NMR) and analytical task to be performed (e.g., pre-processing, pre-treatment, statistical analysis). For example, CROP (Kouřil et al. 2020) and Decon2LS (Jaitly et al. 2009) are commonly used for pre-processing MS raw data (Table 4). Similarly, MNova NMR (Bruker) and TopSpin (Bruker) are widely used for the pre-processing of NMR data (Table 4). There is also software dedicated to statistical analysis (ROBPCA) and metabolite annotation (e.g., GNPS, MetFrag (Ruttkies et al. 2016)). Comprehensive platforms, such as MetaboAnalyst (Pang et al. 2024), MetaboScape (Bruker), Progenesis QI (Nonlinear Dynamics), and XCMS Online (Tautenhahn et al. 2012), offer end-to-end solutions for all data mining processes, making them popular choices in plant metabolomics workflows (Table 4). These tools enhance the accuracy and efficiency of metabolomics analysis, which is essential for ensuring robust, reproducible data processing and meaningful biological interpretation.

Table 4 Available software tools for data mining processes, including data pre-processing, pre-treatment, statistical analysis, and metabolite annotation

Software name	Category	Analytical technique	Website	Reference
CROP (correlation-based removal of multiplicities)	Pre-processing	LC–MS	www.github.com/rendju/CROP (R-script)	Kouřil et al. (2020)
Decon2LS	Pre-processing	LC–MS	https://ncrr.pnl.gov/software/ (software installation)	Jaitly et al. (2009)
Galaxy-M	Pre-processing, pre-treatment, statistical analysis	LC–MS	https://github.com/Viant-Metabolomics/ (R script, Pythons)	Davidson et al. (2016)
GNPS (global natural products social molecular network)	Metabolite annotation	LC–MS	http://gnps.ucsd.edu/ (web-based)	Wang et al. (2016)
MetaboAnalyst 6.0	All[a]	LCMS, NMR	http://metaboanalyst.ca/ (web-based)	Pang et al. (2024)
MetaboScape	All	LC–MS	http://www.bruker.com/ (commercial)	Bruker
MetFrag	Metabolite annotation	LC–MS	http://www.msbi.ipb-halle.de or https://ipb-halle.github.io/MetFrag/	Ruttkies et al. (2016)
MNova NMR	Pre-processing,	NMR	http://www.bruker.com/ (commercial)	Bruker
MZMine 3	All	LC–MS, GC–MS	http://www.mzmine.org (software installation)	Schmid et al. (2023)
MS-DIAL 4	Pre-processing	LC–MS, GC–MS	http://systemsomicslab.github.io/en/ (software installation)	Tsugawa et al. (2020)
OpenMS 3.3.0	Pre-processing, pre-treatment	LC–MS	http://www.openms.de (software installation)	Sturm et al. (2008)
Progenesis QI	All	LC–MS	http://www.nonlinear.com/ (commercial)	Nonlinear dynamics
ROBPCA	Statistical analysis		https://github.com/MLGroup15/ROBPCA/	Hubert et al. (2005)
SIMCA-P	Pre-processing, pre-treatment, statistical analysis	LC–MS, NMR	http://www.sartorius.com/ (commercial)	Umetrics
TopSpin 4.5.0	Pre-processing	NMR	https://www.bruker.com/ (commercial)	Bruker
XCMS online	All	LC–MS, GC–MS	https://xcmsonline.scripps.edu (web-based)	Tautenhahn et al. (2012)

[a]All refers to all data mining processes: data pre-processing, data pre-treatment, statistical analysis, and metabolite annotation

2.4.1 Data Pre-processing

The primary objective of data pre-processing is to reduce the complexity of the raw metabolomics data and extract the most relevant features for downstream analysis (Castillo et al. 2011). This step is pivotal for ensuring analytical reproducibility and biological comparability across samples. Key pre-processing procedures include noise filtering, peak detection, deconvolution, deisotoping, chromatographic alignment, and baseline correction (Castillo et al. 2011; Liland 2011). Each of this step may vary according to the type of analytical technique being used.

Noise filtering is a critical pre-processing step aimed at distinguishing true analytical signals originating from biologically relevant compounds from background signal coming from chemical or instrumental interferences that are not of biological interest (Castillo et al. 2011; Schiffman et al. 2019). To effectively perform noise filtering, an appropriate signal-to-noise ratio (S/N) threshold is recommended to be set, as overly stringent thresholds may potentially eliminate low-abundance yet biologically significant metabolites (Gadara et al. 2021). Peak detection, on the other hand, involves identifying the correct signal shapes that potentially correspond to real compounds and estimating their intensities (Castillo et al. 2011). This step is essential as peak lists often include a high density of artifacts or false positives that do not represent actual chemical analytes (Stancliffe and Patti 2023). While peak detection and noise filtering are closely related, the latter must be performed before peak detection to enhance the accuracy and reliability of the identified features. Furthermore, deconvolution is a computational step that separates co-eluting compounds and reconstructs a pure mass spectrum for each individual component (Du and Zeisel 2013). This process is essential for resolving overlapping signals and accurately estimating the contribution of each compound to the total signal. Another important pre-processing step is deisotoping, which combines multiple signals from the same molecule that appear with different charge levels into one clear signal, usually the monoisotopic peak (Smith et al. 2014). This is important when searching a database, as only the single-charge *m/z* ratio and sometimes the retention time are used.

Working on metabolomics involves the analysis of hundreds to thousands of samples, which may take several days or weeks to be completed. Due to the large data volume and inherent variability introduced during sample preparation and instrument performance, robust pre-processing steps such as chromatographic alignment, baseline correction, scaling, and normalization are essential to ensure data quality and reliability. Chromatographic alignment is used to match peaks corresponding to the same compound across all samples, accounting for retention time shifts during data acquisition (Castillo et al. 2011). Baseline correction, on the other hand, involves removing background signal and noise from chromatograms, which may arise from instrument fluctuations or within- and between-batch variations in signal intensity (Brunius et al. 2016).

2.4.2 Data Pre-treatment

Data pre-treatment is the critical step that follows data pre-processing and precedes statistical analysis. It involves refining the structured datasets generated during pre-processing to enhance the reliability, comparability, and interpretability of downstream analyses (Sun and Xia 2024). A key objective of pre-treatment is to reduce the impact of unwanted systematic variation, such as instrumental drift, sampling inconsistencies, or experimental noise, so that biological differences can be more accurately detected. Normalization, a core pre-treatment technique, aims to correct for these technical biases by adjusting the data across samples, thereby ensuring that observed differences reflect true biological variability rather than artifacts of the experimental process (Sun and Xia 2024). It is a strategy to reduce sample-to-sample variability due to differences in the amounts or concentrations of samples (Misra 2020). Without proper normalization, the data generated will be erroneous, unreliable, and misleading. Another important technique is scaling, which standardizes each metabolite by dividing it by a specific scaling factor. This helps make metabolites more comparable, especially when they vary greatly in concentration, and ensures each contributes equally to statistical models (van den Berg et al. 2006). Furthermore, transformation is also used to improve statistical performance by addressing issues such as heteroscedasticity (non-constant variance) and non-normal distributions (van den Berg et al. 2006; Sun and Xia 2024). If some metabolites are much larger or smaller than others, the data might be hard to compare or may not follow a normal distribution. Common transformations in metabolomics include log transformation, which reduces data heteroscedasticity; power transformation, which minimizes data skewness; and variance stabilization normalization (VSN), which reduces sample-to-sample variation and adjusts the variance of different metabolites (Sun and Xia 2024).

2.4.3 Statistical Analysis

Given the large number of metabolites generated in plant metabolomics (as well as other metabolomic studies), robust statistical analysis is critical for uncovering meaningful patterns, comparing experimental groups, and identifying potential biomarkers. Two main types of statistical approaches are commonly employed: supervised and unsupervised statistical methods (Ren et al. 2015). Supervised methods are guided by known class labels and aim to reveal relationships between metabolite profiles and predefined sample categories (Bartel et al. 2013). In contrast, unsupervised methods explore the data without prior knowledge of class labels, allowing for the discovery of intrinsic patterns or clustering the dataset (Gal et al. 2020).

In plant metabolomics, unsupervised statistical methods play a critical role in data exploration, pattern recognition, and hypothesis generation. Commonly used techniques include principal component analysis (PCA), hierarchical cluster analysis (HCA), K-means clustering, and correlation analysis. Among these, PCA is the most widely adopted multivariate analysis tool for metabolomics (Worley and

Powers 2013). It is a non-parametric method that reduces data dimensionality by projecting complex NMR or MS spectra into a lower-dimensional space, thereby revealing inherent data structure and trends (Nyamundanda et al. 2010). PCA is primarily used to visualize the distribution and variability of metabolite profiles post-dimensional reduction (Jiang et al. 2022). Similarly, HCA offers a way to group samples based on their metabolic similarity using a dendrogram, allowing researchers to identify patterns or outliers within the dataset (Caesar et al. 2018). Like PCA, it provides an overview of how samples or variables relate to one another, but in a hierarchical manner. On the other hand, K-means clustering partitions the dataset into a pre-defined number of clusters, making it useful for identifying metabolic groupings or conditions with shared profiles in a large dataset (Ikotun et al. 2023). Lastly, correlation analysis is employed to explore the relationships between one dependent variable and one or more variables in a big or high-dimensional dataset (Xiao et al. 2016). In plant systems, this could explore the relationships between metabolites or between metabolites and phenotypic traits, offering insights into metabolic network interactions and functional associations.

Supervised statistical methods are essential in plant metabolomics, particularly for classification, prediction, and the identification of discriminant metabolites. Commonly used methods include partial least squares discriminant analysis (PLS–DA), orthogonal partial least squares discriminant analysis (OPLS–DA), random forest (RF), and volcano plots. PLS–DA is a supervised multivariate technique that reduces data dimensionality while incorporating class label information, making it especially useful for identifying key variables that distinguish between experimental conditions (Ruiz-Perez et al. 2020). It supports both predictive and descriptive modeling and aids in selecting significant variables with high discriminant power (Lee et al. 2018). An extension of this, OPLS–DA, enhances model interpretability by isolating variation that is orthogonal (i.e., unrelated) to the class distinction, effectively separating relevant predictive variation from noise in complex metabolomics datasets (Boccard and Rutledge 2013). In plant metabolomics, OPLS–DA is particularly valuable for understanding plant responses to different treatments or environmental conditions, as it reveals changes in metabolic profiles. Furthermore, random forest (RF) is a type of machine learning method that builds multiple decision trees and aggregates their results for classification or regression tasks (Acharjee et al. 2020). In plant metabolomics, RF can be used to classify samples (e.g., by species, treatment, or stress response) and to rank metabolites by group separation, facilitating biomarker discovery (Acharjee et al. 2020). The volcano plot combines univariate statistical results, particularly p-values from t-tests and fold-change values, to visualize and identify significantly different metabolites between two groups. It is especially useful in screening for differential metabolites, while also controlling for false discovery rate (Kumar et al. 2018).

Following statistical analysis, metabolite annotation is performed. In LC–MS-based metabolomics, this process involves assigning identities to detected peak features, which are typically defined as a peak or a set of aligned peaks across samples with a unique *m/z* and a specific retention time (Domingo-Almenara et al. 2018). Accurate annotation is essential for the biological interpretation of metabolomics data (Chaleckis et al. 2019). The most common approach to annotating these features is to search the exact mass of the precursor ion and MS/MS against established spectral libraries. The METLIN database is a web-based data repository that provides an extensive, annotated list of known metabolite structures, which can be readily cross-referenced with its catalog of high-resolution Fourier transforms mass spectrometry (FTMS), MS/MS, and LC–MS data. This database was developed to assist metabolite identification via mass analysis (Smith et al. 2005). Similarly, MassBank serves as a public repository of mass spectra for small chemical compounds (i.e., < 3000 Da) relevant to the life sciences (Horai et al. 2010). It operates as a distributed database, with contributors preparing and sharing their data via individual servers. GNPS is an open-access platform designed to facilitate community-wide sharing and analysis of raw, processed, or annotated MS/MS data. It enables users to analyze their datasets and compare them against all publicly available data, thus fostering collaborative metabolite identification (Wang et al. 2016). PubChem is another widely used public repository that provides comprehensive information on chemical substances and their biological activities (Kim et al. 2015). Furthermore, a unique plant database called KNapSAck (Nakamura et al. 2013) contains information on plant metabolites and their corresponding plant species. It is regularly updated to include newly identified metabolites, their associated biological functions, and their geographical distribution. In plant metabolomics research, scientists are not restricted to a single database. Instead, they often utilize multiple complementary databases to improve the accuracy and reliability of metabolite annotation.

However, due to limitations of spectral databases and the vast chemical diversity in plant metabolomes, many detected features remain unidentified. To address this, bioinformatic tools such as MetFrag (Ruttkies et al. 2016) and SIRIUS (Dührkop et al. 2019) have been developed to predict in silico MS/MS spectra or molecular fingerprints and to match them with candidate structures (Zhou et al. 2022). MetFrag is an in silico fragmentation predictor tool for small-molecule identification (Ruttkies et al. 2016). It ranks candidate compounds from various databases by comparing predicted fragments to experimental spectra. MetFrag then produces a list of "candidate compounds" that can be scored and filtered based on different criteria, such as elemental composition, thereby refining metabolite selection. Similarly, SIRIUS is another tool that enables rapid molecular structure identification by combining high-resolution isotope pattern analysis with fragmentation tree algorithms. It integrates these features to deliver a unified and efficient structural elucidation of compounds from MS/MS data (Dührkop et al. 2019). Both tools have been successfully employed in plant metabolomics (Bhattarai et al. 2022; Molina et al. 2022).

3 Applications of Metabolomics in Plants

Metabolomics has revolutionized plant research. Due to the vast diversity and complexity of metabolites present in plants, metabolomics offers a powerful approach to comprehensively analyze these biochemical signatures and link them to biological functions. As a result, metabolomics has been widely applied across multiple areas of plant science, including research on plant development, physiology and stress response, plant–microbe interactions, agricultural applications, and crop improvement, as well as bioactive plant metabolites and drug discovery.

3.1 Plant Development, Physiology, and Stress Response

Plants undergo complex developmental stages, including germination, vegetative growth, flowering, and senescence, which are tightly regulated by intrinsic genetic mechanisms and environmental cues (Guiboileau et al. 2010; Coen and Prusinkiewicz 2024). Throughout their life cycle, plants are continuously exposed to a multitude of abiotic and biotic stresses, which can vary in intensity, duration and severity (Fig. 4). Biotic stress is triggered by microorganisms, such as bacteria, fungi, viruses and parasites, and insect herbivory (Fig. 4). As plants accumulate nutrients and organic compounds essential for growth and reproduction, they often become targets for microbial colonization, leading to either deleterious or beneficial plant–microbe interactions (Anzano et al. 2022). Abiotic stress, on the other hand, is induced by salinity, heat (drought), cold, heavy metal, and air pollution (Fig. 4). In the context of climate change, the impact of these abiotic stresses continues to pose significant global challenges for plant survival and productivity. As sessile organisms, plants cannot avoid these adverse conditions. Instead, plants evolved sophisticated physiological and biochemical strategies to respond to and mitigate stress impacts (Du et al. 2024).

Metabolomics, therefore, is a powerful approach to understanding these complex plant responses at the biomolecular level. For example, in a study on the mangrove *Avicennia marina* subsp. *australasica*, the levels of non-structural carbohydrates (NSCs), which are known to influence stress tolerance, were experimentally manipulated, followed by a controlled drought and salinity treatment (Ravi et al. 2020). The results showed that low-NSC plants exhibited increased concentrations of amino acids (i.e., leucine and valine) under drought and high salinity stress. In contrast, under combined drought and salinity conditions, high-NSC plants displayed a reduction in α-ketoglutarate and an accumulation of soluble sugars. These findings underscore the functional importance of specific metabolites in conferring salinity and drought stress tolerance in *A. marina*, providing valuable insights into mangrove resilience under changing environmental conditions.

Metabolomics has also been employed to investigate the metabolic responses of radish (*Raphanus sativus* L.) under heavy metal stress conditions, specifically,

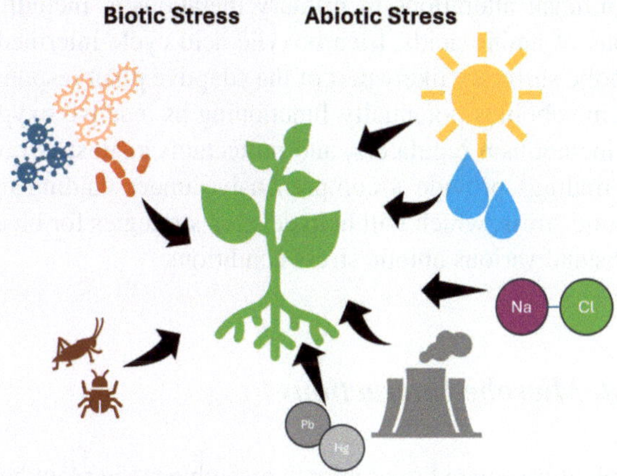

Fig. 4 Various abiotic (e.g., drought, flood, salinity, heavy metals, air pollution) and biotic (e.g., microbial pathogens, insects) stresses that affect plants

exposure to lead (Pb) and cadmium (Cd) (Wang et al. 2015). Both Pb and Cd stresses induced substantial alterations in metabolite profiles, particularly affecting sugars, amino acids, and organic acids. Additionally, Pb stress was found to disrupt carbohydrate metabolism, energy metabolism, and glutathione metabolism. In contrast, Cd stress significantly impacted pathways related to energy production, amino acid metabolism, and oxidative phosphorylation. These findings provided valuable insights into the underlying mechanisms of heavy metal accumulation or tolerance in plants, thereby developing effective management of heavy metal contamination in crops through genetic manipulation.

Furthermore, plant–insect interactions were also investigated using metabolomics. Using GC–MS-based metabolomics, Peng et al. (2016) analyzed both the metabolic profiles of rice plants and brown planthopper (BPH) *Nilaparvata luges* Stål insects during infestation. The study analyzed both the metabolic profiles of resistant (YHY15) and susceptible (TN1) rice cultivars, where YHY15 carries the resistance gene *BPH15*, as well as changes in the honeydew metabolites of BPH Biotypes 1 and Y after feeding on these cultivars. The study revealed that TN1 and YHY15 activated distinct metabolic pathways in response to BPH attack. Furthermore, BPH insects feeding on the resistant YHY15 plants exhibited enhanced amino acid absorption and upregulated cerebral ammonia metabolism, potentially aiding in the detoxification of plant secondary metabolites. These findings highlight the complex metabolic interaction between plant defense and insect adaptation.

The impact of air pollution on plants has also been explored using metabolomics. Metabolite fingerprinting using one- and two-dimensional NMR was performed on the leaves of *Bougainvillea spectabilis* (a plant known to tolerate various abiotic stresses) exposed to prolonged vehicular emissions, compared to plants grown under controlled conditions without such exposure (Mishra et al. 2019). The results revealed significant alterations in primary metabolism, including changes in the concentrations of amino acids, tricarboxylic acid cycle intermediates, and sugars. These metabolic shifts are likely part of the adaptive plant response to air pollution, with altered metabolites potentially functioning as reactive oxygen species (ROS) scavengers, metabolism regulators, and protectants against ultraviolet (UV) radiation. These findings provide a comprehensive understanding of the response of plants to abiotic stress, which will help develop strategies for bioengineering plants that can withstand various abiotic stress conditions.

3.2 Plant–Microbe Interactions

The role of plant-associated microbiome assemblages in plant health has also been investigated through metabolomics. For example, the metabolites produced by soil-borne and endophytic microbiota (i.e., bacteria and fungi) in healthy and infected ginger plants (*Zingiber officinale* L. Roscoe) (rhizome rot disease) were profiled using untargeted metabolomics approach (Wang et al. 2024). It was revealed that lipids and lipid-like molecules were the most abundant metabolites strongly associated with plant health compared to those linked with rhizome rot disease. These findings highlight the importance of metabolome-driven microbiome assembly in maintaining plant health and offer valuable insights for developing microbiome-based strategies in crop protection. Similarly, a metabolomics approach was employed to investigate disease suppression via intercropping of chrysanthemum (*Chrysanthemum morifolium*) with ginger (*Zingiber officinale*). It was discovered that *Fusarium* wilt incidence was successfully suppressed by the modulation of various metabolites produced by the microbiome found in the rhizosphere (Zhu et al. 2025). Ginger root exudates triggered the release of specific metabolites from chrysanthemum roots, which in turn promoted the growth and biofilm formation of *Burkholeria* sp. These findings provide valuable insights into the role of root-exudate-mediated plant–microbe interactions in disease suppression and highlight potential strategies for enhancing crop health and productivity through intercropping systems. Furthermore, metabolomics was also employed to investigate the pathogenesis of *Rhizoctonia solani* and the role of its phytotoxin, phenylacetic acid (PAA), on maize leaves and sheaths (Hu et al. 2018). The analysis revealed distinct metabolite profiles between the two tissue types, suggesting a higher susceptibility to *R. solani* and providing insights into the differential mechanisms of host–pathogen interactions. These studies emphasize the essential role of metabolomics in elucidating the complex dynamics of plant–microbe interactions.

3.3 Agricultural Applications and Crop Improvement

Metabolomics is widely applied in agriculture to improve crop yield and quality. For example, untargeted metabolomics combined with agro-meteorological data was used to monitor the development of sorghum (*Sorghum bicolor* (L.)) across 12 crop fields in Italy (Mandrone et al. 2021). The study identified dhurrin, a cyanogenic glucoside, as a biomarker of crop quality and development, and provided new insights into its turnover during plant maturation. Additionally, *p*-glucosyloxy-2-hydroxyphenylacetic acid was detected as a major metabolite that accumulated during ripening. The findings also demonstrated that metabolomic profiles varied in response to fertilization, biotic and abiotic stresses, water supply, and soil properties. By profiling the metabolites, a more informed crop monitoring is made possible to enhance management practices for crop improvement. Furthermore, metabolite profiling has also been applied to vegetables to evaluate their health-promoting properties (Tao et al. 2023). A total of 403 flavonoids (e.g., flavonoid, flavonols, anthocyanins, isoflavones, flavonoid carbonoside, dihydroflavone, chalcones, tannin, proanthocyanidins) from 20 commonly consumed vegetables have been profiled, with bean sprouts and hot peppers exhibiting the highest flavonoid abundance. Notably, bean sprouts and hot pepper also demonstrated significant antioxidant enzyme activity, highlighting their potential nutritional and functional value. These findings contribute to the establishment of a comprehensive database on major flavonoid contents in vegetables, which can inform the selection and breeding of vegetable varieties with enhanced health-promoting properties.

Analyzing the metabolomes of plants grown under stress can help identify metabolites associated with stress resistance, which are helpful for plant breeders in selecting resilient and high-performing phenotypes (Abdelrahman et al. 2018). For example, a metabolomics approach was applied to profile 210 recombinant inbred lines of rice (*Oryza sativa*) to support hybrid breeding, which is a key strategy to grow more rice. This study demonstrated that selecting the top 10 hybrids based on metabolite data could lead to an estimated 30% increase in yield (Xu et al. 2016). It was then further revealed that metabolomic prediction (i.e., prediction based on plant metabolites) outperformed genomic prediction, highlighting the superior potential of metabolomics in enhancing crop yield. Similarly, metabolomics was also applied to *Ocimum basilicum* (basil) to optimize the production of bioactive metabolites with nutraceutical and antibacterial properties. Treatments with humic substances and compost teas derived from artichoke (ART) and coffee grounds (COF) enhanced the metabolism of basil plants (Verrillo et al. 2022). The findings revealed 149 basil metabolites with antioxidant and antibacterial properties. The application of ART and COF extracts promoted the biosynthesis of aromatic amino acids and saccharides involved in hydroxycinnamic acid production, thereby stimulating overall plant metabolism. Additionally, there was a notable increase in the production of phenolic compounds (i.e., caffeic acid, naringenin, resveratrol, and rosmarinic acid), which are potent antioxidants known for their health-promoting properties. These findings highlight the potential of metabolomics as a powerful tool for improving crop performance and contributing to global food security through the development of high-yielding, stress-resilient, and nutritionally enriched cultivars.

3.4 Bioactive Plant Metabolites and Drug Discovery

Plants are rich sources of bioactive secondary metabolites, which have important pharmaceutical applications. Metabolomics has emerged as a powerful tool for identifying these bioactive compounds within the complex plant metabolome. For example, crude extracts of the biennial herb *Ammi visnaga* L. demonstrated inhibitory activity against colon, breast, and hepatocellular carcinoma cells (Ahmed et al. 2022). Metabolomics analysis revealed several key bioactive compounds, such as junipediol A 4-*O*-glucoside, junipediol A 8-*O*-glucoside, acacetin, and apiumetin-*O*-glucoside, which exhibited strong binding affinities for the epidermal growth factor receptor (EGFR) tyrosine kinase, suggesting their potential role in targeted cancer therapy. Furthermore, three *Artemisia* species native to Tibet were profiled for their metabolites with antibacterial and antioxidants properties (Liu et al. 2023). A total of 1109 metabolites were identified, including lipids, amino acids, nucleotides, flavonoids, terpenes, coumarins, organic acids, and phenolic compounds. All three *Artemisia* species demonstrated strong antibacterial and antioxidant activities, highlighting their potential as sources of bioactive compounds for pharmaceutical and nutraceutical applications. Similarly, metabolomics was applied to six lettuce (*Lactuca sativa* L.) varieties (i.e., four green and two red varieties), resulting in the identification of 195 metabolites (Otify et al. 2023a, 2023b). Crude extracts from these varieties exhibited strong antibacterial and antihemolytic activities against *Staphylococcus aureus*, highlighting their potential as sources of natural antimicrobial agents. A metabolomics study was also carried out on tree peonies (*Paeonia suffruticosa*), which are important pharmaceutical and oilseed crops in China (Bai et al. 2024). The findings revealed that the flower buds contain bioactive compounds, including paeniflorin, albiflorin, cynaroside, and pentagalloylglucose, which exhibited strong inhibitory activity against *Listeria monocytogenes.* These studies emphasize that metabolomics is an effective approach for identifying biologically active metabolites from natural sources, supporting its potential applications in the pharmaceutical industry.

4 Challenges and Future Directions

Metabolomics has revolutionized plant science research by offering a powerful platform for both fundamental understanding and applied innovation. It has enabled comprehensive insights into plant physiology, stress responses, and biochemical pathways that were previously difficult to elucidate using traditional tools, such as chromatography and spectroscopy, enzymatic assays, and bioassay-guided fractionation. Through targeted and untargeted metabolomics, researchers can now explore complex plant datasets exposed to various environmental, developmental, and pathological conditions, leading to the discovery of distinct biomarkers and regulatory mechanisms. Metabolomics has also contributed to the comprehensive understanding of plant–microbe interactions, guided the optimization of agricultural practices, and has facilitated discovery of novel bioactive metabolites with pharmaceutical and nutraceutical applications. As such, metabolomics has answered important scientific questions, which were previously unattainable through conventional approaches.

Despite its significant advancements and contributions to plant science research, metabolomics research is not without challenges. Key issues include the inherent complexity of plant metabolomes, the limited availability of annotated metabolites and comprehensive databases, and the scarcity of suitable internal standards. Additionally, the lack of standardized protocols for sample preparation, metabolite extraction, data mining, and analysis across diverse plant species poses a barrier to reproducibility and cross-study comparisons. The sensitivity and specificity of current analytical techniques are also limiting factors, particularly when detecting low-abundance or highly unstable compounds. To address these challenges, the field is moving toward integrative and collaborative solutions. Researchers are adopting integrative approaches, such as combining complementary analytical techniques (e.g., LC–MS, GC–MS, CE–MS, and NMR) to capture a broader and more representative spectrum of plant metabolites. At the same time, community-driven initiatives are expanding metabolite reference libraries and developing user-friendly, publicly accessible bioinformatics platforms to support data mining, analysis, and interpretation. Standardization efforts are also underway, with guidelines such as the metabolomics standards initiative (MSI) and minimum reporting standards for plant metabolomics aiming to reduce variability, promote reproducibility, and facilitate cross-study comparison. Furthermore, the integration of metabolomics with other OMICS technologies, such as genomics, transcriptomics, and proteomics, provides more holistic insights into plant biology.

The future of plant metabolomics depends on the continuous advancement of high-resolution analytical platforms and robust bioinformatics tools. As environmental pressures such as climate change, soil degradation, and emerging plant pathogens intensify, these technological advancements are essential in elucidating plant stress response, uncovering novel plant metabolites, and improving crop resilience. Furthermore, extending metabolomics research beyond model plant species and into underexplored plant species will pave the way for the discovery of novel metabolites and adaptive traits. As global priorities shift towards food security, climate-resilient agriculture, and biodiversity conservation, metabolomics will play a critical role in real-time monitoring of plant health and in developing sustainable agricultural practices. In the coming years, plant metabolomics will evolve from a predominantly research-focused tool to a practical component of plant science, driving impactful discoveries that will benefit humans and the environment alike. Additionally, the integration of artificial intelligence (AI), spatial and single-cell metabolomics, and the development of field-deployable metabolomics (e.g., hand-held sensors) in plant research will expand the reach of the field, enabling more sophisticated methodologies and in situ visualization of metabolites within plant tissues. These advancements offer new ways of studying plant growth and development, as well as for developing solution-based strategies in agriculture and medicine. Metabolomics will therefore continue to be a transformative force for scientific discovery, shaping the next generation of plant science research across both fundamental and applied aspects and supporting global efforts towards a more sustainable future.

Acknowledgements Refined with the assistance of artificial intelligence (AI) tools developed by OpenAI, ChatGPT (GPT-5.1), 2025. Every section was carefully reviewed, revised, and approved by the author/s to ensure that the text is appropriate concerning content and scientific correctness and maintains the integrity of the original content.

Conflict of Interest Disclosure The authors declare that they have no conflict of interest.

References

Abdelhafez OH, Othman EM, Fahim JR, Desoukey SY, Pimentel-Elardo SM, Nodwell JR et al (2019) Metabolomics analysis and biological investigation of three Malvaceae plants. Phytochem Anal 31(2):204–214. https://doi.org/10.1002/pca.2883

Abdelrahman M, Burritt DJ, Tran LSP (2018) The use of metabolomic quantitative trait locus mapping and osmotic adjustment traits for the improvement of crop yields under environmental stresses. Semin Cell Dev Biol 83:86–94. https://doi.org/10.1016/j.semcdb.2017.06.020

Acharjee A, Larkman J, Xu Y, Cardoso VR, Gkoutos GV (2020) A random forest based biomarker discovery and power analysis framework for diagnostic research. BMC Med Genet 13:178. https://doi.org/10.1186/s12920-020-00826-6

Ahmed SST, Fahim JR, Youssif KA, AboulMagd AM, Amin MN, Abdelmohsen UR, Hamed ANE (2022) Metabolomics of the secondary metabolites of *Ammi visnaga* L. roots (family Apiaceae) and evaluation of their biological potential. S Afr J Bot 149:860–869. https://doi.org/10.1016/j.sajb.2022.01.011

Albert NW, Lafferty DJ, Moss SMA, Davies KM (2023) Flavonoids—flowers, fruit, forage and the future. J R Soc N Z 53(3):304–331. https://doi.org/10.1080/03036758.2022.2034654

Albrecht CF, Stander MA, Grobbelaar MC, Colling J, Kossmann J, Hills PN, Makunga NP (2012) LC-MS-based metabolomics assists with quality assessment and traceability of wild and cultivated plants of *Sutherlandia frutescens* (Fabaceae). S Afr J Bot 82:33–45. https://doi.org/10.1016/j.sajb.2012.07.018

Alseek S, Aharoni A, Brotman Y, Contrepois K, D'Auria J, Ewald J et al (2021) Mass spectrometry-based metabolomics: a guide for annotation, quantification and best reporting practices. Nat Methods 18:747–756. https://doi.org/10.1038/s41592-021-01197-1

Anzano A, Bonanomi G, Mazzoleni S, Lanzotti V (2022) Plant metabolomics in biotic and abiotic stress: a critical overview. Phytochem Rev 21:503–524. https://doi.org/10.1007/s11101-021-09786

Awad H, Khamis MM, El-Aneed A (2015) Mass spectrometry, review of the basics: ionization. Appl Spectrosc Rev 50:158–175. https://doi.org/10.1080/05704928.2014.954046

Azizan KA, Ghani NHA, Nawawi MF (2015) GC-MS based metabolomics and multivariate statistical analysis of *Wedelia trilobata* extracts for the identification of potential phytochemical properties. Plant Omics 8(6):537–543

Bai ZZ, Li YJ, Yu R, Li ZN, Qi RY, Ji XT et al (2024) LC-MS-based metabolomics for the profiling of bioactive compounds in tree peony flower buds with multiple bioactive potentials. Ind Crop Prod 217:118873. https://doi.org/10.1016/j.indcrop.2024.118873

Bartel J, Krumsiek J, Theis FJ (2013) Statistical methods for the analysis of high-throughput metabolomics data. Comput Struct Biotechnol J 4:e201301009. https://doi.org/10.5936/csbj.201301009

Beale DJ, Pinu FR, Kouremenos KA, Poojary MM, Narayana VK, Boughton BA et al (2018) Review of recent developments in GC-MS approaches to metabolomics-based research. Metabolomics 14:152. https://doi.org/10.1007/s11306-018-1449-2

Benamar OA, Craquelin M, Herfurth D, Molinié R, Fontaine JX, Srifa A et al (2024) NMR and LC-MS-based metabolomics to study the effect of surfactin on the metabolome of flax. Appl Sci 14:11999. https://doi.org/10.3390/app142411999

Bhattarai K, Paudel B, Dahal S, Yadav P, Aryal N, Baral B, Bhattarai HD (2022) Bioprospecting the metabolome of plant *Urtica dioica* L.: a fast dereplication and annotation workflow in plant metabolomics. Evid Based Complement Alternat Med 1:3710791. https://doi.org/10.1155/2022/3710791

Bijttebier S, Van der Auwera A, Foubert K, Voorspoels S, Pieters L, Apers S (2016) Bridging the gap between comprehensive extraction protocols in plant metabolomics studies and method validation. Anal Chim Acta 935:136–150. https://doi.org/10.1016/j.aca.2016.06.047

Boccard J, Rutledge DN (2013) A consensus orthogonal partial least squares discriminant analysis (OPLS-DA) strategy for multiblock omics data fusion. Anal Chim Acta 769:30–39. https://doi.org/10.1016/j.aca.2013.01.022

Bragagnolo FS, Funari CS, Ibáñez E, Cifuentes A (2021) Metabolomics as a tool to study underused soy parts: in search of bioactive compounds. Foods 10:1308. https://doi.org/10.3390/foods10061308

Brunius C, Shi L, Landberg R (2016) Large-scale untargeted LC-MS metabolomics data correction using between-batch feature alignment and cluster-based within-bacth signal intensity drift correction. Metabolomics 12:173. https://doi.org/10.1007/s11306-016-1124-4

Caesar LK, Kvalheim OM, Cech NB (2018) Hierarchical cluster analysis of technical replicates to identify interferents in untargeted mass spectrometry metabolomics. Anal Chim Acta 1021:69–77. https://doi.org/10.1016/j.aca.2018.03.013

Castillo S, Gopalacharyulu P, Yetukuri L, Orešič M (2011) Algorithms and tools for preprocessing of LC-MS metabolomics data. Chemom Intell Lab Syst 108(1):23–32. https://doi.org/10.1016/j.chemolab.2011.03.010

Chaleckis R, Meister I, Zhang P, Wheelock CE (2019) Challenges, progress and promises of metabolite annotation for LC-MS-based metabolomics. Curr Opin Biotechnol 55:44–50. https://doi.org/10.1016/j.copbio.2018.07.010

Coen E, Prusinkiewicz P (2024) Developmental timing in plants. Nat Commun 15:2674. https://doi.org/10.1038/s41467-024-46941-1

Crescenzi MA, Cerulli A, Masullo M, Montoro P, Piacente S (2025) Comparison of vegetable waste byproducts of selected cultivars of *Foeniculum vulgare* mill. by an integrated LC-(HR) MS and ^{1}H-NMR-based metabolomics approach. Phytochem Anal 36:1017–1030. https://doi.org/10.1002/pca.3488

Davidson RL, Weber RJM, Liu H, Sharma-Oates A, Viant MR (2016) Galaxy-M: a galaxy workflow for processing and analyzing direct infusion and liquid chromatography mass spectrometry-based metabolomics data. GigaScience 5:10. https://doi.org/10.1186/s13742-016-0115-8

de Hoffman E (2006) Tandem mass spectrometry: fundamentals and instrumentation. In: Encyclopedia of analytical chemistry: applications, theory and instrumentation. Wiley. https://doi.org/10.1002/9780470027318.a6017

Domingo-Almenara X, Montenegro-Burke JR, Benton HP, Siuzdak G (2018) Annotation: a computational solution for streamlining metabolomics analysis. Anal Chem 90(1):480–489. https://doi.org/10.1021/acs.analchem.7b03929

Du X, Zeisel SH (2013) Spectral deconvolution for gas chromatography mass spectrometry-based metabolomics: current status and future perspectives. Comput Struct Biotechnol J 4:5. https://doi.org/10.5936/csbj.201301013

Du B, Haensch R, Alfarraj S, Rennenberg H (2024) Strategies of plants to overcome abiotic and biotic stresses. Biol Rev 99(4):1524–1536. https://doi.org/10.1111/brv.13079

Dührkop K, Fleischauer M, Ludwig M, Aksenov AA, Melnik AV, Meusel M et al (2019) SIRIUS 4: a rapid tool for turning tandem mass spectra into metabolite structure information. Nat Methods 16:299–302. https://doi.org/10.1038/s41592-019-0344-8

Emam Y, Abdou RM, Younis IY, Farag MA (2025) Metabolomics in guava: a comprehensive review of quality, agricultural traits, composition, and nutraceutical applications of *Psidium guajava* fruits and byproducts. Phytocohem Rev. https://doi.org/10.1007/s11101-025-10153-2

Fernie AR, Aharoni A, Willmitzer L, Stitt M, Tohge T, Kopka J et al (2011) Recommendations for reporting metabolite data. Plant Cell 23(7):2477–2482. https://doi.org/10.1105/tpc.111.086272

Fiehn O, Kopka J, Dörmann P, Altmann T, Trethewey RN, Willmitzer L (2000) Metabolite profiling for plant functional genomics. Nat Biotechnol 18:1157–1161. https://doi.org/10.1038/81137

Fiehn O, Sumner LW, Rhee SY, Ward J, Dickerson J, Lange BM et al (2007) Minimum reporting standards for plant biology context information in metabolomics studies. Metabolomics 3:195–201. https://doi.org/10.1007/s11306-007-0068-0

Gadara D, Coufalikova K, Bosak J, Smajs D, Spacil Z (2021) Systematic feature filtering in exploratory metabolomics: application toward biomarker discovery. Anal Chem 93:9103–9110. https://doi.org/10.1021/acs.analchem.1c00816

Gal J, Bailleux C, Chardin D, Pourcher T, Gilhodes J, Jing L et al (2020) Comparison of unsupervised machine-learning methods to identify metabolomics signatures in patients with localized breast cancer. Comput Struct Biotechnol J 18:1509–1524. https://doi.org/10.1016/j.csbj.2020.05.021

Gautam V, Garg V, Meena N, Kumari S, Patel S, Mukesh, Singh H, Sing RK (2025) Harnessing NMR technology for enhancing field crop improvement: applications, challenges, and future perspectives. Metabolomics 21:27. https://doi.org/10.1007/s11306-025-02229-z

Gika HG, Wilson ID, Theodoridis GA (2014) LC-MS-based holistic metabolic profiling. Problems, limitations, advantages, and future perspectives. J Chromatogr B 966:1–6. https://doi.org/10.1016/j.jchromb.2014.01.054

Glauser G, Veyrat N, Rochat B, Wolfender JL, Turlings TCJ (2013) Ultra-high pressure liquid chromatography-mass spectrometry for plant metabolomics: a systematic comparison of high-resolution quadrupole-time-of-flight and single stage orbitrap mass spectrometers. J Chromatogr A 1292:151–159. https://doi.org/10.1016/j.chroma.2012.12.009

Gonulalan EM, Nemutlu E, Bayazeid O, Koçak E, Yalçın FN, Demirezer LO (2020) Metabolomics and proteomics profiles of some medicinal plants and correlation with BDNF activity. Phytomedicine 74:152920. https://doi.org/10.1016/j.phymed.2019.152920

Guiboileau A, Sormani R, Meyer C, Masclaux-Daubresse C (2010) Senescence and death of plant organs: nutrient recycling and developmental regulation. C R Biol 333(4):382–391. https://doi.org/10.1016/j.crvi.2010.01.016

Hamade K, Fliniaux O, Fontaine JX, Molinié R, Petit L, Mathiron D et al (2024) NMR and LC-MS-based metabolomics to investigate the efficacy of a commercial bio stimulant for the treatment of wheat (*Triticum aestivum*). Metabolomics 20:58. https://doi.org/10.1007/s11306-024-02131-0

Harada K, Fukusaki E (2009) Profiling of primary metabolite by means of capillary electrophoresis-mass spectrometry and its application for plant science. Plant Biotechnol 26:47–52. https://doi.org/10.5511/plantbiotechnology.26.47

Hartung J, Wagener J, Ruser R, Piepho HP (2019) Blocking and re-arrangement of pots in greenhouse experiments: which approach is more effective? Plant Methods 15:143. https://doi.org/10.1186/s13007-019-0527-4

Harzing AW (2024) Publish or Perish version 23.3.0 [Software]. Available at: https://harzing.com/resources/publish-or-perish. Accessed: 5 Mar 2025

Heiles S (2021) Advanced tandem mass spectrometry in metabolomics and lipidomics—methods and applications. Anal Bioanal Chem 413:5927–5948. https://doi.org/10.1007/s00216-021-03425-1

Hein JA, Sherrard ME, Manfredi KP, Abebe T (2016) The fifth leaf and spike organs of barley (*Hordeum vulgare* L.) display different physiological and metabolic responses to drought stress. BMC Plant Biol 16:248. https://doi.org/10.1186/s12870-016-0922-1

Himmelsbach M (2012) 10 years of MS instrumental developments—impact on LC-MS/MS in clinical chemistry. J Chromatogr B 883-884:3–17. https://doi.org/10.1016/j.jchromb.2011.11.038

Horai H, Arita M, Kanaya S, Nihei Y, Ikeda T, Suwa K et al (2010) MassBank: a public repository for sharing mass spectral data for life sciences. J Mass Spectrom 45(7):703–714. https://doi.org/10.1002/jms.1777

Hu W, Pan X, Li F, Dong W (2018) UPLC-QTOF-MS metabolomics analysis revealed the contributions of metabolites to the pathogenesis of *rhizoctonia solani* strain AG-1-IA. PLoS One 13(2):e0192486. https://doi.org/10.1371/journal.pone.0192486

Hubert M, Rousseeuw PJ, Branden KV (2005) ROBPCA: a new approach to robust principal component analysis. Technometrics 47:1. https://doi.org/10.1198/004017004000000563

Huberty M, Steinauer K, Heinen R, Jongen R, Hannula SE, Choi YH, Bezemer TM (2022) Temporal changes in plant-soil feedback effects on microbial networks, leaf metabolomics and plant-interactions. J Ecol 110(6):1328–1343. https://doi.org/10.1111/1365-2745.13872

Ibrahim RM, El-Shiekh RA, Mohamed OG, Al-Karmalawy AA, Tripathi A, Abdel-Baki PM (2024) LC/MS-based metabolomics reveals chemical variations of two broccoli varieties in relation to their anticholinesterase activity: *in vitro* and *in silico* studies. Plant Food Hum Nutr 79:359–366. https://doi.org/10.1007/s11130-024-01161-2

Ikotun AM, Ezugwu AE, Abualigah L, Abuhaija B, Heming J (2023) K-means clustering algorithms: a comprehensive review, variants analysis, and advances in the era of big data. Inf Sci 622:178–210. https://doi.org/10.1016/j.ins.2022.11.139

Ito J, Herter T, Baidoo EEK, Lao J, Vega-Sánchez ME, Smith-Moritz AM et al (2014) Analysis of plant nucleoside sugars by hydrophilic interaction liquid chromatography and tandem mass spectrometry. Anal Biochem 448:14–22. https://doi.org/10.1016/j.ab.2013.11.026

Jacyna J, Kordalewska M, Markuszewski MJ (2019) Design of experiments in metabolomics-related studies: an overview. J Pharm Biomed Anal 164:598–606. https://doi.org/10.1016/j.jpba.2018.11.027

Jaitly N, Mayampurath A, Littlefield K, Adkin JN, Anderson GA, Smith RD (2009) Decon2LS: an open-source software package for automated processing and visualization of high resolution mass spectrometry data. BMC Bioinf 10:87. https://doi.org/10.1186/1471-2105-10-87

Jha UC, Warburton ML, Nayyar H, Chatti D, Ghatak A, Siddique KHM, Prasad PVV (2025) Heat stress-induced metabolomic shifts in chickpea (*Cicer arietinum* L.) flowers insights from contrasting genotypes. Sci Rep 15:34821. https://doi.org/10.1038/s41598-025-21697-w

Jiang L, Sullivan H, Wang B (2022) Principal component analysis (PCA) loading and statistical tests for nuclear magnetic resonance (NMR) metabolomics involving multiple study groups. Anal Lett 55(10):1648–1662. https://doi.org/10.1080/00032719.2021.2019758

Jiang M, Sha Y, Zou Y, Xu X, Ding M, Lian X et al (2025) Integration of deep neural network modelling and LC-MS-based pseudo-targeted metabolomics to discriminate easily confused ginseng species. J Pharm Anal 15:101116. https://doi.org/10.1016/j.jpha.2024.101116

Jiao Y, Bai Z, Xu J, Zhao M, Khan Y, Hu Y, Shi L (2018) Metabolomics and its physiological regulation process reveal the salt-tolerant mechanism in *Glycine soja* seedling roots. Plant Physiol Biochem 126:187–196. https://doi.org/10.1016/j.plaphy.2018.03.002

Jing H, Xue X, Zhang X, Xu X, Tang Y, Wang H et al (2025) Metabolomics and microbiome analysis elucidate the detoxification mechanisms of *Hemarthria compressa*, a low cadmium accumulating plant, in response to cadmium stress. J Hazard Mater 487:137226. https://doi.org/10.1016/j.jhazmat.2025.137226

Jorge TF, Rodrigues JA, Caldana C, Schmidt R, van Dongen JT, Thomas-Oates J, António C (2015) Mass spectrometry-based plant metabolomics: metabolite responses to abiotic stress. Mass Spectrom Rev 35:620–649. https://doi.org/10.1002/mas.21449

Jorge TF, Rodrigues JA, Caldana C et al (2016) Mass spectrometry-based plant metabolomics: metabolite responses to abiotic stress. Mass Spectrom Rev 35:620–649. https://doi.org/10.1002/mas.21449

Jorge TF, Mata AT, António C (2016) Mass spectrometry as a quantitative tool in plant metabolomics. Philos Trans A 374:20150370. https://doi.org/10.1098/rsta.2015.0370

Kang Z, Babar MA, Khan N, Guo J, Khan J, Islam S, Shrestha S, Shahi D (2019) Comparative metabolomic profiling in the roots and leaves in contrasting genotypes reveals complex mechanisms involved in post-anthesis drought tolerance in wheat. PLoS One 14(3):e0213502. https://doi.org/10.1371/journal.pone.0213502

Kim S, Thiessen PA, Bolton EE, Chen J, Fu G, Gindulyte A, Han L et al (2015) PubChem substance and compound database. Nucleic Acids Res 44(D1):D1202–D1213. https://doi.org/10.1093/nar/gkv951

Kouřil S, de Sousa J, Václavik J, Friedecký D, Adam T (2020) CROP: correlation-based reduction of feature multiplicities in untargeted metabolomic data. Bioinformatics 36(9):2941–2942. https://doi.org/10.1093/bioinformatics/btaa012

Kruve A, Rebane R, Kipper K, Oldekop ML, Evard H, Herodes K, Ravio P, Leito I (2015) Tutorial review on validation of liquid chromatography-mass spectrometry methods: part I. Anal Chem Acta 870:29–44. https://doi.org/10.1016/j.aca.2015.02.017

Kumar N, Hoque MA, Sugimoto M (2018) Robust volcano plot: identification of differential metabolites in the presence of outliers. BMC Bioinf 19:128. https://doi.org/10.1186/s12859-018-2117-2

Lavergne FD, Broeckling CD, Cockrell DM, Haley SD, Peairs FB, Jahn CE, Heuberger AL (2018) GC-MS metabolomics to evaluate the composition of plant cuticular waxes for four *Triticum aestivum* cultivars. Int J Mol Sci 19(2):249. https://doi.org/10.3390/ijms19020249

Lee LC, Liong CY, Jemain AA (2018) Partial least squares-discriminant analysis (PLS-DA) for classification of high-dimensional (HD) data: a review of contemporary practice strategies and knowledge gaps. Analyst 143:3526–3539. https://doi.org/10.1039/C8AN00599K

Li L, Long W, Wan X, Ding Q, Zhang F, Wan D (2015) Studies on quantitative determination of total alkaloids and berberine in five origins of crude medicine "Sankezhen". J Chromatogr Sci 53(2):307–311. https://doi.org/10.1093/chromsci/bmu060

Li Y, He Q, Du S, Guo S, Geng Z, Deng Z (2018) Study of methanol extracts from different parts of *Peganum harmala* L. using ^{1}H-NMR plant metabolomics. J Anal Methods Chem 6532789. https://doi.org/10.1155/6532789

Li Y, Miao S, Tan J, Zhang Q, Chen DDY (2024) Capillary electrophoresis: a three-year literature review. Anal Chem 96:7799–7816. https://doi.org/10.1021/acs.analchem.4c00857

Liebelt DJ, Jordan JT, Doherty CJ (2019) Only a matter of time: the impact of daily and seasonal rhythms on phytochemicals. Phytochem Rev 18:1409–1433. https://doi.org/10.1007/s11101-019-09617-z

Liland KH (2011) Multivariate methods in metabolomics—from pre-processing to dimension reduction and statistical analysis. TrAC Trends Anal Chem 30(6):827–841. https://doi.org/10.1016/j.trac.2011.02.007

Lin ZH, Chen CS, Zhong QS, Ruan QC, Chen ZH, You XM, Shan RY, Li XL (2021) The GC-TOF/MS-based metabolomic analysis reveals altered metabolic profiles in nitrogen-deficient leaves and roots of tea plants (*Camellia sinensis*). BMC Plant Biol 21:506. https://doi.org/10.1186/s12870-021-03285-y

Lisec J, Schauer N, Kopka J, Willmitzer L, Fernie AR (2006) Gas chromatography mass spectrometry-based metabolite profiling in plants. Nat Protoc 1:387–396. https://doi.org/10.1038/nprot.2006.59

Liu X, Renzengwangdui, Tang S, Zhu Y, Wang M, Cao B, Wang J, Zhao B, Lu H (2023) Metabolomic analysis and antibacterial and antioxidant activities of three species of *Artemisia* plants in Tibet. BMC Plant Biol 23:208. https://doi.org/10.1186/s12870-023-04219-6

Liu X, Zhou F, Wen M, Jiang S, Long P, Ke JP et al (2024) LC-MS and GC-MS based metabolomics analysis revealed the impact of tea trichomes on chemical and flavor characteristics of white tea. Food Res Int 191:114740. https://doi.org/10.1016/j.foodres.2024.114740

Liu Q, Gu X, Zhang Y, Zhang T, Wang Y, Parkash O et al (2025) Molecular and metabolomics analysis reveals new insight into the mechanism underlying *Iris halophila* Pall. *IhCHSI*-mediated regulation of plant salt tolerance. Environ Exp Bot 229:106080. https://doi.org/10.1016/j.envexpbot.2024.106080

Lu W, Bennett BD, Rabinowitz JD (2008) Analytical strategies for LC-MS-based targeted metabolomics. J Chromatogr B Analyt Technol Biomed Life Sci 871:236–42.

Mais E, Alolga RN, Wang SL, Linus LO, Yin X, Qi LW (2018) A comparative UPLC-Q/TOF-MS-based metabolomics approach for distinguishing *Zingiber officinale* Roscoe of two geographical origins. Food Chem 240:239–244. https://doi.org/10.1016/j.foodchem.2017.07.106

Mandrone M, Chiocchio I, Barbanti L, Tomasi P, Tacchini M, Poli F (2021) Metabolomic study of sorghum (*Sorghum bicolor*) to interpret plant behavior under variable field conditions in view of smart agriculture applications. J Agric Food Chem 69(3):1132–1145. https://doi.org/10.1021/acs.jafc.0c06533

Martin AC, Pawlus AD, Jewett EM, Wyse DL, Angerhofer CK, Hegeman AD (2014) Evaluating solvent extraction systems using metabolomics approaches. RSC Adv 4:26325–26334. https://doi.org/10.1039/C4RA02731K

Mishra S, Gogna N, Dorai K (2019) NMR-based investigation of the altered metabolic response of *Bougainvillea spectabilis* leaves exposed to air pollution stress during the circadian cycle. Environ Exp Bot 164:58–70. https://doi.org/10.1016/j.envexpbot.2019.04.019

Misra BB (2020) Data normalization strategies in metabolomics: current challenges, approaches, and tools. Eur J Mass Spectrom 26:3. https://doi.org/10.1177/1469066720918446

Misra BB (2021) New software tools, databases, and resources in metabolomics: updates from 2020. Metabolomics 17:49. https://doi.org/10.1007/s11306-021-01796-1

Mlynárik V (2017) Introduction to nuclear magnetic resonance. Anal Biochem 529:4–9. https://doi.org/10.1016/j.ab.2016.05.006

Molina J, Nikolic D, Jeevarathanam JR, Abzalimov R, Park EJ, Pedales R et al (2022) Living with a giant, flowering parasite: metabolic differences between *Tetrastigma loheri* Gagnep. (Vitaceae) shoots uninfected and infected with *Rafflesia* (Rafflesiaceae) and potential applications for propagation. Planta 255:4. https://doi.org/10.1007/s00425-021-03787

Mönchgesang S, Strehmel N, Trutschel D, Westphal L, Neumann S, Scheel D (2016) Plant-to-plant variability in root metabolite profiles of 19 *Arabidopsis thaliana* accessions is substance-class-dependent. Int J Mol Sci 17(9):1565. https://doi.org/10.3390/ijms17091565

Mondello L, Cordero C, Janssen HG, Synovec RE, Zoccali M, Tranchida PQ (2025) Comprehensive two-dimensional gas chromatography-mass spectrometry. Nat Rev Methods Prim 5:7. https://doi.org/10.1038/s43586-024-00379-3

Moros G, Chatziioannou AC, Gika HG, Raikos N, Theodoridis G (2016) Investigation of the derivatization conditions for GC-MS metabolomics of biological samples. Bioanalysis 9:53–65. https://doi.org/10.4155/bio-2016-0224

Mota MFS, Waktola HD, Nolvachai Y, Marriott PJ (2021) Gas chromatography—mass spectrometry for characterisation, assessment of quality and authentication of seed and vegetable oils. TrAC Trends Anal Chem 138:116238. https://doi.org/10.1016/j.trac.2021.116238

Mozumder NHMR, Lee YR, Hwang KH, Lee MS, Kim EH, Hong YS (2020) Characterization of tea leaf metabolites dependent on tea (*Camellia sinensis*) plant age through ^{1}H NMR-based metabolomics. Appl Biol Chem 63:10. https://doi.org/10.1186/s13765-020-0492-7

Munjal Y, Tonk RK, Sharma R (2022) Analytical techniques used in metabolomics: a review. Syst Rev Pharm 13(5):550–556. https://doi.org/10.31858/0975-8453.13.5.550-556

Mushtaq MY, Choi YH, Verpoorte R, Wilson EG (2014) Extraction for metabolomics: access to the metabolome. Phytochem Anal 25(4):291–306. https://doi.org/10.1002/pca.2505

Nakamura K, Shimura N, Otabi Y, Hirai-Morita A, Nakamura Y, Ono N, Ul-Amin MA, Kanaya S (2013) KNApSAck-3D: a three-dimensional structure database of plant metabolites. Plant Cell Physiol 54(2):e4. https://doi.org/10.1093/pcp/pcs186

Nurazah Z, Idris AS, Din AM, Manaf MAA, Othman A, Ramli US (2021) Metabolite fingerprinting of oil palm (*Elaeis guineensis* Jacq.) root for the identification of altered metabolic pathways associated with basal stem rot (BSR) disease. *Physiological and molecular*. Plant Physiol 115:101647. https://doi.org/10.1016/j.pmpp.2021.101647

Nyamundanda G, Brennan L, Gormley IC (2010) Probabilistic principal component analysis for metabolomic data. BMC Bioinf 11:571. https://doi.org/10.1186/1471-2105-11-571

Omar J, Alonso I, Olivares M, Vallejo A, Etxebarria N (2012) Optimization of comprehensive two-dimensional gas chromatography (GC × GC) mass spectrometry for the determination of essential oils. Talanta 88:145–151. https://doi.org/10.1016/j.talanta.2011.10.023

Otify AM, ElBanna SA, Eltanany BM, Pont L, Benavente F, Ibrahim RM (2023a) A comprehensive analytical framework integrating liquid chromatography-tandem mass spectrometry metabolomics with chemometrics for metabolite profiling of lettuce varieties and discovery of antibacterial agents. Food Res Int 172:113178. https://doi.org/10.1016/j.foodres.2023.113178

Otify AM, Mohamed OG, El-Amier YA, Saber FR, Tripathi A, Younis IY (2023b) Bioherbicidal activity and metabolic profiling of allelopathic metabolites of three *Cassia* species using UPLC-qTOF-MS/MS and molecular networking. Metabolomics 19:16. https://doi.org/10.1007/s11306-023-01980-5

Pan H, Yao C, Yao S, Yang W, Wu W, Guo D (2020) A metabolomics strategy for authentication of plant medicines with multiple botanical origins, a case study of Uncaria Rammulus Cum Uncis. J Sep Sci 43:1043–1050. https://doi.org/10.1002/jssc.201901064

Pang Z, Lu Y, Zhou G, Hui F, Xu L, Viau C et al (2024) MetaboAnalyst 6.0: towards a unified platform for metabolomics data processing, analysis and interpretation. Nucleic Acids Res 52(W1):W398–W406. https://doi.org/10.1093/nar/gkae253

Patel RS, Roy M, Dutta GK (2012) Mass spectrometry—a review. Vet World 5(3):185–192. https://doi.org/10.5455/vetworld.2012.185-192

Peng L, Zhao Y, Wang H, Zhang J, Song C, Shangguan X, Zhu L, He G (2016) Comparative metabolomics of the interaction between rice and the brown planthopper. Metabolomics 12:132. https://doi.org/10.1007/s11306-016-1077-7

Pinheiro GP, Silva AAR, de Melo Porcari A, Sawaya ACHF (2025) Effects of shading and water availability on *Kalanchoe pinnata* (Lam.) Pers. (Crassulaceae) metabolite profile via a LC-MS-based untargeted metabolomics approach. J Plant Interact 20:2448116. https://doi.org/10.1080/17429145.2024.2448116

Pontes JGM, Ohashi WY, Brasil AJM, Filgueiras PR, Espíndola APDM, Silva JS, Poppi RJ, Coletta-Filho HD, Tasic L (2016) Metabolomics by NMR spectroscopy in plant disease diagnostic: Huanglongbing as a case study. ChemistrySelect 6:1176–1178. https://doi.org/10.1002/slct.201600064

Putri SP, Ikram MMM, Sato A, Dahlan HA, Rahmawati D, Ohto Y, Fukusaki E (2022) Application of gas chromatography-mass spectrometry-based metabolomics in food science and technology. J Biosci Bioeng 133(5):425–435. https://doi.org/10.1016/j.jbiosc.2022.01.011

Ramautar R, Somsen GW, de Jong GJ (2008) CE-MS in metabolomics. Electrophoresis 30:276–291. https://doi.org/10.1002/elps.200800512

Ravi S, Young T, Macinnis-Ng C, Nyugen TV, Duxbury M, Alfaro AC, Leuzinger S (2020) Untargeted metabolomics in halophytes: the role of different metabolites in New Zealand mangroves under multi-factorial abiotic stress conditions. Environ Exp Bot 173:103993. https://doi.org/10.1016/j.envexpbot.2020.103993

Ren S, Hinzman AA, Kang EL, Szczesniak RD, Lu LJ (2015) Computational and statistical analysis of metabolomics data. Metabolomics 11:1492–1513. https://doi.org/10.1007/s11306-015-0823-6

Ren Z, Fang M, Muhae-Ud-Din G, Gao H, Yang Y, Liu T, Chen W, Gao L (2021) Metabolomics analysis of grains of wheat infected and non-infected with *Tilletia controversa* Kühn. Sci Rep 11:18876. https://doi.org/10.1038/s41598-021-98283-3

Ribbenstedt A, Ziarrusta H, Benskin JP (2018) Development, characterization and comparisons of targeted and non-targeted metabolomics methods. PLoS One 13(11):e0207082. https://doi.org/10.1371/journal.pone.0207082

Rivas-Ubach A, Aranda I, Sardans J, Liu Y, de Quijano MD, Paša-Tolić L et al (2025) An outline on the chemical phenotype flexibility of forest species: an eco-metabolomics study of *Pinus uncinate* along an altitudinal gradient. Perspect Plant Ecol Evol Syst 66:125844. https://doi.org/10.1016/j.ppees.2024.125844

Rodrigues AM, Ribeiro-Barros AI, António C (2019) Experimental design and sample preparation in forest tree metabolomics. Meta 9:285. https://doi.org/10.3390/metabo9120285

Rohloff J (2015) Analysis of phenolic and cyclic compounds in plants using derivatization techniques in combination with GC-MS-based metabolite profiling. Molecules 20(2):3431–3462. https://doi.org/10.3390/molecules20023431

Ruiz-Perez D, Guan H, Madhivanan P, Mathee K, Narasimhan G (2020) So you think you can PLS-DA? BMC Bioinf 21:2. https://doi.org/10.1186/s12859-019-3310-7

Ruttkies C, Schymanski EL, Wolf S, Hollender J, Neumann S (2016) MetFrag relaunched: incorporating strategies beyond *in silico* fragmentation. J Chemom 8:3. https://doi.org/10.1186/s13321-016-0115-9

Salem MA, de Souza LP, Serag A, Fernie AR, Farag MA, Ezzat SM, Alseekh S (2020) Metabolomics in the contect of plant natural products research: from sample preparation to metabolite analysis. Meta 10(1):37. https://doi.org/10.3390/metabo10010037

Sampaio BL, Edrada-Ebel R, Da Costa FB (2016) Effect of the environment on the secondary metabolic profile of *Tithonia diversifolia*: a model for environmental metabolomics of plants. Sci Rep 6:29265. https://doi.org/10.1038/srep29265

Sarkar J, Singh R, Chandel S (2025) Understanding LC/MS-based metabolomics: a detailed reference for natural product analysis. PROTEOMICS—Clin Appl 19:e202400048. https://doi.org/10.1002/prca.202400048

Schiffman C, Petrick L, Perttula K, Yano Y, Carlsson H, Whitehead T et al (2019) Filtering procedures for untargeted LC-MS metabolomics data. BMC Bioinf 20:334. https://doi.org/10.1186/s12859-019-2871-9

Schmid R, Heuckeroth S, Korf A, Smirnov A, Myers O, Dyrlund TS et al (2023) Integrative analysis of multimodal mass spectrometry data in MZmine 3. Nat Biotechnol 41:447–449. https://doi.org/10.1038/s41587-023-01690-2

Schripsema J (2010) Application of NMR in plant metabolomics: techniques, problems and prospects. Phytochem Anal 21:14–21. https://doi.org/10.1002/pca.1185

Serrano-Garcia I, Dominguez-Garcia J, Hurtado-Fernández E, González-Fernández JJ, Hormaza JI, Beiro-Valenzuela MG, Monasterio R, Pedreschi R, Olmo-Garcia L, Carrasco-Pancorbo A (2023) Assessing the RP-LC-MS-based metabolic profile of *Hass* avocados marketed in Europe from different geographical origins (Peru, Chile, and Spain) over the whole season. Plants 12(16):3004. https://doi.org/10.3390/plants12163004

Shen S, Zhan C, Yang C, Fernie AR, Luo J (2023) Metabolomics-centered mining of plant metabolic diversity and function: past decade and future perspectives. Mol Plant 16:43–63. https://doi.org/10.1016/j.molp.2022.09.007

Sikandar A, Rao W, He H, Chen B, Xu X, Wu H (2024) Metabolomics and histopathological analysis of two tomato cultivar after co-infection with soil-borne pathogens (Southern root-knot nematode and fusarium wilt fungus). Plant Physiol Biochem 215:108983. https://doi.org/10.1016/j.plaphy.2024.108983

Singh S, Pathak N, Fatima E, Negi AS (2021) Plant isoquinoline alkaloids: advances in the chemistry and biology of berberine. Eur J Med Chem 226(15):113839. https://doi.org/10.1016/j.ejmech.2021.113839

Smith CA, Maille GO, Want EJ, Qin C, Trauger SA, Brandon TR et al (2005) METLIN a metabolite mass spectral database. Ther Drug Monit 27(6):747–751. https://doi.org/10.1097/01.ftd.0000179845.53213.39

Smith R, Mathis AD, Prince JT (2014) Proteomics, lipidomics, metabolomics: a mass spectrometry tutorial from a computer scientist's point of view. BMC Bioinf 15:S9. https://doi.org/10.1186/1471-2105-15-S7-S9

Stancliffe E, Patti GJ (2023) PeakDetective: a semisupervised deep learning-basede approach for peak curation in untargeted metabolomics. Anal Chem 95(25):9397–9403. https://doi.org/10.1021/acs.analchem.3c00764

Sturm M, Bertsch A, Gröpl C, Hildebrandt A, Hussong R, Lange E et al (2008) OpenMS—an open-source software framework for mass spectrometry. BMC Bioinf 9:163. https://doi.org/10.1186/1471-2105-9-163

Sumner LW, Amberg A, Barrett D, Beale MH, Beger R, Daykin CA et al (2007) Proposed minimum reporting standards for chemical analysis chemical analysis working group (CAWG) metabolomics standards initiative (MSI). Metabolomics 3(3):211–221. https://doi.org/10.1007/s11306-007-0082-2

Sun J, Xia Y (2024) Pretreating and normalizing metabolomics data for statistical analysis. Genes Dis 11:100979. https://doi.org/10.1016/j.gendis.2023.04.018

Tang DQ, Zou L, Yin XX, Ong CN (2014) HILIC-MS for metabolomics: an attractive and complementary approach to RPLC-MS. Mass Spectrom Rev 35:574–600. https://doi.org/10.1002/mas.21445

Tang S, Wei K, Huang H, Li X, Min Y, Tai J et al (2024) Effect of soil factors on flavonoid metabolites in *Striga asiatica* using LC-MS based on untargeted metabolomics. Chem Biol Technol Agric 11:89. https://doi.org/10.1186/s40538-024-00614-9

Tao H, Zhao Y, Li L, He Y, Zhang X, Zhu Y, Hong G (2023) Comparative metabolomics of flavonoids in twenty vegetables reveal their nutritional diversity and potential health benefits. Food Res Int 164:112384. https://doi.org/10.1016/j.foodres.2022.112384

Tautenhahn R, Patti GJ, Rinehart D, Siuzdak G (2012) XCMS online: a web-based platform to process untargeted metabolomic data. Anal Chem 84(11):5035–5039. https://doi.org/10.1021/ac300698

Tsugawa H, Ikeda K, Takahashi M, Satoh A, Mori Y, Uchino H et al (2020) A lipidome atlas in MS-DIAL4. Nat Biotechnol 38:1159–1163. https://doi.org/10.1038/s41587-020-0531-2

Tugizimana F, Steenkamp PA, Piater LA, Dubery IA (2016) A conversation on data mining strategies in LC-MS untargeted metabolomics: pre-processing and pre-treatment steps. Meta 6:40. https://doi.org/10.3390/metabo6040040

Uthe H, van Dam NM, Hervé MR, Sorokina M, Peters K, Weinhold A (2021) Chapter six—a practical guide to implementing metabolomics in plant ecology and biodiversity research. Adv Bot Res 98:163–203. https://doi.org/10.1016/bs.abr.2020.09.019

van den Berg RA, Hoefsloot HCJ, Westerhuis JA, Smilde AK, van der Werf MJ (2006) Centering, scaling, and transformations: improving the biological information content of metabolomics data. BMC Genomics 7:142. https://doi.org/10.1186/1471-2164-7-142

Veenstra TD (2021) Omics in systems biology: current progress and future outlook. Proteomics 21:2000235. https://doi.org/10.1002/pmic.202000235

Verrillo M, Koellensperger G, Puehringer M, Cozzolino V, Spaccini R, Rampler E (2022) Evaluation of sustainable recycled products to increase the production of nutraceutical and antibacterial molecules in basil plants by a combined metabolomic approach. Plants 12(3):513. https://doi.org/10.3390/plants12030513

Wang Y, Xu L, Shen H, Wang J, Liu W, Zhu X et al (2015) Metabolomic analysis with GC-MS to reveal potential metabolites and biological pathways involved in Pb & Cd stress response of radish roots. Sci Rep 5:18296. https://doi.org/10.1038/srep18296

Wang M, Carver JJ, Phelan VV, Sanchez LM, Garg N, Peng Y et al (2016) Sharing and community curation of mass spectrometry data with global natural products social molecular networking. Nat Biotechnol 34:828–837. https://doi.org/10.1038/nbt.3597

Wang W, Portal-Gonzalez N, Wang X, Li J, Li H, Portieles R, Borras-Hidalgo O, He W et al (2024) Metabolome-driven microbiome assembly determining the health of ginger crop (*Zingiber officinale* L. Roscoe) against rhizome rot. Microbiome 12:167. https://doi.org/10.1186/s40168-024-01885-y

Waris M, Koçak E, Gonulalan EM, Demirezer LO, Kır S, Nemutlu E (2022) Metabolomics analysis insight into medicinal plant science. TrAC Trends Anal Chem 157:116795. https://doi.org/10.1016/j.trac.2022.116795

Wei Z, Zhongbing C, Xiuqin Y, Luying S, Huan M, Sixi Z (2023) Integrated transcriptomics and metabolomics reveal key metabolic pathway responses in *Pistia stratiotes* under Cd stress. J Hazard Mater 452:131214. https://doi.org/10.1016/j.jhazmat.2023.131214

Wong YF, Perlmutter P, Marriott PJ (2017) Untargeted metabolic profiling of *Eucalyptus* spp. leaf oils using comprehensive two-dimensional gas chromatography with high resolution mass spectrometry: expanding the metabolic coverage. Metabolomics 13:46. https://doi.org/10.1007/s11306-017-1173-3

Worley B, Powers R (2013) Multivariate analysis in metabolomics. Curr Metabolomics 1(1):92–107. https://doi.org/10.2174/2213235X11301010092

Wu X, Li N, Li H, Tang H (2014) An optimized method for NMR-based plant seed metabolomic analysis with maximized polar metabolite extraction efficiency, signal-to-noise ratio, and chemical shift consistence. Analyst 139:1769–1778. https://doi.org/10.1039/C3AN02100A

Xiao C, Ye J, Esteves RM, Rong C (2016) Using Spearman's correlation coefficients for exploratory data analysis on big dataset. Concurrency Comput Pract Exp 28:3866–3878. https://doi.org/10.1002/cpe.3745

Xiao Q, Mu X, Liu J, Li B, Liu H, Zhang B, Xiao P (2022) Plant metabolomics: a new strategy and tool for quality evaluation of Chinese medicinal materials. Chin Med 17:45. https://doi.org/10.1186/s13020-022-00601-y

Xiao X, Huang Q, Rao B, Wu Z, Wang Y (2025) Integrating LC-MS based molecular networks and comparative metabolomics for variety identification and origin tracing of *Zanthoxylum nitidum* (Roxb.) DC. Ind Crop Prod 226:120633. https://doi.org/10.1016/j.indcrop.2025.120633

Xu S, Xu Y, Gong L, Zhang Q (2016) Metabolomic prediction of yield in hybrid rice. Plant J 88(2):219–227. https://doi.org/10.1111/tpj.13242

Yan S, Bhawal S, Yin Z, Thannhauser TW, Zhang S (2022) Recent advances in proteomics and metabolomics in plants. Mol Hortic 2:17. https://doi.org/10.1186/s43897-022-00038-9

Yang Z, Li L, Chen CH, Zhang YY, Yang Y, Zhang P, Bao GH (2022) Chemical composition and antibacterial activity of 12 medicinal plant ethyl acetate extracts using LC-MS feature-based molecular networking. Phytochem Anal 33(3):473–489. https://doi.org/10.1002/pca.3103

Zhang W, Ramautar R (2020) CE-MS for metabolomics: developments and applications in the period 2018-2020. Electrophoresis 42(4):381–401. https://doi.org/10.1002/elps.202000203

Zhang A, Sun H, Wang P, Han Y, Wang X (2012) Modern analytical techniques in metabolomics analysis. Analyst 137:293–300. https://doi.org/10.1039/C1AN15605E

Zhang H, Lu L, Zhao X, Zhao S, Gu X, Du W et al (2019) Metabolomics reveals the "invisible" responses of spinach plants exposed to CeO_2 nanoparticles. Environ Sci Technol 53(10):6007–6017. https://doi.org/10.1021/acs.est.9b00593

Zhao Y, Zhao J, Zhao C, Zhou H, Li Y, Zhang J et al (2015) A metabolomics study delineating geographical location-associated primary metabolic changes in leaves of growing tobacco plants by GC-MS and CE-MS. Sci Rep 5:16346. https://doi.org/10.1038/srep16346

Zhao J, Xu J, Li H, An W, Yin Y, Wang B et al (2024) Metabolite-based genome-wide association studies enable the dissection of the genetic bases of flavonoids, betaine and spermidine in wolfberry (*Lycium*). Plant Biotechnol J 22:1435–1452. https://doi.org/10.1111/pbi.14278

Zhou Z, Luo M, Zhang H, Yin Y, Cai Y, Zhu ZJ (2022) Metabolite annotation from knowns to unknowns through knowledge-guided multi-layer metabolic networking. Nat Commun 13:6656. https://doi.org/10.1038/s41467-022-34537-6

Zhu M, Geng S, Chakravorty D, Guan Q, Chen S, Assmann SM (2019) Metabolomics of red-light induced stomatal opening in *Arabidopsis thaliana*: coupling with abscisic acid and jasmonic acid metabolism. Plant J 101(6):1331–1348. https://doi.org/10.1111/tpj.14594

Zhu L, Zhou W, Wang J, Guo J, Zhou C (2025) Root exudate-mediated assemblage of rhizo-microbiome enhances *Fusarium* wilt suppression in chrysanthemum. Microbiol Res 292:128031. https://doi.org/10.1016/j.micres.2024.128031

Proteomics for Crop Improvement

Lucas Rodrigues Xavier, Laura Eliza de Oliveira Alves, Carlos Eduardo Assis da Silva, Daniel Dastan Rezabala Pacheco, Roberta Pena da Paschoa, Claudete Santa-Catarina, and Vanildo Silveira

Abstract Mass spectrometry (MS) has emerged as an indispensable tool in plant biology, offering exceptional versatility for investigating protein-level responses to environmental stimuli. This chapter provides an overview of quantitative proteomics applied to plant breeding, covering the experimental workflow, sample preparation, and data analysis. Special attention is given to experimental design, emphasizing the importance of an adequate number of biological replicates and choosing the appropriate proteomics approach according to the comparisons to be analyzed. Although pairwise comparisons are common, time-resolved and dose-dependent designs are increasingly preferred for their robustness in revealing co-accumulated proteins and reducing uncertainty in determining biomarkers. This chapter highlights the power of proteomics to elucidate plant responses to abiotic and biotic stresses, and heterosis, demonstrating its ability to uncover signaling pathways, regulatory proteins, and stress- and trait-specific biomarkers. An outlook on future directions is provided, highlighting the transformative potential of proteomics in plant science and crop improvement.

Keywords Plant breeding · Sample preparation · Proteomics workflow · Mass spectrometry

---

L. R. Xavier · L. E. de Oliveira Alves · C. E. A. da Silva · D. D. R. Pacheco · R. P. da Paschoa
V. Silveira (✉)
Laboratório de Biotecnologia, Centro de Biociências e Biotecnologia (CBB), Universidade Estadual do Norte Fluminense Darcy Ribeiro (UENF), Campos dos Goytacazes, Rio de Janeiro, Brazil

Setor de Genômica e Proteômica, Unidade de Biologia Integrativa, Universidade Estadual do Norte Fluminense Darcy Ribeiro (UENF), Campos dos Goytacazes, Rio de Janeiro, Brazil
e-mail: vanildo@uenf.br

C. Santa-Catarina
Laboratório de Biologia Celular e Tecidual, CBB-UENF, Campos dos Goytacazes, Rio de Janeiro, Brazil

L. M. Vaschetto (ed.), *Omics Technologies in Plant Science*, Interdisciplinary Biotechnological Advances, https://doi.org/10.1007/978-981-92-0533-2_6

Abbreviations

ABA	Abscisic acid
ABC	ATP-binding cassette
ADH	Alcohol dehydrogenase
APX	Ascorbate peroxidase
BLAST	Basic Local Alignment Search Tool
Bph3	Brown planthopper resistance gene 3
bZIP	Basic leucine zipper (transcription factor family)
CMV	Cucumber mosaic virus
DAP	Differentially accumulated protein
DDA	Data-dependent acquisition
DIA	Data-independent acquisition
EAS	Egg-associated secretions
ETI	Effector-triggered immunity
FASP	Filter-aided sample preparation
GA_3	Gibberellin A_3 or gibberellic acid
GeLC	SDS-PAGE prefractionation-MS/MS
GST	Glutathione S-transferase
HSP	Heat shock protein
IGD	In-gel digestion
ISD	In-solution digestion
KEGG	Kyoto Encyclopedia of Genes and Genomes
LC-MS/MS	Liquid chromatography-MS/MS
m/z	Mass-to-charge ratio
MALDI-TOF	Matrix-assisted laser desorption/ionization time-of-flight
MAPK	Mitogen-activated protein kinase
MS	Mass spectrometry
MS/MS	Tandem mass spectrometry
MS1	First stage of MS (precursor ion scan)
MS2	Second stage of MS (fragment ion scan)
nanoLC-MS/MS	Nano-flow LC-MS/MS
PAL	Phenylalanine ammonia-lyase
PAMP	Pathogen-associated molecular pattern
PCA	Principal component analysis
PER12	Peroxidase 12
PR	Pathogenesis-related protein
PSI	Photosystem I
PSII	Photosystem II
PTI	Pattern-triggered immunity
PTM	Posttranslational modification
ROS	Reactive oxygen species
Rubisco	Ribulose-1,5-bisphosphate carboxylase/oxygenase
SDS-PAGE	Sodium dodecyl sulfate-polyacrylamide gel electrophoresis
SOD	Superoxide dismutase

STRING	Search Tool for the Retrieval of Interacting Genes/Proteins
TIC	Total ion chromatogram
UPS	Ubiquitin–proteasome system
ZIP	Zinc-regulated/iron-regulated-like protein
2DE	Two-dimensional gel electrophoresis

1 Introduction

1.1 Proteomics Toolbox for Crop Breeding

It was clear from the sequencing of the first genomes that the next big challenge for biotechnology would be to discover the functions of each gene encoded in the DNA (Holtorf et al. 2002). In this context, high-throughput approaches applied to investigate multiple layers of regulation of gene expression have evolved and become increasingly popular, with genomics, transcriptomics, proteomics, and metabolomics receiving a lot of focus at the beginning (Yan et al. 2022). Since then, omics tools have been rapidly incorporated into plant research and have been employed to elucidate biological processes such as stress, embryogenesis, photosynthesis, flowering, and reproduction (Mergner and Kuster 2022). Large-scale proteomic analyses initially relied on two-dimensional gel electrophoresis (2DE) coupled with matrix-assisted laser desorption/ionization–time-of-flight (MALDI-TOF) mass spectrometry for the identification of differentially abundant proteins (DAPs) (Görg et al. 2004). Although this 2DE platform was fundamental for early proteome mapping, the advent of gel-free strategies, next-generation mass spectrometers, and nano-flow liquid chromatography systems has dramatically expanded the depth and resolution of proteome coverage (Aydoğan 2024). While early proteomic studies using 2DE typically identified only a few dozen proteins, modern bottom-up approaches can now detect and quantify thousands of proteins in a single experiment.

Top-down and bottom-up proteomics refer to the characterization of proteins based on intact proteins or peptides generated through proteolytic digestion, respectively. When applied to complex protein mixtures, the bottom-up strategy is known as shotgun proteomics, which provides an indirect measurement of protein abundance and identity through the detection of their corresponding peptides (Zhang et al. 2013). Bottom-up proteomics can be performed using gel-free or gel-based workflows, both widely applied in shotgun proteomics experiments (Fig. 1). In the gel-free approach, proteins are digested, commonly via in-solution digestion (ISD) or filter-aided sample preparation (FASP), and the resulting complex peptide mixture is directly analyzed by nanoLC-MS/MS. In the shotgun gel-based workflow, also called GeLC, protein extracts are first separated by one-dimensional sodium dodecyl sulfate-polyacrylamide gel electrophoresis (SDS-PAGE), followed by in-gel digestion (IGD) of each gel lane (Balbuena et al. 2012; Piersma et al. 2013). Peptides from all gel slices are subsequently analyzed by nanoLC-MS/MS, and the resulting data are merged to provide comprehensive identification and

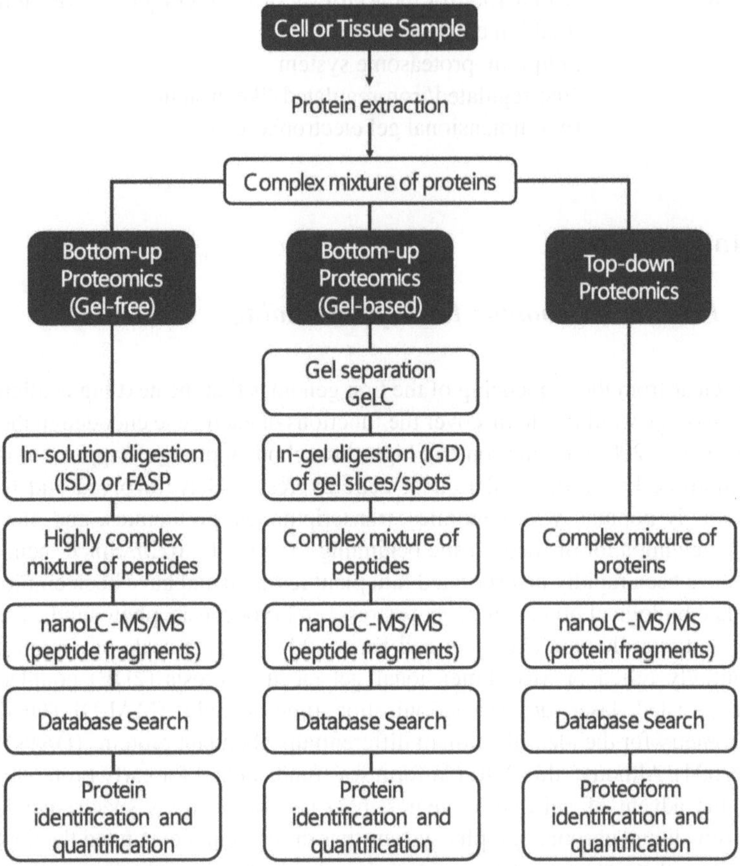

Fig. 1 Overview of the main proteomics workflows. The figure illustrates the three principal strategies used in mass spectrometry-based proteomics: bottom-up gel-free, bottom-up gel-based (GeLC), and top-down proteomics. Each workflow is depicted from sample preparation through to peptide or proteoform identification and quantification, highlighting key steps such as in-solution digestion (ISD/FASP), SDS-PAGE separation, in-gel digestion (IGD), and nanoLC-MS/MS analysis

quantification (Piersma et al. 2013). Alongside bottom-up proteomics, top-down proteomics has emerged as a complementary and increasingly common workflow (Fig. 1). The top-down approach involves the direct analysis of intact proteins, offering distinct advantages for the identification and quantification of proteoforms, including precise detection of post-translational modifications (PTMs) and protein isoforms, with notable success in various biological systems (Zhang et al. 2013).

The advances in proteomics have established proteomics as a fundamental approach to the study of gene function and have provided advances in both basic science and biotechnology (Mergner and Kuster 2022). The application of proteomic tools in plant science has made it possible to study the regulation of metabolic pathways, transcription factors, and protein kinases/phosphatases, protein–protein interaction networks, post-translational modifications, and the cellular localization of proteins during different biological processes (Yan et al. 2022). In this sense, working with plant cells, tissues, and organs usually involves challenges, such as the presence of cellulose, chlorophylls, and very abundant proteins, such as ribulose-1,5-bisphosphate carboxylase/oxygenase (Rubisco), where an important factor contributing to the advancement of plant proteomics has been the development of methods for collecting and preparing samples (Mikulášek et al. 2021; Varnavides et al. 2022). In addition, the development of databases and bioinformatics tools specific to plant organisms has resulted in an increase in the depth of proteomic data analysis, leading to an expanded interpretation of changes in the proteomic profile and hypothesis testing (Tyanova et al. 2016b; Zheng et al. 2016; Szklarczyk et al. 2022). On the other hand, aspects such as the identification of post-translational modifications and the quantification of low-abundance proteins still represent challenges for plant proteomics (Picotti 2015). The existence of cells with different functions in samples of plant tissues and organs (such as leaves and roots) also creates a background that generates doubts when interpreting data. Recently, advances in enrichment techniques for low-abundance peptides, targeted proteomics, and single-cell analysis can help overcome these challenges (Balotf et al. 2022). This chapter aims to provide an accessible and comprehensive review of how bottom-up proteomics is applied to drive plant breeding, highlighting experimental designs, methodological advances, and applications in elucidating molecular responses to abiotic and biotic stresses and heterosis.

1.2 A Workflow for Plant Bottom-Up Gel-Free Proteomics

Bottom-up proteomics is the most widely used strategy for large-scale protein identification and quantification, comprising a multistep workflow that includes sample selection and protein extraction, proteolytic digestion, peptide chromatographic separation, tandem mass spectrometry (MS/MS) analysis, data processing, bioinformatic analysis, and functional annotation. Careful optimization of each step is essential to ensure reproducibility and maximize proteome coverage. A general overview of the main steps involved in a bottom-up proteomics workflow for plant samples is presented in Fig. 2.

1.2.1 Experiment Design

Several factors influence the identification of proteins by mass spectrometry, including separation technology, instrument type, search software, and database selection (Taylor et al. 2007). On the other hand, a fundamental step during research into the regulation of the proteome consists of selecting the most appropriate experimental design to answer a biological question. In this sense, during experimental planning, key aspects must be clearly defined, including the research questions, comparison

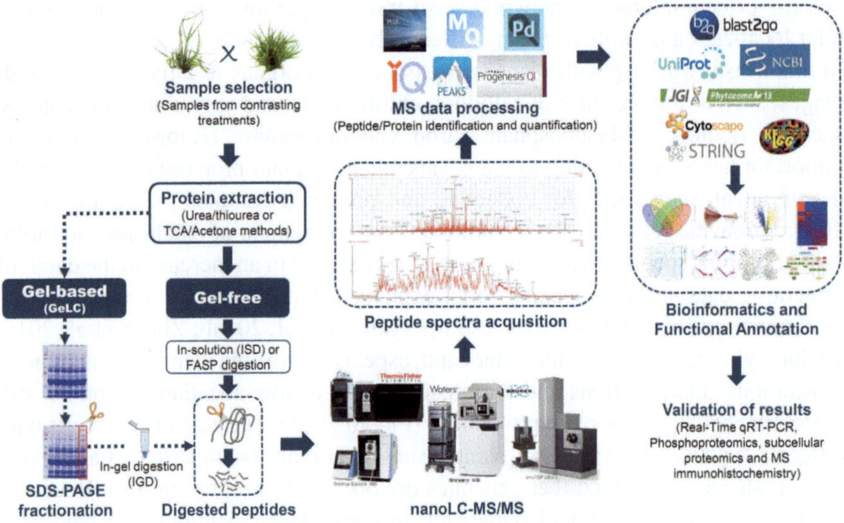

Fig. 2 Bottom-up proteomics workflow applied to plant science. After selecting contrasting biological conditions, protein extraction is performed using chaotropic agents (e.g., urea/thiourea) or precipitation methods (e.g., TCA/acetone). Two main digestion strategies are commonly employed: gel-based (GeLC), where proteins are separated by SDS-PAGE and each band is subjected to in-gel digestion (IGD), or gel-free approaches, including in solution digestion (ISD) or filter-aided sample preparation (FASP). Peptides are analyzed by high-resolution mass spectrometers (e.g., Orbitrap, timsTOF, or Synapt G2-Si) coupled to nanoLC systems. Peptide spectra are processed using dedicated software for identification and quantification of proteins. Subsequently, the results are explored through bioinformatics tools for functional annotation, pathway analysis, and visualization of protein-protein interaction networks. The workflow concludes with the validation of selected targets using complementary approaches such as qRT-PCR, phosphoproteomics, or subcellular proteomics

design (e.g., simple comparison, time series, or dose-dependent analysis), sampling strategies (which organ or tissue to collect, when, and how), the mass spectrometry data acquisition approach (DDA or DIA), and quantification strategy (labeled or label-free) (Balotf et al. 2022; Yan et al. 2022; Aydoğan 2024).

The most widely accepted experimental design in comparative proteomics involves a minimum of three biological replicates and a direct comparison between two conditions or treatments (e.g., control and water stress) (Corrêa et al. 2025). In this context, the inclusion of a control condition is essential for identifying DAPs in response to a specific stimulus. However, such pairwise comparisons are susceptible to background noise caused by stochastic fluctuations in protein abundance, which may increase the rate of false-positive identifications. As an alternative, time-resolved and dose-dependent experimental designs have become increasingly prevalent (Nikonorova et al. 2021; Xavier et al. 2023). These approaches enable the identification of co-accumulated proteins under defined conditions, thereby facilitating the discovery of potential regulatory proteins. Time-course and dose-dependent analyses involve the collection of samples at three or more distinct time points or treatment conditions, representing robust strategies for the more accurate identification of molecular targets.

1.2.2 Sample Preparation Methods

The reliability of proteomics data is directly related to the quality of the samples, and each tissue, organ, or cell studied presents unique challenges (Isaacson et al. 2006). The main steps that make up sample preparation for proteomics are material collection, protein extraction and purification, and proteolytic digestion to obtain peptides (Aguilar-Hernández and Loyola-Vargas 2018). In terms of statistical analysis in a comparative study, at least three biological replicates representing the treatments being evaluated are required. The lysis of the cell wall from plant tissue being analyzed can be performed using a sonicator (in the case of cell suspensions), a mortar and pestle with liquid nitrogen, or a grinding mill with beads to automate the process for harder biological materials. In emerging single-cell proteomics work, strategies to isolate cells of a specific type may be necessary, including fluorescent activated cell sorting or laser-mediated capture dissection (Clark et al. 2021).

During the extraction process, the method chosen must be able to capture as many substances as possible in the sample (Heringer et al. 2018). For samples with a high concentration of contaminants, the trichloroacetic acid/acetone precipitation method followed by resuspension in urea/thiourea buffer usually generates good results (Isaacson et al. 2006; Heringer et al. 2018). Alternatively, the phenol-based protein extraction method has more steps and is more laborious, but it is a process that can produce highly purified proteins (Isaacson et al. 2006). In summary, TCA/acetone precipitation and phenol-based extraction are more suitable for plant tissues rich in contaminants, as they promote more efficient removal of polysaccharides, phenolics, and lipids (e.g., leaves and seeds). Although the phenol method is more laborious, it is capable of generating higher-purity protein extracts. In contrast, direct extraction using urea/thiourea buffer is a simpler and faster procedure, suitable for materials with lower chemical complexity, such as embryogenic cultures, although it is less effective at removing contaminants (Almeida et al. 2020).

Once the proteins have been extracted and purified, the following step normally consists of quantifying the samples in order to measure their concentration before running on a gel or digesting with proteases (Varnavides et al. 2022). The quantification stage involves some challenges, such as the limited amount of material available in some cases and the presence of interferents in the sample (e.g., urea, detergents, and reducing agents) (Ni et al. 2017). In this sense, different strategies are available for total protein quantification, including methods with and without precipitation steps. Protein precipitation allows sample concentration, buffer exchange, and removal of interfering substances, although these methods usually require more time and reagents compared with nonprecipitation methods such as the Bradford assay (Olson 2016).

Among the proteases used for proteolytic digestion during proteomic sample preparation, trypsin is the most widely employed and cleaves peptide bonds at the

C-terminal side of lysine (K) and arginine (R) residues (Aguilar-Hernández and Loyola-Vargas 2018). In addition to trypsin, other proteases, such as endoproteinases Lys-C and Glu-C, are also available for use in proteomics analysis, but tryptic digestion provides higher sequence coverage of identified proteins (Giansanti et al. 2016). In addition, trypsin is a relatively low-cost enzyme compared with other proteases and is readily available, factors that contribute to its widespread use in shotgun proteomics sample preparation (Aguilar-Hernández and Loyola-Vargas 2018). Protein digestion can be performed using various strategies, including ISD, filter-aided sample preparation (FASP), IGD, and other protocol variations depending on sample complexity, buffer composition, and downstream analytical requirements.

1.2.3 Data Acquisition and Processing

During the quantification and sequencing of peptides, two different acquisition modes of mass spectra can be used, the data-dependent acquisition (DDA) or data-independent acquisition (DIA) (Hu et al. 2016). The DDA acquisition mode is the most widely used in proteomics experiments to date, but DIA allows the identification of a greater number of proteins and is gaining ground with the development of new processing algorithms and the sensitivity of mass spectrometers (Guan et al. 2020). While the mode using DDA selects the precursors (peptides) with the highest intensity in MS1, DIA selects all precursors within a range of m/z in MS1, eliminating the bias of the most intense precursors (Zhang et al. 2020). Thus, DDA experiments are likely to have problems with missing values and perform better with a high number of biological replicates, whereas DIA relies on the availability of reference proteomes and the development of high-quality spectral libraries (Zhang et al. 2020).

A fundamental step in proteome studies is the selection of an appropriate reference proteome (protein database), together with the choice of a suitable software or algorithm for mass spectra deconvolution, peptide and protein identification and quantification (Schubert et al. 2017). During this process, the selection of the appropriate parameters is a key factor in ensuring the reliability of the data, including the mass tolerance differences of the precursor ion and its fragments (Chen et al. 2020). On the other hand, there are different software packages for performing database searches (DDA/DIA), de novo sequencing (DIA), and also hybrid identification approaches, where de novo sequencing and protein database searches are performed simultaneously (Chen et al. 2020).

After acquiring and processing the mass spectra of the samples, it is important to assess the quality of biological replicates to ensure the reproducibility and representativeness of the proteomic analysis. For this purpose, correlation analysis (e.g., Pearson correlation) and principal component analysis (PCA) are usually used as initial approaches for the data quality control and reliability assessment (Lin et al. 2022).

1.2.4 Bioinformatics Analysis and Statistics

Given the complexity of mass spectrometry data and omics analysis, an important aspect of interpreting and communicating proteomics results is the consistent and accessible presentation of graphs and figures (Fig. 3). In this sense, graphical representations are used in different stages of proteomic analysis, including quality control, identification, quantification, functional annotation, metabolic pathway analysis, interaction network studies, and data exploration in proteome databases (Oveland et al. 2015). In addition, specific types of graphs and visualizations are commonly used to analyze data generated at different stages of the proteomics

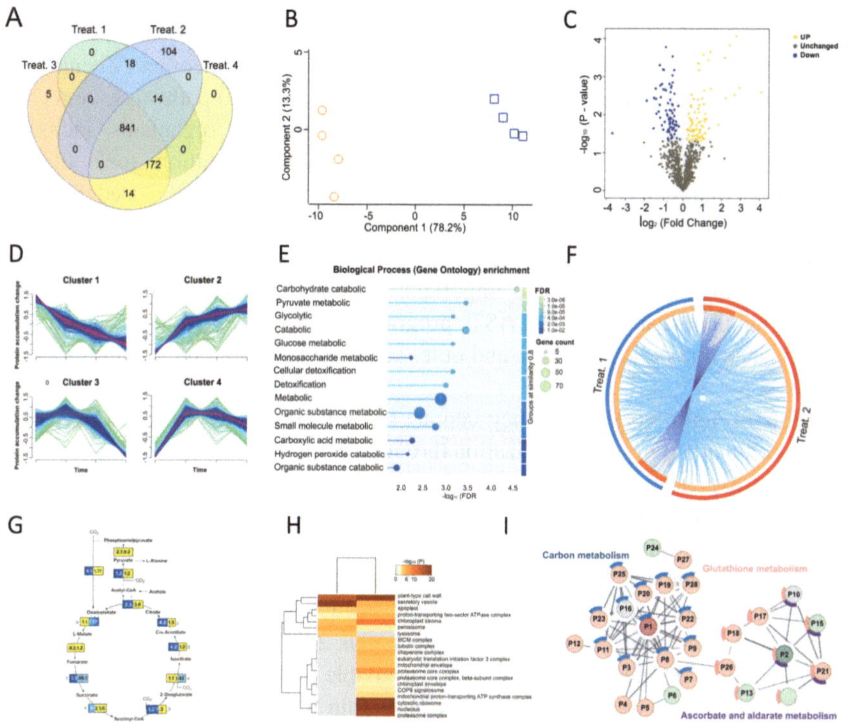

Fig. 3 Representative data visualization strategies commonly employed in large-scale proteomic studies of plants. The Venn diagram (**a**) shows the overlap of identified proteins across multiple experimental conditions; principal component analysis (PCA) (**b**) illustrates the separation of samples based on their proteomic profiles; the volcano plot (**c**) highlights differentially abundant proteins, with upregulated (yellow) and downregulated (blue) proteins; temporal clustering analysis (**d**) groups proteins with similar abundance patterns over time; Gene Ontology enrichment (**e**) reveals the most significantly enriched biological processes; the chord diagram (**f**) represents changes in protein–protein associations between treatment and control; pathway mapping (**g**) integrates identified proteins into the KEGG pathways. The heatmap (**h**) displays the enrichment of specific cellular components and protein complexes; protein–protein interactions (**i**) illustrate interaction networks derived from significantly enriched protein modules. (The figures are presented for illustrative purposes only)

workflows (Schessner et al. 2022). These include total ion chromatogram (TIC) for MS1 data, fragmentation spectrum with second stage of mass spectrometry (MS2) data, lollipop plots for visualizing post-translational modifications, correlation analysis based on protein intensity data, volcano plots for illustrating differential accumulation, and multivariate approaches such as PCA and clustering for multidimensional experiments. Thus, several software tools are available for proteomics data visualization, including Perseus, OmicScope, PatternLab, ProVision, DEP, protti, Eatomics, amica, MSstats, and FragPipe-Analyst (Carvalho et al. 2016; Hsiao et al. 2024; Reis-de-Oliveira et al. 2024).

In order to provide biological meaning and interpretation to proteomics data, tools for functional analysis and identification of regulatory proteins (e.g., transcription factors, kinases, and phosphatases) are commonly used, enabling broader, more reproducible, and more detailed data exploration. For the identification of regulatory proteins, iTAK represents a specific option for plant proteomes (Zheng et al. 2016). Kyoto Encyclopedia of Genes and Genomes (KEGG) pathways are a common option for studying metabolic pathways in different organisms, including plants (Kanehisa et al. 2016). For protein functional annotation, tools such as g:Profiler, Metascape, OmicsBox, and Mercator4 are commonly used for gene ontology analysis, enrichment analysis, and protein domains characterization (Conesa and Götz 2008; Schwacke et al. 2019; Zhou et al. 2019; Kolberg et al. 2023). During this process, it is important to define parameters such as the basic local alignment search tool (BLAST) program, the reference database (taxon), the significance threshold (e-value), and other settings that ensure the reliability of the results.

Proteins don't work alone; they interact to form complexes with other proteins, shape cellular activity and localization through post-translational changes, and control gene expression, signaling cascades, and other biological processes (Van de Graaf et al. 2024). In this context, the Search Tool for the Retrieval of Interacting Genes/Proteins (STRING) is a database that allows you to create and analyze protein–protein interaction networks from sources such as automated text mining of the scientific literature, co-expression, genomic context, interaction experiments, and complexes from curated sources (Szklarczyk et al. 2022). The use of network analysis tools fed with interaction information available in databases is a necessary step for the biological interpretation of results within a systems biology perspective. When analyzing thousands of proteins simultaneously, it is essential to classify biological complexes through clustering analyses, determine interaction hubs, and include post-translational modification information, functional enrichment, and differential accumulation information. The use of STRING with auxiliary tools such as Cytoscape allows the customization of interaction networks and the use of plugins to identify clusters and integrate with experimental data from PTMs (Shannon et al. 2003). Databases such as Reactome, Pathway Commons, and Omnipath can also be used to study interaction networks.

A fundamental step during proteomic analysis is data mining in complex tables with thousands of values, variables, and information to identify DAPs (Aguilan et al. 2020). At this point, the processing of the data prior to statistical analysis is fundamental and includes stages of normalization, cleaning, imputation, and the handling of missing values (Tyanova and Cox 2018). In most pairwise comparisons, DAPs are commonly defined using a combination of statistical significance (Student's t-test, $p < 0.05$) and arbitrary fold-change cutoffs, typically ranging from 1.41 to 2.0, which may also be expressed as ± $\log_2$ fold change values (e.g., > 0.5 or < −0.5) to classify proteins as up- or down-accumulated. However, there are bottlenecks in the analysis of proteomics data that mask potentially important discoveries due to the superficiality of functional and bioinformatics analyses downstream of data acquisition (Tyanova et al. 2016a). For this reason, there are already tutorials to help novice scientists with the statistical analyses involved in comparative bottom-up proteomics (Aguilan et al. 2020). In addition, different software can be used to process tables and generate visualizations for exploratory and functional analysis of proteomics data (Love et al. 2014; Tyanova et al. 2016b).

2 Proteomics in Crop Breeding

2.1 Proteomics of Plant Responses to Abiotic Stress

Plants are constantly exposed to abiotic stress factors such as drought, salinity, temperature, waterlogging, and nutrient imbalances, all of which can negatively impact growth, development, and productivity. These stresses influence plant physiology at multiple levels, leading to reduced photosynthesis, altered metabolism, and impaired reproductive success (Zhang et al. 2023). The increasing frequency and intensity of abiotic stresses due to climate change have become a major concern for global agriculture, according to the Food and Agriculture Organization (FAO) (Holleman et al. 2020). By analyzing changes in protein abundances, modifications, and interactions, proteomics allows researchers to identify key candidate proteins involved in stress responses and their potential roles in maintaining cellular homeostasis under adverse conditions. Since proteins are direct effectors of physiological processes, investigating their dynamics under stress conditions offers a more comprehensive understanding of how plants adjust their metabolism and signaling pathways to cope with environmental challenges (Wang et al. 2004).

Proteomic approaches have greatly contributed to the identification of enriched biological processes and candidate proteins associated with tolerance to various abiotic stresses in different plant species (Table 1). In this section, we summarize the studies on changes in the plant proteome in response to the impact of several abiotic stress factors, including water (drought, flooding/waterlogging), temperature extremes (cold and heat), salinity, and enhanced concentrations of heavy metals. Also, we provide updated information about the contribution of proteomic studies for unraveling the mechanisms underlying abiotic stress responses in plants. The role of proteomics in plant stress is detailed in chapter "Plant Leaf Proteomics and Drought Stress Tolerance".

Table 1 Summary of proteomic studies addressing abiotic stress responses in plants

Plant species	Abiotic stress	Highlighted proteins	Reference
Sorghum bicolor	Drought stress, heat stress, combined drought and heat stress	APX (ascorbate peroxidase), SOD (superoxide dismutase), RuBisCO (Ribulose-1,5-bisphosphate carboxylase/oxygenase)	Ali et al. (2025)
Zea mays	Drought stress	SS (sucrose synthase), SOD (superoxide dismutase), and RuBisCO (Ribulose-1,5-bisphosphate carboxylase/oxygenase)	Schmitt et al. (2024)
Setaria italica	Drought stress and recovery	MPK6 (mitogen-activated protein kinase 6), PAL (phenylalanine ammonia-lyase), SOD (superoxide dismutase), PER12 (peroxidase 12)	Gao et al. (2023)
Solanum lycopersicum	Drought stress	AREB1 (ABA-response element binding factor 1), GSH-Px-1 (glutathione peroxidase 1), Hsp (heat shock protein)	Liu et al. (2023)
Glycine max	PEG-induced drought stress	GmGSH (glutathione synthetase), GmGST1 (glutathione S-transferase 1), GmCAT (catalase), Gm6PGD (6-Phosphogluconate dehydrogenase)	Zhou et al. (2022)
Z. mays	Drought stress	HSP20 (heat shock protein 20), COR410 (cold-regulated 410), RuBisCO (Ribulose-1,5-bisphosphate carboxylase/oxygenase)	Ren et al. (2022)
Nicotiana benthamiana	Drought stress and re-watering	CA1 (carbonic anhydrase 1), PGK (phosphoglycerate kinase), GSA (Glutamate-1-semialdehyde 2,1-aminomutase)	Li et al. (2020)
Saccharum spp.	Drought stress	bZIP (leucine-rich transcription factor), C2H2 (zinc finger protein), HSF (heat shock factor), ARF (auxin response factor)	Salvato et al. (2019)
Solanum melongena	Waterlogging stress (6, 12, and 24 h)	FBA (fructose-bisphosphate aldolase), ADH (alcohol dehydrogenase), 14–3-3-like protein A	Yang et al. (2023)
Triticum aestivum	Waterlogging in contrasting genotypes	Fe-S cluster assembly factor, HSC70 (heat shock cognate 70), SAR1A (GTP-binding protein SAR1A-like), CBS domain-containing protein	Yang et al. (2022)
G. max	Flooding stress	CNX (Calnexin), CRT (Calreticulin), ADH (alcohol dehydrogenase)	Komatsu et al. (2021)
Vitis vinifera	Waterlogging stress (0, 10, and 20 days)	SOD (superoxide dismutase), POD (peroxidase), CAT (catalase), APX (ascorbate peroxidase)	Wang et al. (2021b)

(continued)

Table 1 (continued)

Plant species	Abiotic stress	Highlighted proteins	Reference
Arachis hypogaea	Waterlogging	LDH (L-lactate dehydrogenase), NAD-ME (NAD^+-dependent malic enzyme), AspAT (aspartate aminotransferase), GDH (glutamate dehydrogenase)	Liu et al. (2021)
Z. mays	Heat stress	ZmMKK2 (mitogen-activated protein kinase kinase 2), ZmMPK7 (mitogen-activated protein kinase 7), HSP70 (heat shock protein 70), HSP90 (heat shock protein 90)	Ma et al. (2025)
Nicotiana tabacum	Heat stress	GluTR (Glutamyl-tRNA reductase), MOAC (monomethyl ester aerobic oxidative cyclase), PSII-Cyt b559 (photosystem II cytochrome b559), FBPase (Fructose-1,6-bisphosphatase), GBSS (granule-bound starch synthase)	Song et al. (2025)
Gossypium hirsutum	Heat stress (35 °C control vs. 45 °C stress)	HSP70 (heat shock protein 70), APX (ascorbate peroxidase), RuBisCO (Ribulose-1,5-bisphosphate carboxylase/oxygenase)	Perveen et al. (2025)
O. sativa	Heat stress during grain filling	GH3.4 (Indole-3-acetic acid-amido synthetase), GH3.8 (Indole-3-acetic acid-amido synthetase), SPC4 (cationic peroxidase), ETHE1 (Persulfide dioxygenase), RA5 (seed allergy protein)	Zhao et al. (2023)
Solanum tuberosum	Heat stress and drought stress	SOD (superoxide dismutase), CAT (catalase), RuBisCO (Ribulose-1,5-bisphosphate carboxylase/oxygenase)	Boguszewska-Mańkowska et al. (2020)
G. max	Cold stress	ATAF1 (transcription factor), MEblue module proteins, MEgreen module proteins	Bukhari et al. (2025)
O. sativa	Chilling stress during seed germination	OsJ_16716, OsPAL1, RINO1	Zhang et al. (2024)
O. sativa subsp. japonica	Cold stress (8 °C for 24, 48, and 72 h)	14-3-3 proteins and cold-regulated proteins	Qing et al. (2022)
Saccharum spontaneum	Cold stress	SOD (superoxide dismutase), CAT (catalase), POD (peroxidase), MDA (malondialdehyde)	Zhang et al. (2022a)
T. aestivum	Salt stress (150 mM NaCl)	30S and 50S ribosomal proteins, ATP-dependent caseinolytic protease, filamentous temperature-sensitive H proteases, lipoxygenases, Allene oxide synthases	Wijeweera et al. (2025)

(continued)

Table 1 (continued)

Plant species	Abiotic stress	Highlighted proteins	Reference
Avena sativa	Salt stress (control, moderate, and severe)	SOD (superoxide dismutase), CAT (catalase), POD (peroxidase), ADH (alcohol dehydrogenase), PDC (pyruvate decarboxylase)	Chen et al. (2025)
G. hirsutum	Salt stress (three-leaf stage, 72-h exposure)	SOD (superoxide dismutase), POD (peroxidase), CAT (catalase), APX (ascorbate peroxidase)	Li et al. (2025b)
S. lycopersicum	Saline conditions (moderate salinity)	Photosystem I proteins, Fructose-1,6-bisphosphatase, superoxide dismutase (SOD), Acylamino acid-releasing enzymes	Barba-Espín et al. (2025)
G. max	Salt stress (300 mM NaCl)	Calreticulin (CRT), Calnexin (CNX), Annexin 4 (ANN4)	Fu et al. (2024a)
M. sativa	Salt stress (150 mM NaCl for 2 and 6 h)	GSTL3 (glutathione S-transferase L3), GSTU3 (glutathione S-transferase U3), PER5 (peroxidase 5), PER10 (peroxidase 10), APR (5′-Adenylylsulfate reductase), COMT (Caffeic acid O-methyltransferase)	Hang et al. (2024)
M. sativa	Salt stress (100 mM and 200 mM NaCl for 3 weeks)	Ascorbate peroxidase, POD (peroxidase), CBS protein, PR-10, HSP70 (heat shock protein 70), TCP-1, GroES, Calmodulin, Annexin (MsANN2)	Gao et al. (2024b)
Cannabis sativa	Salt stress (140 mM NaCl)	UDP-glycosyltransferases, Germacrene B synthase, Secoisolariciresinol dehydrogenase-like protein, auxin response factor 6-like protein	Yang et al. (2024)
Pfaffia glomerata	Salt stress (50 mM NaCl) and epigenetic modification (25 μM 5-azacytidine)	SOD (superoxide dismutase), CAT (catalase), POD (peroxidase), Jasmonate biosynthesis-related proteins	Fortini et al. (2023)
Saccharum spp.	Salt stress (180 mM NaCl for 20 days)	Calcium-dependent protein kinase, photosystem I, phospholipase D, Glyceraldehyde-3-phosphate dehydrogenase	Passamani et al. (2017)
Brassica napus	Heavy metal stress (Cadmium, lead, arsenic)	Peroxidase, ABC transporters, Abscisic acid-related proteins	Li et al. (2025a)
Pogonatherum crinitum	Lead (Pb) stress at various concentrations (400 mg/kg, 800 mg/kg, 1000 mg/kg)	PcCAT3 (catalase), PcCht1 (Chitinase), PcSPS5 (sucrose phosphate synthase), PcGME-1 (GDP-mannose 3,5-epimerase), PcHSP26.7 (heat shock protein)	Meng et al. (2024)

(continued)

Table 1 (continued)

Plant species	Abiotic stress	Highlighted proteins	Reference
O. sativa subsp. indica and *japonica*	Cadmium (Cd) toxicity stress (3 mg/kg Cd)	psbO, PetH, LHCA, ASCAB9, PEPC, OsAld, OsRCCR1, SE5, CATC, GLO4, APX8, OsGSTU13	Zhu et al. (2024)
B. napus	Cadmium (Cd) toxicity stress with citric acid (CA) supplementation	Superoxide dismutase, L-ascorbate peroxidase 3, glutamine synthetase, glutathione S-transferase DHAR1, cysteine synthase, S-adenosylmethionine synthase 2, thiosulfate/3-mercaptopyruvate sulfur transferase 2	Mittra et al. (2023)
Z. mays	Cadmium (Cd) toxicity stress	ZmHMA3 (heavy metal ATPase 3), glutamine synthetase, Peroxiredoxin, Chitinase chem5	Guo et al. (2023)
N. tabacum	Zinc deficiency and excess toxicity	ZIPs (zinc-regulated/iron-regulated-like proteins), NAS3s (Nicotianamine synthases), HMA1s (heavy metal ATPases), SODs (superoxide dismutases), APX1s (ascorbate peroxidases), GPX6 (glutathione peroxidase), GSTs (glutathione S-transferases)	Lu et al. (2023)
O. sativa	Zinc (Zn) stress under low cadmium (Cd) conditions	Plasma membrane ATPase, ABC transport proteins, P1B-ATPase, V-type ATPase	Zhang et al. (2022b)
Z. mays var. everta	Aluminum (Al) toxicity stress	Sucrose synthase, citrate synthase, Glucose-6-phosphate isomerase, fructose-bisphosphate aldolase	Pinto et al. (2021)
Z. mays	Chromium (Cr) toxicity stress (100 μM Cr)	Glutathione peroxidase, glutathione S-transferases, pathogenesis-related proteins, glyoxalases, superoxide dismutase	Terzi and Yıldız (2021)
Pisum sativum	Copper (Cu) toxicity stress (200 μM $CuCl_2$) with calcium (Ca) or citrate supplementation	Glutathione reductase (GR), glutathione-S-transferase (GST), glutathione peroxidase (GPX), Thioredoxin (Trx), ferredoxin (Fdx)	Massoud et al. (2022)
Vitis labruscana × *Vitis vinifera*	Low degree of excess copper stress (low-ECS) with $CuSO_4$ treatment	HMA5, ABC transporters, PMM, GME, DHAR, MDHAR, ARGs, ARPs	Chen et al. (2021)
S. lycopersicum	Cadmium (Cd) toxicity stress (35 μM $CdCl_2$ for 4 days)	Superoxide dismutase (SOD), glutathione-S-transferase (GST), peroxidase (POD), Chitinase, ATP-sulfurylase	Borges et al. (2019)
Populus deltoides	Arsenic (As) toxicity stress	PdDRT102 (DNA repair/toleration protein), superoxide dismutase (SOD), catalase (CAT), peroxidase (POD)	Liu et al. (2017)

For each species, the applied abiotic stress, the main stress-responsive candidate proteins, and corresponding references are listed

2.1.1 Water Stress

Water is particularly vital, as it is the most abundant compound in the cytosol where critical biochemical reactions occur (Zahedi et al. 2025). Furthermore, its hydraulic properties drive cell expansion and contribute to structural support. However, from an agricultural standpoint, water availability represents a major abiotic factor limiting crop productivity worldwide (Fu et al. 2024b). Water stress in plants arises under two contrasting conditions: water deficit (drought) and excess water (waterlogging). Different plant species exhibit varying levels of tolerance or sensitivity to these extremes, depending on their ecological adaptations (Fu et al. 2024b).

Drought, characterized by limited water availability and cellular dehydration, is the most widespread abiotic stress globally and poses a severe threat to agricultural yields, affecting plants at morphological, physiological, biochemical, and molecular levels (Seleiman et al. 2021). Osmotic adjustment in response to drought involves metabolic reprogramming, which is often reflected in changes in the proteome (Lyon et al. 2016). Over recent decades, root-specific proteomic analyses have been instrumental in linking functional protein groups to drought response, underscoring the central role of roots in detecting and responding to soil water deficits (Schachtman and Goodger 2008; Aroca et al. 2011; Prinsi et al. 2018). Since proteins represent the active products of gene expression, proteomic approaches are particularly informative for revealing stress-related mechanisms that may not be evident through DNA or RNA analyses.

Using a comparative proteomic approach in leaves and roots of *Setaria italica*, tissue-specific responses to drought stress were studied, highlighting the complexity of plant adaptation mechanisms (Gao et al. 2023). In leaves, DAPs were mainly associated with photosynthesis and carbohydrate metabolism, whereas roots showed enrichment in proteins related to cellular homeostasis and hormone biosynthesis (Gao et al. 2023). Notably, proteins such as mitogen-activated protein kinase 6 (MPK6) and phenylalanine ammonia-lyase (PAL) were up-accumulated during drought stress, while peroxidase 12 (PER12) was more abundant during the recovery phase (Gao et al. 2023). These findings suggest complementary strategies between organs: leaves prioritize the maintenance of photosynthetic capacity, while roots contribute to stress sensing and structural reinforcement (Gao et al. 2023). In *Zea mays var. everta* (popcorn), proteomic analysis revealed significant alterations in protein profiles in response to drought stress of the drought-tolerant line (L71) and the drought-sensitive line (L61) (Schmitt et al. 2024). Key proteomic differences between the genotypes involved energy metabolism, photosynthesis, oxidative stress response, and protein synthesis pathways (Schmitt et al. 2024).

Multiple studies have utilized root proteomics to investigate mechanisms of drought tolerance. In this context, a proteomic approach has been applied to analyze *Glycine max* roots subjected to PEG-simulated drought stress (Zhou et al. 2022). The analysis revealed significant alterations in proteins involved in energy metabolism, antioxidant defense, and signal transduction. Among these, components of the glutathione metabolism pathway, such as glutathione synthetase (GmGSH), glutathione S-transferase 1 (GmGST1), and catalase (GmCAT), were notably upregulated, suggesting a central role in maintaining redox balance under stress (Zhou et al. 2022). Likewise, the use of proteomics to investigate drought stress responses in maize seedlings identified key drought-responsive proteins involved in redox homeostasis, cell wall organization, photosynthesis, abscisic acid (ABA) biosynthesis, and carbohydrate metabolism (Ren et al. 2022). Molecular chaperones, heat shock proteins (HSPs), and enzymes associated with primary metabolism exhibited significant differential abundance, emphasizing their roles in maintaining cellular function and enhancing drought resilience (Ren et al. 2022).

A broader integrative analysis combined transcriptomic and proteomic approaches to investigate drought responses in *Solanum lycopersicum* (Liu et al. 2023). Among the up-accumulated proteins, those involved in antioxidant defense mechanisms, particularly HSPs, and osmolyte metabolism, such as enzymes associated with proline biosynthesis, played pivotal roles in enhancing drought tolerance. Additionally, proteins related to signal transduction, including ABA-responsive transcription factors, were identified as key regulators of the drought response (Liu et al. 2023). These results suggested that drought stress led to elevated ABA levels, which in turn increased the expression of *ABA-response element binding factor* genes and respective protein synthesis. These transcription factors regulate the synthesis of HSPs and other protective proteins, thereby promoting adaptive responses to water deficit conditions (Liu et al. 2023).

Proteomic responses to drought were also explored in shoot tissues. In *Saccharum* spp., a comprehensive proteomic analysis of sugarcane stems under drought stress was conducted (Salvato et al. 2019). Proteins involved in chromatin remodeling, RNA processing, and transcriptional regulation were highlighted, including transcription factors with domains like basic leucine zipper (bZIP), NAC, and Myb-related, as well as components of the spliceosome and THO/TREX complex. These results suggest a significant role of transcriptional reprogramming and post-transcriptional regulation in sugarcane drought tolerance (Salvato et al. 2019). Furthermore, in *Nicotiana benthamiana*, chloroplast proteomics under drought stress identified 1087 DAPs, revealing extensive reprogramming of the chloroplast proteome (Li et al. 2020). Notably, carbonic anhydrase 1 (CA1) showed a significant decrease in chloroplasts but a notable increase in leaves under drought conditions (Li et al. 2020). This contrasting distribution suggests that CA1 and similar proteins may be translocated from chloroplasts to the cytosol, acting as retrograde signal messengers to regulate nuclear gene expression (Li et al. 2020). These results underscore the systemic role of chloroplast proteins in coordinating drought responses and emphasize the importance of organelle-specific proteomics in dissecting the intracellular signaling networks involved in plant stress adaptation.

In contrast to drought, waterlogging results from excess water in the soil surpassing its absorption capacity, leading to surface saturation and oxygen deprivation in the root zone. The resulting hypoxic or anoxic conditions shift root metabolism from aerobic respiration to anaerobic pathways, such as glycolysis and fermentation, thereby reducing energy efficiency (Manghwar et al. 2024). This stress, most of the time triggered by heavy or prolonged rainfall in poorly drained soils, represents one of the most severe environmental stresses, especially during early plant development (Manghwar et al. 2024).

Some proteomic studies have investigated the physiological effects and metabolic pathways associated with low oxygen conditions, such as those occurring during waterlogging, contributing to a better understanding of plant adaptive mechanisms to flooding-induced water stress. In *Solanum melongena*, a proteomic investigation revealed that short-term waterlogging induced early and dynamic changes in root protein profiles (Yang et al. 2023). The majority of DAPs were associated with energy metabolism, amino acid biosynthesis, signal transduction, and nitrogen metabolism. Notably, enzymes involved in glycolysis and fermentation, such as fructose-bisphosphate aldolase and alcohol dehydrogenase (ADH), were up-accumulated in a time-dependent manner, reflecting metabolic adjustment to maintain cellular activity and flexibility under hypoxic conditions (Yang et al. 2023). In this context, *G. max* mutant lines generated through gamma irradiation demonstrated enhanced tolerance to flooding (Komatsu et al. 2021). Notably, the tolerant mutant lines displayed increased levels of glycoprotein-folding enzymes, such as calnexin and calreticulin (Komatsu et al. 2021). In addition, fermentation-related proteins like ADH were identified, although the tolerant mutant lines showed more moderate activity of ADH compared to wild-type plants, suggesting a balanced approach to anaerobic energy production (Komatsu et al. 2021). Proteins such as exostosin domain-containing, which are crucial for secondary cell-wall biogenesis, also showed increased abundance in tolerant mutant lines under flooding stress, in contrast to their reduced abundance in wild-type plants (Komatsu et al. 2021). This improved regulation of stress-mitigation processes likely underpins the suppression of cell death observed in tolerant mutant lines, enabling a more efficient flooding tolerance mechanism (Komatsu et al. 2021).

Genotypic differences in waterlogging tolerance have also been elucidated through proteomic approaches, as demonstrated in *Vitis* spp. (Wang et al. 2021b), *Triticum aestivum* L. (Yang et al. 2022), and *Arachis hypogaea* (Liu et al. 2021). These findings underscore the complexity and plasticity of plant responses to water-related stress. Proteomic studies consistently reveal that successful adaptation involves coordinated changes in energy metabolism, redox balance, protein turnover, and signal transduction, pathways that are central to survival under both drought and waterlogging conditions.

2.1.2 Temperature Stress

Temperature is one of the most critical environmental variables influencing plant development and productivity (Zhao et al. 2025a). Both high and low temperature extremes, commonly referred to as heat and cold stress, can significantly alter plant metabolism, cellular structures, and overall physiological functions (Distéfano et al.

2024). Plants have evolved various mechanisms to perceive, respond to, and survive such temperature fluctuations, many of which are reflected at the proteomic level (Zhao et al. 2025a). Among temperature-related stresses, heat stress stands out as one of the most extensively studied at the molecular level, particularly through proteomic analyses.

Proteomic studies have provided valuable insights into how different species, organs, and genotypes modulate their protein profiles in response to heat, shedding light on key components involved in stress perception, signaling, and tolerance (Kausar et al. 2022). For instance, a proteomic study in *T. aestivum* investigated the molecular basis of thermotolerance by comparing heat-tolerant and heat-sensitive genotypes (Ma et al. 2025). In the heat-tolerant genotype, there was a significant enrichment of proteins involved in translational and post-translational modifications, protein turnover, and chaperone binding, particularly within the mitogen-activated protein kinase (MAPK) signaling pathway. Moreover, the accumulation of heat shock proteins, notably HSP70, along with transcription factors such as bZIP and MYB, suggested enhanced regulation of stress-responsive pathways and improved cellular adaptation (Ma et al. 2025).

In *N. tabacum*, integrated proteomic and physiological analyses of heat-stressed leaves treated with melatonin revealed significant modulation of photosynthesis-related proteins (Song et al. 2025). Among the up-accumulated proteins were glutamyl-tRNA reductase and monomethyl ester aerobic oxidative cyclase, both involved in chlorophyll biosynthesis, as well as photosystem II (PSII) cytochrome b559 and fructose-1,6-bisphosphatase, which are associated with the PSII photochemical activity and the Calvin cycle, respectively (Song et al. 2025). The study highlighted the use of melatonin for enhancing photosynthetic efficiency and mitigating photoinhibition, emphasizing its potential in improving plant thermotolerance.

Proteomic analysis of *Solanum tuberosum* (Boguszewska-Mańkowska et al. 2020) roots under drought and high temperature stress revealed distinct molecular responses between tolerant (cv. Gwiazda) and sensitive (cv. Oberon) cultivars. In the tolerant cultivar Gwiazda, proteins associated with energy and carbohydrate metabolism, such as enolase and malate dehydrogenase, were significantly up-accumulated under both stress conditions, indicating a capacity for enhanced metabolic flexibility and sustained energy production (Boguszewska-Mańkowska et al. 2020). Conversely, the sensitive Oberon cultivar displayed a marked increase in proteins related to defense and detoxification, suggesting a reactive response rather than an adaptive strategy. Moreover, elevated temperatures led to a decrease in proteins involved in cellular energy metabolism in the sensitive cultivar, while the tolerant cultivar maintained higher levels of these proteins, suggesting a more robust energy production system (Boguszewska-Mańkowska et al. 2020).

In *Gossypium hirsutum*, a proteomic analysis under heat stress identified that up-accumulated proteins in the tolerant genotype included those involved in photosynthesis, oxidoreductase activity, and carbohydrate metabolism, reflecting enhanced stress adaptation. In contrast, the susceptible genotype showed significant down-accumulation of proteins linked to these pathways, indicating impaired stress responses (Perveen et al. 2025). In *Oryza sativa*, elevated temperatures during the grain-filling stage modulate the abundance of stimulus-responsive proteins involved in storage substance accumulation and stress adaptation (Zhao et al. 2023). Proteomic analysis identified key proteins such as alpha-amylase isozyme 3E and seed allergy protein RA5, which are involved in regulating starch and protein accumulation under high-temperature conditions. Additionally, proteins related to reactive oxygen species (ROS) metabolism, including cationic peroxidase SPC4 and persulfide dioxygenase ETHE1 homolog, were up-accumulated, contributing to oxidative stress mitigation and physiological homeostasis (Zhao et al. 2023). These findings highlight the molecular mechanisms underlying heat tolerance, providing potential targets for breeding programs.

In contrast, cold stress, triggered by chilling (<20 °C) or freezing (<0 °C) temperatures, causes a different set of physiological challenges (Chinnusamy et al. 2007). The reduction in temperature slows down enzyme-catalyzed reactions, disrupting metabolic balance and often resulting in oxidative stress due to an accumulation of ROS. Moreover, low temperatures impair water uptake, leading to cellular dehydration, which further exacerbates stress (Chinnusamy et al. 2007). To counteract these adverse effects, plants activate adaptive responses, including the stimulation of primary metabolic pathways to increase energy production, up-accumulation of antioxidant systems and chaperone proteins, and modifications to membrane composition that help maintain osmotic balance (Sharma et al. 2005). These responses are reflected in profound changes in the plant proteome under cold stress, underscoring the importance of proteomic studies in understanding cold acclimation and tolerance.

In *O. sativa*, a crop particularly sensitive to chilling stress, proteomic analysis revealed the differential accumulation of proteins related to antioxidant defense, phenylpropanoid biosynthesis, and inositol phosphate metabolism under low-temperature conditions (Zhang et al. 2024). In seeds primed with gibberellin A_3 (GA_3) and brassinolide, key proteins such as peroxidase, phenylalanine ammonia-lyase, and inositol phosphate synthase were up-accumulated (Zhang et al. 2024). In a study with *O. sativa* subsp. *Japonica*, cold stress elicited significant proteomic changes, resulting in the up-accumulation of proteins involved in photosynthesis, metabolic pathways, secondary metabolites biosynthesis, and carbon metabolism (Qing et al. 2022). Similarly, in *G. max*, low-temperature stress elicited contrasting proteomic and transcriptomic responses in cold-tolerant (HX3) and cold-sensitive (HH43) varieties (Bukhari et al. 2025). Proteomic analysis revealed that the cold-tolerant variety exhibited broader and sustained upregulation of proteins associated with metabolic activity and cellular homeostasis, while the cold-sensitive variety showed limited proteomic responses and degradation of early responsive proteins. Key findings included the identification of transcription factors such as *Arabidopsis* transcription activation factor 1 (ATAF1) and their potential targets, which likely contribute to the enhanced cold tolerance in HX3 (Bukhari et al. 2025). These findings highlight the role of enhanced antioxidant capacity and metabolic regulation in promoting germination and resilience under chilling stress. Other proteomic studies have also investigated cold stress responses in *T. aestivum* (Wang et al. 2023) and sugarcane (*Saccharum* spp.) (Zhang et al. 2022a).

These studies reflect a highly integrated network of metabolic, homeostatic, and signaling processes, whose precise regulation underpins a plant's ability to endure thermal extremes. Beyond changes in protein abundance, thermotolerance and cold acclimation are also shaped by PTMs, organelle-specific responses, and spatially coordinated signaling events that collectively safeguard cellular function and developmental integrity under heat and cold stress.

2.1.3 Salt Stress

Salt stress, primarily caused by the excessive accumulation of ions such as sodium (Na^+), chloride (Cl^-), calcium (Ca^{+2}), potassium (K^+), carbonate (CO_3^{-2}), nitrate (NO_3^-), and sulfate (SO_4^{-2}) in the soil, leads to a reduction in soil water potential, decreasing water uptake by plant roots, initiating osmotic stress (Kamran et al. 2020; Stavi et al. 2021). Salt stress is a major constraint on agricultural productivity, especially in hot, dry, and semiarid regions where crop cultivation often relies on irrigation. Globally, salinity affects at least 20% of irrigated lands, significantly reducing crop yields (Tarolli et al. 2024). Roots, being the first organs to sense salinity in the soil, play a central role in stress perception and response. In addition to stress adaptation responses, proteomic tools have been important in elucidating the dynamics of the proteome of primary roots, root hairs, and lateral roots, as well as the effects of heterosis during development (Hochholdinger et al. 2018). In this sense, advances in proteomic approaches have provided valuable insights into the molecular mechanisms underlying root adaptation to salt stress.

In *T. aestivum*, salt stress disrupted chloroplast protein homeostasis, particularly affecting plastid-localized protein synthesis machinery, including 30S and 50S ribosomal proteins and elongation factors (Wijeweera et al. 2025). Proteomic analysis revealed significant reductions in these proteins, correlating with altered K^+/Na^+ ratios along the leaf gradient. Additionally, decreased abundance of plastid proteases, such as adenosine triphosphate (ATP)-dependent caseinolytic protease, highlighted impaired protein turnover (Wijeweera et al. 2025). These findings underscore the critical role of chloroplast protein homeostasis in maintaining *T. aestivum* resilience under salinity stress (Wijeweera et al. 2025).

In *Avena sativa*, proteomic analysis under salt stress revealed significant alterations in root respiratory metabolism. Key findings included differential accumulation of proteins involved in glycolysis, the tricarboxylic acid cycle, and the pentose phosphate pathway (Chen et al. 2025). Notably, salt-tolerant cultivars exhibited enhanced antioxidant enzyme activity and pyruvate-ethanol fermentation, while salt-sensitive cultivars relied more on pyruvate-lactic acid fermentation. These results highlight distinct metabolic strategies employed by oat genotypes to adapt to salinity stress (Chen et al. 2025).

In *Saccharum* spp., proteomic analysis under salt stress revealed significant alterations in protein abundance in micropropagated sugarcane shoots (Passamani et al. 2017). The salt-tolerant cultivar RB855536 exhibited a higher abundance of proteins involved in nonenzymatic antioxidant mechanisms, ion transport, and photosynthesis, compared to susceptible CB38-22. Notably, proteins such as calcium-dependent protein kinase, photosystem I (PSI), phospholipase D, and glyceraldehyde-3-phosphate dehydrogenase were more abundant in RB855536 under salt stress, suggesting their role in maintaining ionic and osmotic homeostasis (Passamani et al. 2017).

In *Medicago sativa*, integrated transcriptomic and proteomic analyses have revealed genotype-specific mechanisms of salt stress response (Hang et al. 2024). In a comparative study between salt-tolerant (ZM-1) and salt-sensitive (AT) genotypes, proteomic profiling identified DAPs involved in key metabolic pathways, including glutathione metabolism, biosynthesis of secondary metabolites, gluconeogenesis, carbon fixation, and photosynthesis (Hang et al. 2024). In a complementary root-specific analysis, proteins such as ascorbate peroxidase (APX), pathogenesis-related protein 10 (PR-10), and cystathionine beta-synthase (CBS) were upregulated, enhancing defense mechanisms (Gao et al. 2024b). Additionally, the study emphasized the role of unfolded protein response (UPR) and ubiquitin–proteasome system (UPS) in maintaining protein homeostasis under stress (Gao et al. 2024b). Proteomic approaches have also been applied to investigate plant responses to salt stress in various species, such as *G. hirsutum* (Li et al. 2025b), *Pfaffia glomerata* (Fortini et al. 2023), *S. lycopersicum* (Barba-Espín et al. 2025), and *G. max* (Fu et al. 2024a).

These studies reinforce the view that the proteomic response to salinity stress extends beyond basic ionic regulation. It involves the integration of metabolic, antioxidant, and signaling processes, whose fine coordination determines a plant's ability to maintain cellular function under saline conditions. The regulation of salt stress responses involves not only changes in protein abundance but also protein turnover, redox control, and ion compartmentalization mechanisms critical for sustaining homeostasis.

2.1.4 Heavy Metal Stress

The extensive and uncontrolled use of chemicals over recent decades has resulted in a widespread accumulation of environmental pollutants (Peralta-Videa et al. 2009). Among these, heavy metals such as cadmium (Cd), arsenic (As), aluminum (Al), and lead (Pb) are particularly concerning due to their persistence and frequent detection in contaminated soils and groundwater (Pinto et al. 2021). As sessile organisms, plants are continuously exposed to these toxic elements, making the investigation of their physiological and molecular responses to heavy metal stress a critical area of study (Peralta-Videa et al. 2009). Heavy metal ions can bind to sulfhydryl groups in proteins, inactivate enzymes, displace essential cations from specific binding sites, and promote the generation of ROS, leading to oxidative damage to lipids, proteins, and nucleic acids (Hossain and Komatsu 2013). These harmful effects not only reduce crop productivity but also pose serious ecological risks to plants and animals alike (Ahsan et al. 2009).

In *Z. mays*, multiple studies have demonstrated genotype-specific and compound-specific responses to heavy metal stress (Terzi and Yıldız 2021). In *Z. mays var. everta* (popcorn), proteomic analysis revealed key metabolic pathways associated with aluminum (Al) tolerance in roots (Pinto et al. 2021). Conserved tolerance-associated metabolic pathways, particularly starch and sucrose metabolism and glycolysis/gluconeogenesis, were significantly regulated, suggesting that these pathways contribute to popcorn tolerance to Al stress (Pinto et al. 2021). Additionally, citrate synthase accumulation in the Al-tolerant inbred line points to its role in detoxification processes (Pinto et al. 2021). Another investigation demonstrated that exogenous application of cysteine (Cys) mitigated chromium (Cr) toxicity by limiting Cr translocation to aerial tissues and enhancing its sequestration in roots (Terzi and Yıldız 2021). Proteins involved in oxidative stress response, energy and carbohydrate metabolism, protein turnover, and amino acid metabolism were affected by Cr exposure (Terzi and Yıldız 2021). Proteins such as glutathione peroxidase, glutathione S-transferase, glyoxalase, and superoxide dismutase (SOD) were regulated by Cys, contributing to Cr detoxification (Terzi and Yıldız 2021).

Proteomic studies in *Brassica napus* under Cd, Pb, and As stresses revealed enhanced activity of antioxidant enzymes, such as peroxidase, and increased abundance of transporters, particularly ATP-binding cassette (ABC) proteins, highlighting their central roles in mitigating oxidative stress and facilitating metal detoxification (Li et al. 2025a). ABA signaling was also implicated in modulating these responses, underscoring the importance of hormonal regulation in metal stress adaptation (Li et al. 2025a). Additionally, another study demonstrated that citric acid supplementation altered global protein abundance under Cd stress. Key DAPs were involved in energy metabolism, photosynthesis, signal transduction, protein turnover, and antioxidant defense (Mittra et al. 2023). Among them, enzymes such as SOD, APX, dehydroascorbate reductase 1 (DHAR1), cysteine synthase, and S-adenosylmethionine synthase 2 played central roles in redox regulation and sulfur assimilation, contributing to improved stress resilience (Mittra et al. 2023).

In *N. tabacum*, combined transcriptomic and proteomic analyses under varying zinc (Zn) availability identified molecular responses critical for Zn homeostasis and oxidative stress mitigation (Lu et al. 2023). Notably, transport proteins such as zinc-regulated/iron-regulated-like proteins (ZIPs), nicotianamine synthases (NAS3s), and heavy metal ATPases (HMA1s) were up-accumulated under Zn deficiency or limitation, while antioxidant enzymes like SODs, APX1s, glutathione peroxidase (GPX6), and glutathione S-transferases (GSTs) were induced to counter oxidative damage (Lu et al. 2023). Proteomic analyses have also been applied to investigate plant responses to heavy metal stress, such as in *Pogonatherum crinitum* under lead (Pb) exposure (Meng et al. 2024) and *Populus deltoides* under arsenate stress (Liu et al. 2017).

Collectively, these studies underscore the multifaceted nature of plant responses to heavy metal stress. Proteomic investigations reveal that successful tolerance involves a coordinated network of detoxification mechanisms, antioxidant defenses, metabolic adjustments, and signal transduction pathways. As heavy metal contamination remains a major agricultural and environmental concern, proteomics provides a valuable framework for uncovering adaptive strategies and identifying candidate proteins for crop improvement programs focused on enhancing stress tolerance and phytoremediation potential.

2.2 *Proteomics of Plant Responses to Biotic Stress*

The proteomic response of plants to biotic stress is highly dynamic and multifaceted, reflecting the complex signaling networks that coordinate defense, metabolism, and cellular homeostasis (Liu et al. 2019). Recent studies have shown that this response goes far beyond the mere induction of defense proteins, integrating metabolic pathways, PTMs, hormonal regulation, and intra- and intercellular signaling processes.

Biotic stress in plants is caused by a diverse range of agents, including viruses, fungi, bacteria, oomycetes, insects, mites, and nematodes, each capable of activating distinct molecular defense responses. The application of proteomics to plant–pathogen and plant–pest interactions has enabled the identification of key molecular targets associated with resistance mechanisms in various crop species (Table 2).

2.2.1 Fungal Infections

Tropical fruit crops have significantly contributed to the understanding of proteomic responses to fungal infections. Studies in crops such as *Mangifera indica* have revealed that infection by *Alternaria alternata* induces the accumulation of DAPs associated with phenylpropanoid, flavonoid, and MAPK pathways (Xie et al. 2025). The accumulation of H_2O_2, confirmed by histochemical staining, and the increase in enzymes such as SOD, peroxidases, glucanases, and chitinases highlight antioxidant responses and cell wall reinforcement (Xie et al. 2025). In *Carica papaya* under infection by *Fusarium equiseti*, proteins such as 9S-lipoxygenase and ribosomal components were associated with translational reprogramming and defense signaling (Karunamoorthy et al. 2025). In *Hylocereus polyrhizus*, the presence of *Neoscytalidium dimidiatum* triggered activation of proteins related to photosynthesis, redox homeostasis, and hormonal signaling, including ABP and ARP4, indicating a coordinated response defense and metabolic regulation response (Pan et al. 2021).

In *O. sativa* L. infected by *Magnaporthe oryzae*, contrasting genotypes displayed distinct temporal responses, with 18 modulated proteins related to defense (e.g., Pik-1, Argonaute proteins), jasmonate biosynthesis (AOS), and photosynthesis regulation (Rubisco, cytochrome b6-f), highlighting genotype-specific accumulation profiles (Keerthana et al. 2025). The resistance to the fungus *M. oryzae* was associated with reduced glycosylation of the peroxidases Prx4 and Prx13 in the rbb1

Table 2 Proteins associated with plant defense against pathogens identified by proteomic approaches

Plant species	Pathogen/pest species	Highlighted proteins	References
S. lycopersicum	*Ralstonia solanacearum*—Bacteria	Kinase proteins (RipU), mitochondrial and chloroplast proteins (RipB), endomembrane system proteins (RipD)	De Ryck et al. (2025)
T. aestivum	*Wheat yellow mosaic virus (WYMV)*	Peroxidase (TaPOD), Thioredoxin reductase (TaTrxR)	Zhao et al. (2025b)
N. tabacum	*Cucumber mosaic virus (CMV)*—Virus	Catalase, disulfide isomerase, Nt4CL	Shen et al. (2024)
S. lycopersicum	*Tomato spotted wilt orthotospovirus (TSWV)*—Virus	RLKs, SABP, PR proteins, antioxidant proteins, PCD proteins, protease inhibitors	Feng et al. (2024)
Carica papaya	*Papaya meleira virus complex (PMeV-complex)*—Virus	PsbQ-like2, PsbP, β-1,4-glucanase, UDP-glucose 6-dehydrogenase (UGD)	Rodrigues et al. (2024)
Citrus limon	*Candidatus liberibacter*—Bacteria	ACC oxidase, JAZ proteins, β-1,3-glucanase, ferritin, chitinase, SAM synthetase	Ramsey et al. (2020)
Arabidopsis thaliana	*Pseudomonas syringae*—Bacteria	Ex-proteasomes, PR5, PR2, SBT1.7, chitinases (PR1/PR3-like), BGALs	Zahedi et al. (2025)
A. thaliana	*Pseudomonas syringae pv. tomato DC3000 (Pst DC3000)*—Bacteria	Lhcb1, Lhcb2, Lhcb5, Lhcb6, GSTs, ribosomal proteins L2 and S subunits	Ding et al. (2024)
A. thaliana	*Pseudomonas syringae pv. maculicola (Psm)*—Bacteria	NAC90, NAC61, NAC36	Cai et al. (2024)
O. sativa L	*Xanthomonas oryzae pv. oryzae (Xoo) and Magnaporthe oryzae (M. oryzae)*—Bacteria	Prx4, Prx13	Zhang et al. (2025a)
Solanum tuberosum	*Streptomyces scabies*—Bacteria	C9ZCX9 (pyochelin), C9ZCY1 (pyochelin), C9YUJ7 (nitrite/sulfite reductase), C9ZEV0, C9YVQ3 (oxidative stress), C9YY94 (adaptation)	Giroux et al. (2024)
Actinidia chinensis	*Pseudomonas syringae pv. actinidiae (Psa)*—Bacteria	Phenylalanine ammonia-lyase (PAL), MAPK cascade proteins, pathogenesis-related proteins (PRs), pectinase	Qu et al. (2023)
Hevea brasiliensis	*Phytophthora palmivora*—Oomycete	Oxidative phosphorylation proteins, ATP synthesis, energy regulation	Kruaweangmol et al. (2025)
Solanum tuberosum	*Phytophthora infestans*—Oomycete	Parakletos	Zahid et al. (2024)

(continued)

Table 2 (continued)

Plant species	Pathogen/pest species	Highlighted proteins	References
Capsicum annuum	*Phytophthora capsici*—Oomycete	Pectin methylesterase (CA03g36980), PR protein (CA04g13070, CA01g31060), late blight R protein (CA11g02470), APX, aquaporins, β-1,3-glucanase, peroxidases	Shi et al. (2024)
B. napus	*Plasmodiophora brassicae Woronin*—Oomycete	Laccase, peroxidases, plant invertase/pectin methylesterase inhibitors (PInv/PMEI)	Adhikary et al. (2024)
G. max	*Phytophthora sojae*—Oomycete	PsCAP1 (secreted CAP protein)	Wu et al. (2023)
Z. mays L. var everta	*Bipolaris maydis*—Fungus	PSB27-1, RPP0B, DHAR3, APXs	Corrêa et al. (2025)
Z. mays L	*Bipolaris maydis*—Fungus	APX1	Zhang et al. (2022c)
O. sativa (Benugopal and Kalikati 2)	*Magnaporthe oryzae*—Fungus	Pik-1, Allene oxide synthase 1, hypersensitive-induced response protein 1, Phytochrome B	Keerthana et al. (2025)
S. lycopersicum	*Moniliophthora perniciosa*—Fungus	PR2, PR3, PR10	Alcântara et al. (2025)
T. aestivum L	*Fusarium pseudograminearum*—Fungus	TaHDA9, TaP5CS1	Zhang et al. (2025b)
Mangifera indica	*Alternaria alternata*—Fungus	Cellulase, β-glucosidase, MAPK pathway proteins, phenylpropanoid-related enzymes	Xie et al. (2025)
C. papaya	*Fusarium equiseti*—Fungus	Linoleate 9S-lipoxygenase, 30S ribosomal protein S3, heat shock-related protein	Karunamoorthy et al. (2025)
B. napus	*Sclerotinia sclerotiorum*—Fungus	CERK (ceramide kinase), lipase (GDSL motif)	Ouyang et al. (2024)
Cajanus cajan	*Fusarium udum*—Fungus	HMG1/2-like, DEAD-box helicase 3, lipoxygenase 1, Annexin D1	Karmakar et al. (2024)
Saccharum spp.	*Colletotrichum falcatum*—Fungus	DRP-1 to DRP-4, SUT-1, SUT-5 to SUT-8	Gujjar et al. (2024)
Malus domestica	*Penicillium expansum*—Fungus	Catalase (CAT), DAO, PEX16, MAPK-pathway proteins, PR-proteins	Wang et al. (2024a)
O. sativa L, T. aestivum L, Hordeum vulgare	*Magnaporthe oryzae*—Fungus	Peroxidases, glycosidases, serine proteases	Dong et al. (2024)

(continued)

Table 2 (continued)

Plant species	Pathogen/pest species	Highlighted proteins	References
S. lycopersicum	*Botrytis cinerea*—Fungus	SlCOP1-1, SlOpaque2	Gao et al. (2024a,)
Prunus persica	*Rhizopus stolonifera*—Fungus	PpVOZ1, PpVOZ2, PpTCP2	Wang et al. (2025)
Rose spp.	*Botrytis cinerea*—Fungus	PR10 proteins, membrane aquaporin, PTI-related kinases	Li et al. (2023)
Cucurbita pepo	*Podosphaera xanthii*—Fungus	Starch branching enzyme, V-type proton ATPase (V-ATPase), Glutaredoxin, carbonic anhydrase	Theologidis et al. (2024)
T. aestivum	*Blumeria graminis f.* sp. *tritici*—Fungus	TaCRT3 (Calreticulin 3)	Ren et al. (2024)
Poa pratensis	*Blumeria graminis*—Fungus	Lignin (phenylpropanoid) pathway proteins	Dong et al. (2024)
Arachis hypogaea	*Fusarium oxysporum*—Fungus	PAL, CHI, C4H	Wang et al. (2024b)
S. lycopersicum	*Cladosporium fulvum*—Fungus	Transporters, transcription factors, monooxygenases (e.g., cytochrome P450)	Zaccaron et al. (2024)
Hordeum vulgare	*Ramularia collo-cygni (Rcc)*—Fungus	Xylanase inhibitor (HORVU3Hr1G006440), late stress response proteins	Lemcke et al. (2024)
Arachis hypogaea	*Cylindrocladium parasiticum*—Fungus	CYP73A100, SA, JA, MAPK, brassinosteroid-related proteins	Chen et al. (2024)
O. sativa	*Magnaporthe oryzae*—Fungus	PTAC16; phosphorylated kinases (e.g., casein kinase II, ERK1/2 targets); ubiquitinated stress-related proteins	Lee et al. (2023b)
Elaeis guineensis	*Ganoderma boninense, G. tornatum*—Fungus	Oxidoreductases, transferases, hydrolases, GAPDH	Dzulkafli et al. (2023)
Solanum sisymbriifolium	*Verticillium dahliae*—Fungus	Phenylalanine ammonia lyase (PAL), superoxide dismutase (SOD), phenylpropanoid pathway proteins, cell wall proteins, plant-pathogen interaction proteins	Wu et al. (2023)
Malus domestica	*Penicillium expansum*—Fungus	PR proteins, ROS-related proteins	Xu et al. (2022)
Hylocereus spp.	*Neoscytalidium dimidiatum*—Fungus	ABP, ARP4, ASP2, proteins associated with ROS homeostasis	Pan et al. (2021)
Juglans regia L.	*Colletotrichum gloeosporioides*	TMV resistance protein N-like, RPP13-like, PR1, CDPK, FLS2, SNF1-related kinase KIN10	Fang et al. (2021)

(continued)

Table 2 (continued)

Plant species	Pathogen/pest species	Highlighted proteins	References
S. lycopersicum	*Botrytis cinerea*—Fungus	APX1, SOD, chaperonin, malate dehydrogenase, ethylene-related enzyme	Tzortzakis (2019)
S. lycopersicum	*Alternaria solani*—Fungus	Phenylalanine ammonia-lyase (PAL), phenylpropanoid pathway proteins	Nikonorova et al. (2021)
Prunus persica (L.) Batsch	*Monilinia fructicola or Monilinia laxa*—Fungus	Thaumatin, formate dehydrogenase, HSP70, GST, major latex allergen, ACC deaminase	Papavasileiou et al. (2020)
Actinidia deliciosa "Hayward"	*Botrytis cinerea*—Fungus	PR proteins, myosin 10, cell wall proteins, MAPK cascade proteins	Liu et al. (2018)
S. lycopersicum	*Botrytis cinerea*—Fungus	Chitinases, β-1,3-glucanases, osmotin-like, PGIP, thaumatin, peroxidases, protease inhibitors	Shah et al. (2012)
Pinus koraiensis	*Bursaphelenchus xylophilus*—Nematode	Receptor-like serine/threonine kinases, pectin methylation modulators, pinosylvin O-methyltransferase, arabinogalactan/proline-rich proteins	Yu et al. (2025)
Elaeis guineensis Jacq.	*Bursaphelenchus cocophilus*—Nematode	Chitinase, L-ascorbate peroxidase, disulfide-isomerase, UDP-glucosyltransferase, Thaumatin	de Oliveira Braga et al. (2024)
G. max	*Tetranychus urticae Koch*—Mite	Proteins associated with photorespiration, chlorophyll synthesis, amino acid metabolism, Krebs cycle	Ruffatto et al. (2024)
O. sativa	*Nilaparvata lugens (Stål)*—Insect	Bph3-regulated proteins involved in MAPK, calcium, and hormonal pathways	Qing et al. (2024)
T. aestivum L. Var. Zhongmai-175	*Metopolophium dirhodum Walker*—Insect	Glucose dehydrogenase-like protein 1, Beta-glucosidase, phospholipase D, superoxide dismutase (SOD), chemosensory protein	Gebrekidan et al. (2024)
O. sativa	*Nilaparvata lugens*—Insect	Egg-associated secretions (EAS) proteins (53 with ≥2 peptides)	Li et al. (2024)

The main pathogen-responsive plant proteins and associated literature are listed

mutant, suggesting a negative role of this modification in immunity (Zhang et al. 2025a). In addition, proteomics revealed early activation of immune pathways upon Pathogen-Associated Molecular Patterns (PAMP) recognition, underscoring the central role of PTMs in pattern-triggered immunity (PTI) and the identification of new molecular targets for resistance against *M. oryzae* (Lee et al. 2023a, b).

In *Z. mays var. everta*, resistance to *Bipolaris maydis* infection was associated with a coordinated proteomic response involving photosystem II components (e.g., PSB27-1), antioxidant enzymes such as dehydroascorbate reductase type-3 (DHAR3), APX, and ribosomal proteins like 60S P0, which collectively support redox balance under stress conditions (Corrêa et al. 2025). In contrast, the susceptible genotype uniquely accumulated GAPC2, a glycolytic enzyme indicative of disrupted energy homeostasis. Furthermore, in *Z. mays*, the cytosolic APX1, encoded within an SLB resistance-associated QTL, has also been proposed as a key contributor to defense against *B. maydis*, reinforcing the central role of ROS-scavenging mechanisms in disease resistance (Zhang et al. 2022c).

Using *S. lycopersicum* as a model to study the infection of *Moniliophthora perniciosa*, a pathogen originally associated with witches' broom disease in *Theobroma cacao*, highlighted the role of the endoplasmic reticulum in protein homeostasis during infection, with accumulation of PR proteins (PR2, PR3, PR10), increased antioxidant enzymes, and reduced hydrogen peroxide (H_2O_2) levels, indicating a protective effect against fungal stress (Alcântara et al. 2025).

These findings highlight the diversity of proteomic responses involved in fungal disease resistance across plant species. Several studies have also employed proteomics to investigate plant–fungal interactions, revealing key components of defense mechanisms, redox regulation, and stress adaptation, summarized in Table 2.

2.2.2 Oomycetes Infections

Proteomic responses to oomycetes have revealed striking patterns of metabolic reprogramming and activation of antioxidant mechanisms in host plants. In *Hevea brasiliensis*, infection by *Phytophthora palmivora*, the causal agent of leaf fall disease, led to substantial changes in the foliar proteomic and metabolic profiles (Kruaweangmol et al. 2025). DAPs were associated with processes like oxidative phosphorylation, ATP synthesis, and amino acid metabolism, particularly alanine, aspartate, and glutamate, indicating a response focused on energy maintenance and primary metabolism adjustment as part of the plant's strategy to contain the pathogen (Kruaweangmol et al. 2025).

In *B. napus*, infection by *Plasmodiophora brassicae*, the etiological agent of clubroot disease, triggers dynamic alterations in the root proteome and metabolome, especially during the early stages of infection (Adhikary et al. 2024). The DAPs proteins included enzymes involved in amino acid, fatty acid, lipid, and glucosinolate metabolism, as well as in cell wall reinforcement, such as laccases, peroxidases, and invertase/pectin methylesterase inhibitors (PInv/PMEI) (Adhikary et al. 2024). Integration with metabolomic data further revealed regulation of nucleotides, sphingolipid compounds, and intermediates of fatty acid metabolism.

Collectively, these studies demonstrate that responses to oomycete infection involve profound reconfiguration of primary and secondary metabolism, with particular emphasis on energy-related, oxidative, and structural pathways.

2.2.3 Bacterial Infections

Plant–bacteria interactions mediated by effectors, hormonal signaling, and systemic resistance have been extensively explored using proteomic approaches. In *S. lycopersicum*, the bacterium *Ralstonia solanacearum* secretes type III effectors such as RipU, RipD, and RipB, which are capable of interacting with proteins localized in the mitochondria, endoplasmic reticulum, and chloroplasts (De Ryck et al. 2025). Using TurboID-mediated proximity labeling, researchers were able to precisely map the intracellular targets of these effectors, revealing distinct mechanisms for the suppression of both PTI and effector-triggered immunity (ETI). The study highlights the ability of bacteria to manipulate subcellular compartments to evade plant defenses (De Ryck et al. 2025). Other example of study, is the infection by *Candidatus liberibacter* in *Citrus limon*, the causal agent of huanglongbing (HLB), that was studied in using a multi-omics approaches and revealed time-dependent, coordinated changes in genes and proteins, identifying potential biomarkers for early diagnosis and reinforcing the role of applied proteomics in the detection and management of systemic plant diseases (Ramsey et al. 2020).

In addition to protein abundance, PTMs and spatial control of cellular signaling emerge as critical elements of this refined regulation. This complexity was further demonstrated in *Actinidia chinensis* (Kiwi), where exogenous application of salicylic acid enhanced the defense response against *Pseudomonas syringae* pv. actinidiae (Psa), the causal agent of bacterial canker (Qu et al. 2023). SA rapidly activated key immune components, such as the MAPK signaling cascade, the biosynthesis of phenolic compounds via the phenylpropanoid pathway, and the induction of PR-proteins (Qu et al. 2023). Moreover, SA was able to restore pathways previously suppressed by infection, highlighting its central role in hormonal reprogramming and in promoting an immunologically primed state for future challenges (Qu et al. 2023).

In *S. tuberosum*, the interaction with the actinobacterium *Streptomyces scabies*, the causal agent of common scab, led to a marked increase in bacterial proteins related to stress responses, secondary metabolism, and siderophore biosynthesis, particularly pyochelin. This metabolic shift coincided with iron depletion in the culture medium and up-accumulation of pyochelin biosynthetic genes, suggesting active iron competition between pathogen and host (Giroux et al. 2024). Additionally, increased nitrite excretion and H_2O_2 accumulation were observed, indicating the activation of redox defense and detoxification pathways in *S. scabies* in response to live plant tissue (Giroux et al. 2024).

These findings highlight how host-derived signals can trigger bacterial virulence through multiple routes, as further illustrated by additional proteomic studies summarized in Table 2.

2.2.4 Viral Infections

Viral infections in plants trigger highly coordinated proteomic responses, particularly involving changes in redox pathways, cellular stability, and reprogramming of vascular transport. In *T. aestivum*, the viral protein P1, encoded by *Wheat yellow mosaic virus* (WYMV), functioned as an efficient suppressor of plant immunity (Zhao et al. 2025b). Overexpression of P1 led to a reduction in ROS production and repression of antioxidant genes such as TaPOD and TaTrxR, thereby facilitating infection. Silencing of these genes restored antiviral resistance, reversing the susceptible phenotype and demonstrating that manipulation of redox pathways is a key strategy in viral virulence (Zhao et al. 2025b). In *N. tabacum*, infection by *Cucumber mosaic virus* (CMV) revealed contrasting proteomic profiles between resistant and susceptible cultivars (Shen et al. 2024). Resistant plants maintained mitochondrial and chloroplast integrity by activating mechanisms such as selective autophagy and the accumulation of antioxidant enzymes, contributing to infection containment (Shen et al. 2024).

In *S. lycopersicum*, proteomic analysis of vascular exudates from plants infected with *Tomato spotted wilt orthotospovirus* (TSWV) identified several DAPs (Feng et al. 2024). Among these were receptor-like kinases, PR defense proteins, and viral proteins N, Gn, and Gc, suggesting that the xylem simultaneously functions as an active route for viral dissemination and as a molecular battleground between the plant and the pathogen (Feng et al. 2024). In *C. papaya*, the response to the *papaya sticky disease viral complex* (PMeV) was characterized by early accumulation of proteins related to photosynthesis and cell wall synthesis in young plants, suggesting a form of tolerance linked to developmental stage (Rodrigues et al. 2024). These differential responses indicate that plant age may significantly influence viral infection dynamics and the effectiveness of activated proteomic responses.

2.2.5 Interactions with Insects, Mites, and Nematodes

Plant–insect, plant–mite, and plant–nematode interactions represent complex challenges to plant defense, often characterized by temporal responses, hormonal modulation, and metabolic reprogramming. In *O. sativa*, infestation by *Nilaparvata lugens* Stål (the brown planthopper) was studied in plants containing the brown planthopper resistance gene 3 (Bph3) (Qing et al. 2024). Proteomic analysis indicated modulation of proteins involved in hormonal pathways, calcium signaling, and phenylpropanoid biosynthesis, highlighting the multifunctionality of Bph3 in activating defense responses against herbivory (Qing et al. 2024). Beyond feeding, *N. lugens* oviposition also significantly influences plant immunity. Egg-associated secretions (EAS), deposited in damaged *O. sativa* tissues during egg-laying, rapidly induced the production of jasmonic acid (JA), jasmonoyl-isoleucine (JA-Ile), and H_2O_2 (Li et al. 2024). These signals activated defensive responses, leading to reduced performance of aphids and whiteflies (Li et al. 2024). Proteomic analysis of the EAS identified some proteins shared with the salivary proteome of herbivorous insects, while others are specific to oviposition and may represent novel molecular targets for inducing plant resistance (Li et al. 2024).

On the other hand, in *G. max*, infestation by the mite *Tetranychus urticae* triggered temporal changes in the plant proteome (Ruffatto et al. 2024). DAPs were identified, affecting processes such as photosynthesis, amino acid metabolism, energy production, and antioxidant responses. Metabolic reprogramming was dynamic and phase-dependent, reflecting the species' continuous adaptation to prolonged herbivore attack (Ruffatto et al. 2024).

In plant–nematode interactions, oil palm (*Elaeis guineensis*) infected by *Bursaphelenchus cocophilus*, the multi-omics data revealed pathways involved in the production of compounds with nematicidal properties, as well as the identification of biomarkers with potential applications in early diagnosis and genetic improvement (de Oliveira Braga et al. 2024). Similarly, in *Pinus koraiensis* trees, wilt caused by the nematode *Bursaphelenchus xylophilus* activated metabolic pathways associated with the biosynthesis of terpenoids, flavonoids, and toxic alkaloids, in addition to inducing proteins involved in pathogen recognition, such as receptor-like kinases, pectin modulators, and proline-rich proteins. These responses, activated even before symptom onset, illustrate an efficient model of systemic defense in conifers (You et al. 2025).

In summary, proteomic studies have significantly advanced our understanding of the complex molecular interactions between plants and a range of biotic stress agents, including pathogens, insects, mites, and nematodes. By uncovering key regulators such as susceptibility factors, defense-related proteins, and effector targets, these analyses provide valuable tools for breeding more resilient cultivars and offer new avenues for biotechnological interventions aimed at disrupting pest and pathogen physiology.

2.3 Proteomic Insights into Heterosis

Heterosis, or hybrid vigor, is a genetic phenomenon in which the F_1 generation exhibits superior performance compared to its parental lines, influencing quantitative traits such as growth rate, fertility, yield, and biomass, as well as qualitative traits like disease resistance and tolerance to environmental stresses (Lippman and Zamir 2007; Mohayeji et al. 2014). Over the years, considerable efforts have been made to understand the genetic and molecular basis of heterosis (Liu et al. 2020). Although the underlying molecular mechanisms of this phenomenon remain not fully elucidated, advances in genomic and proteomic approaches have significantly contributed to deepening our understanding of this process (Xing et al. 2016).

Germination plays a fundamental role in the plant life cycle, marking the onset of postembryonic development through radicle protrusion. This stage is particularly relevant for heterosis studies, as it reflects early hybrid vigor and directly influences seedling establishment. Proteomic analyses during germination have proven effective in identifying DAPs and revealing molecular mechanisms underlying heterosis.

Protein profile patterns reveal the contribution of dominance, partial dominance, and overdominance effects to the DAPs during germination (Fu et al. 2011). Notably, proteins such as 14-3-3, abscisic acid-insensitive 3 (ABI3), abscisic acid-insensitive 5 (ABI5), and PKABA1 were DAPs in hybrids, indicating that hormonal signaling pathways mediated by ABA and gibberellin are strongly associated with the manifestation of heterosis during *Z. mays* germination (Fu et al. 2011).

In a similar manner, the comparative proteomic analysis conducted on hybrid rice embryos of the LYP9 variety (Liang–You–Pei–Jiu) and its parental lines, 9311 and P64S, led to the identification of 54 DAPs spots. The most represented functional categories included proteins related to nutrient reservoir activity and stress response, suggesting an enhanced metabolic and physiological preparedness in the hybrid (Wang et al. 2008). Additionally, proteins involved in core metabolic pathways, such as UDP-glucose pyrophosphorylase, α-amylase, fructose-bisphosphate aldolase, and enolase, were significantly more abundant in LYP9, exhibiting an overdominant accumulation pattern (Wang et al. 2008). This pattern indicates a more efficient activation of metabolic pathways associated with reserve mobilization and the resumption of growth, supporting a potential functional role of these enzymes in promoting hybrid vigor during the germination stage.

Similarly, root development has also emerged as a valuable system for investigating the molecular basis of heterosis through proteomic approaches. Proteomic analysis has revealed distinct patterns of protein accumulation associated with heterosis in *C. papaya* roots (Vale et al. 2016). Specifically, there was a decrease in proteins related to energy-consuming processes such as protein metabolism, alongside an increase in proteins involved in root development, including those associated with auxin polar transport and signaling regulation (Vale et al. 2016). These findings suggest that hybrid vigor in young papaya roots involves the optimization of energy use, with reduced investment in protein metabolism allowing greater allocation of resources toward biomass accumulation (Vale et al. 2016). This interpretation is supported by the differential accumulation of proteins such as Superroot 1(SUR1), MAPK, and ABC transporters, which are associated with increased root number and dry mass (Vale et al. 2016). Additionally, a proteomic study on the primary roots of a *Z. mays* var. *everta* hybrid showed that heterosis is linked to the differential accumulation of proteins involved in energy metabolism, translation, and ribosome biogenesis (Rockenbach et al. 2018). These findings reinforce the idea that early-stage hybrid vigor is associated with enhanced energy production and translational efficiency, leading to faster and more efficient root development (Rockenbach et al. 2018).

Other studies have also demonstrated that heterosis in roots is associated with the nonadditive regulation of proteins involved in key biological processes such as protein metabolism, energy production, signal transduction, and cell growth (Song et al. 2007; Marcon et al. 2013). Comparative proteomic analyses in *Z. mays* and *T. aestivum* hybrids revealed that enhanced root vigor is linked to the coordinated modulation of multiple functional pathways, with particular emphasis on ribosomal proteins and primary metabolism, reinforcing the complex proteomic landscape underlying hybrid performance (Song et al. 2007, Marcon et al. 2013).

Proteomic analyses have revealed that heterosis in leaves is frequently associated with enhanced cell proliferation, regulatory protein activity, and photosynthetic efficiency. In *Z. mays*, studies using the Mo17/B73 hybrid demonstrated that increased leaf size results from a greater number of cells, linked to the differential accumulation of nuclear proteins involved in RNA regulation, cell division, and development (Uchimiya and Takahashi 1973; Guo et al. 2014). Similar analyses in the SC704 hybrid identified proteins related to energy metabolism, regulatory processes, and defense, with stress-related proteins emerging as key contributors to hybrid vigor (Bandehagh et al. 2021). In *O. sativa*, the superhybrid LYP9 exhibited up-accumulation of proteins involved in photosynthesis and glycolysis, such as phosphoglycerate kinase and peroxiredoxin, correlating with enhanced CO_2 assimilation and quantum efficiency (Zhang et al. 2012).

Variability in proteomic responses among hybrids of the same species has also been observed. In *Z. mays*, the hybrid ZD808 showed enrichment of stress-response proteins, while ZD909 displayed a predominance of proteins linked to photosynthesis, indicating distinct molecular bases for heterosis (Wang et al. 2021a). Moreover, studies on ear shoot proteomes and mitochondrial fractions from *Z. mays* hybrids with contrasting heterotic performance revealed that high-heterosis hybrids exhibit specific changes in proteins related to energy metabolism and stress responses, including components of the respiratory chain and redox enzymes (Dahal et al. 2012). These findings underscore the role of coordinated regulation of metabolic and defense pathways in the expression of heterosis across species and genotypes (Uchimiya and Takahashi 1973; Dahal et al. 2012; Zhang et al. 2012; Guo et al. 2014; Bandehagh et al. 2021; Wang et al. 2021a). Together, these proteomic studies highlight the complexity of heterosis and demonstrate that hybrid vigor arises from the coordinated regulation of diverse metabolic, developmental, and stress-responsive pathways, which vary according to species, developmental stage, and tissue type.

3 Future Prospects

The ongoing development of proteomic technologies promises to greatly enhance our understanding of complex biological systems, particularly in plant science. The growing accessibility and precision of proteomic technologies have greatly expanded their potential applications in plant breeding (Fig. 4). Proteomics enables the identification of key proteins and molecular pathways associated with traits of agronomic interest, including responses to biotic and abiotic stresses, growth and development, physiological performance, and yield and breeding potential.

It should be emphasized that, when not integrated, individual omics approaches can be limited in scope and can lead to highly speculative hypotheses regarding biological findings (Roychowdhury et al. 2023). In this sense, proteomics bridges a central gap between transcriptomics and metabolomics by revealing which gene products are actually present, their abundance, their functional state, and their

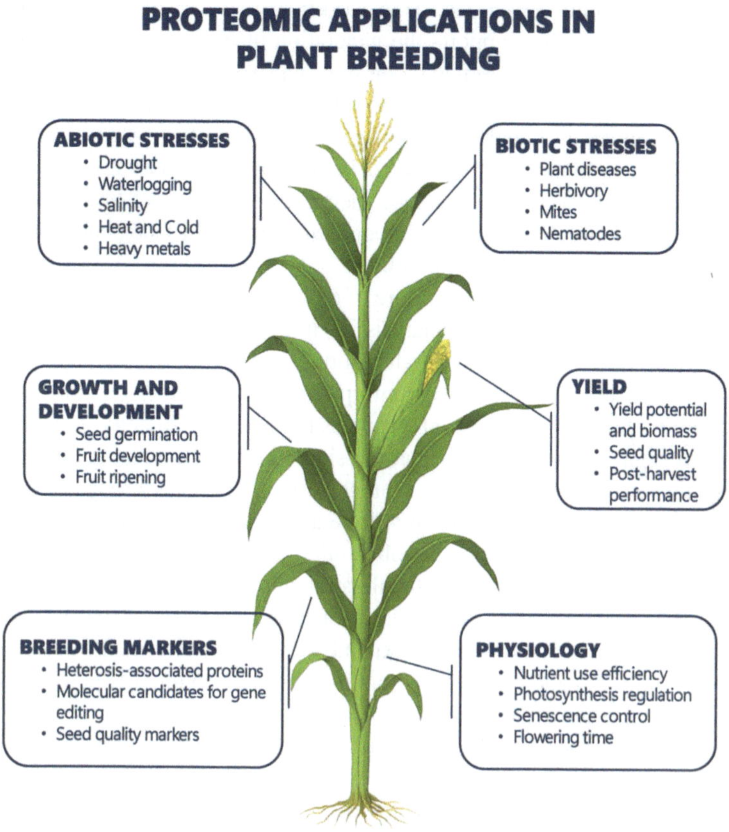

Fig. 4 Proteomic applications in plant breeding. Schematic representation of key areas in which proteomics contributes to plant breeding, including responses to abiotic and biotic stresses, yield and postharvest traits, growth and developmental processes, physiological regulation, and the identification of molecular biomarkers such as those associated with heterosis and gene editing targets

PTMs, information that cannot be inferred directly from RNAs or metabolites alone. On the other hand, metabolomics and transcriptomics complement proteomics by providing information on upstream and downstream layers of cellular regulation, allowing us to interpret not only which proteins are present but also why they are regulated and what functional consequences arise from this regulation. Thus, integration with other omics approaches is a path for the development of plant proteomics and can provide a comprehensive view of biological systems, enhance predictive power, facilitate precision breeding, and facilitate disease management (Cembrowska-Lech et al. 2023).

The gel-free bottom-up proteomics approach has become increasingly detailed, accessible, and effective for comparative analyses in plant biology. With advances in instrumentation and software, this methodology is now adopted by a growing number of laboratories worldwide, enabling high-resolution protein profiling across diverse experimental conditions. This workflow (Fig. 2) offers a practical and streamlined strategy that supports the generation of biologically meaningful data, reinforcing its role as a core tool in modern plant proteomics. A key area of advancement is the refinement of quantitative methods, which are essential for accurate protein quantification. Label-free techniques offer cost-effective solutions for analyzing large sample sets, but continued improvements in data normalization and bias reduction are needed to increase their robustness. Label-based approaches, while providing more precise absolute quantification, still face challenges such as high reagent costs and limited scalability.

An interesting prospect for the future of proteomics is the expansion of top-down approaches, which enable the analysis of intact proteoforms and offer unique insights into post-translational modifications and isoform diversity, areas that remain limited in conventional bottom-up strategies. Another exciting frontier is single-cell proteomics. With recent advances in mass spectrometry, it is now possible to investigate protein dynamics at the single-cell level, offering unprecedented insights into cellular heterogeneity. Nonetheless, issues related to data consistency, sensitivity, and reproducibility must still be addressed in applications before these techniques can be widely adopted. Subcellular proteomics also presents promising opportunities. By mapping protein localization within specific organelles, researchers can gain deeper insights into cellular functions, signaling networks, and stress responses. This field is expected to expand with the development of more refined isolation techniques and high-throughput methods, expanding the potential for discoveries in cell compartmentalization and protein function.

Altogether, as proteomic technologies and methodologies continue to evolve, they are poised to play a central role in unraveling the complexities of plant breeding and advancing broader biotechnological applications.

Acknowledgments This research was supported by funds from the Fundação de Amparo à Pesquisa do Estado do Rio de Janeiro (FAPERJ) (Proc E-26/211.310/2021; Proc E-26/204.185/2024) and Conselho Nacional de Desenvolvimento Científico e Tecnológico (CNPq) (312258/2023-2), awarded to V.S. This study was also financed in part by the Coordenação de Aperfeiçoamento de Pessoal de Nível Superior, Brazil (CAPES), Finance Code 001. The authors would like to thank FAPERJ, CNPq, and CAPES for providing scholarships to graduate students.

Conflict of Interest Disclosure The authors declare that they have no competing interests.

References

Adhikary D et al (2024) Proteome- and metabolome-level changes during early stages of clubroot infection in *Brassica napus* canola. Mol Omics 20(4):265–282

Aguilan JT et al (2020) Guide for protein fold change and p-value calculation for non-experts in proteomics. Mol Omics 16(6):573–582

Aguilar-Hernández V, Loyola-Vargas VM (2018) Advanced proteomic approaches to elucidate somatic embryogenesis. Front Plant Sci 9:1658

Ahsan N et al (2009) Recent developments in the application of proteomics to the analysis of plant responses to heavy metals. Proteomics 9(10):2602–2621
Alcântara G d M et al (2025) Proteomic analysis of plants with binding immunoglobulin protein overexpression reveals mechanisms related to defense against *Moniliophthora perniciosa*. Plants 14(4):503
Ali AEE et al (2025) Proteomic dataset of sorghum leaf and root responses to single and combined drought and heat stress. Sci Data 12(1):97
Almeida FA et al (2020) Label-free quantitative phosphoproteomics reveals signaling dynamics involved in embryogenic competence acquisition in sugarcane. J Proteome Res 19(10):4145–4157
Aroca R et al (2011) Regulation of root water uptake under abiotic stress conditions. J Exp Bot 63(1):43–57
Aydoğan C (2024) Critical review of new advances in food and plant proteomics analyses by nano-LC/MS towards advanced foodomics. TrAC Trends Anal Chem 176:117759
Balbuena TS et al (2012) Large-scale proteome comparative analysis of developing rhizomes of the ancient vascular plant equisetum hyemale. Front Plant Sci 3:131
Balotf S et al (2022) Shotgun proteomics as a powerful tool for the study of the proteomes of plants, their pathogens, and plant–pathogen interactions. Proteomes 10(1):5
Bandehagh A et al (2021) Assessment of heterosis proteins in maize (*Zea mays* L.) leaves by two-dimensional gel electrophoresis. Plant Gene 28:100331
Barba-Espín G et al (2025) Halophyte-based crop managements induce biochemical, metabolomic and proteomic changes in tomato plants under saline conditions. Physiol Plant 177(1):e70060
Boguszewska-Mańkowska D et al (2020) Comparative proteomic analysis of drought and high temperature response in roots of two potato cultivars. Plant Growth Regul 92(2):345–363
Borges KLR et al (2019) Quantitative proteomic analysis of tomato genotypes with differential cadmium tolerance. Environ Sci Pollut Res 26(25):26039–26051
Bukhari B et al (2025) Proteomic, transcriptomic, biochemical and physiological investigations shed light on responses to low temperature stress in two contrasting soybean varieties. Plant Cell Environ. https://doi.org/10.1111/pce.15492
Cai J et al (2024) A NAC triad modulates plant immunity by negatively regulating N-hydroxy pipecolic acid biosynthesis. Nat Commun 15(1):7212
Carvalho PC et al (2016) Integrated analysis of shotgun proteomic data with PatternLab for proteomics 4.0. Nat Protoc 11(1):102–117
Cembrowska-Lech D et al (2023) An integrated multi-omics and artificial intelligence framework for advance plant phenotyping in horticulture. Biology 12(10):1298
Chen C et al (2020) Bioinformatics methods for mass spectrometry-based proteomics data analysis. Int J Mol Sci 21(8):2873
Chen M et al (2021) Multi-omics analyses on the response mechanisms of 'Shine Muscat' grapevine to low degree of excess copper stress (low-ECS). Environ Pollut 286:117278
Chen X et al (2024) Comparative proteomic analysis of the moderately resistant and susceptible peanut cultivars during infestation by *Cylindrocladium parasiticum*. Can J Plant Pathol 46(1):50–64
Chen X et al (2025) Label-free proteomics reveals the response of oat (*Avena sativa* L.) seedling root respiratory metabolism to salt stress. Int J Mol Sci 26(6):2630
Chinnusamy V et al (2007) Cold stress regulation of gene expression in plants. Trends Plant Sci 12(10):444–451
Clark NM et al (2021) To the proteome and beyond: advances in single-cell omics profiling for plant systems. Plant Physiol 188(2):726–737
Conesa A, Götz S (2008) Blast2GO: a comprehensive suite for functional analysis in plant genomics. Int J Plant Genomics 2008(1):619832
Corrêa CCG et al (2025) Comparative proteomic analysis of popcorn genotypes identifies differentially accumulated proteins associated with resistance pathways to southern leaf blight disease. Plants 14(3):426

Dahal D et al (2012) Specific changes in total and mitochondrial proteomes are associated with higher levels of heterosis in maize hybrids. Plant J 72(1):70–83

de Oliveira Braga Í et al (2024) A multi-omics deconstruction of red ring disease in oil palm. J Plant Pathol 108:171

De Ryck J et al (2025) Exploring the tomato root protein network exploited by Core type 3 effectors from the *Ralstonia solanacearum* species complex. J Proteome Res 24(2):696–709

Ding Y et al (2024) Proteomic analysis of lysine acetylation and succinylation to investigate the pathogenicity of virulent *Pseudomonas syringae* pv. tomato DC3000 and avirulent line *Pseudomonas syringae* pv. tomato DC3000 avrRpm1 on *Arabidopsis thaliana*. Genes 15(4):499

Distéfano AM et al (2024) Heat stress in plants: sensing, signalling, and ferroptosis. J Exp Bot 76(5):1357–1369

Dong W et al (2024) Comparative proteomic and metabolomic analysis of resistant and susceptible Kentucky Bluegrass cultivars in response to infection by powdery mildew. BMC Plant Biol 24(1):1195

Dzulkafli SB et al (2023) Understanding the effects of carbon and nitrogen starvation on the comparative secretomes of *Ganoderma boninense* and *Ganoderma tornatum*. Physiol Mol Plant Pathol 127:102084

Fang H et al (2021) Transcriptome and proteome analysis of walnut *(Juglans regia* L.) fruit in response to infection by *Colletotrichum gloeosporioides*. BMC Plant Biol 21(1):249

Feng H et al (2024) Integrated biological experiments and proteomic analyses of *Nicotiana tabacum* xylem sap revealed the host response to tomato spotted wilt orthotospovirus infection. Int J Mol Sci 25(20):10907

Fortini EA et al (2023) Physiological, epigenetic, and proteomic responses in *Pfaffia glomerata* growth in vitro under salt stress and 5-azacytidine. Protoplasma 260(2):467–482

Fu Z et al (2011) Proteomic analysis of heterosis during maize seed germination. Proteomics 11(8):1462–1472

Fu S et al (2024a) Integrated transcriptomic, proteomic, and metabolomic analyses revealed molecular mechanism for salt resistance in soybean (*Glycine max* L.) seedlings. Int J Mol Sci 25(24):13559

Fu Z et al (2024b) Global critical soil moisture thresholds of plant water stress. Nat Commun 15(1):4826

Gao H et al (2023) Proteomic analysis of leaves and roots during drought stress and recovery in *Setaria italica* L. Front Plant Sci 14:1240164

Gao G et al (2024a) E3 ligase SlCOP1-1 stabilizes transcription factor SlOpaque2 and enhances fruit resistance to *Botrytis cinerea* in tomato. Plant Physiol 196(2):1196–1213

Gao Y et al (2024b) Comparative proteomic discovery of salt stress response in alfalfa roots and overexpression of MsANN2 confers salt tolerance. Plant Physiol Biochem 215:109033

Gebrekidan AG et al (2024) A comprehensive transcriptomic and proteomics analysis of candidate secretory proteins in rose grain aphid, *Metopolophium dirhodum* (Walker). Curr Issues Mol Biol 46(12):13383–13404

Giansanti P et al (2016) Six alternative proteases for mass spectrometry–based proteomics beyond trypsin. Nat Protoc 11(5):993–1006

Giroux L et al (2024) Proteomics fingerprinting reveals importance of iron and oxidative stress in *Streptomyces scabies–Solanum tuberosum* interactions. Front Microbiol 15:1466927

Görg A et al (2004) Current two-dimensional electrophoresis technology for proteomics. Proteomics 4(12):3665–3685

Guan S et al (2020) Data Dependent–Independent Acquisition (DDIA) proteomics. J Proteome Res 19(8):3230–3237

Gujjar RS et al (2024) *Colletotrichum falcatum* infection influences the abundance of sucrose transporters and disease resistant proteins in sugarcane stalk. J Plant Biochem Biotechnol 33(3):367–373

Guo B et al (2014) Maize (*Zea mays* L.) seedling leaf nuclear proteome and differentially expressed proteins between a hybrid and its parental lines. Proteomics 14(9):1071–1087

Guo H et al (2023) Phenotypic and proteomic insights into differential cadmium accumulation in maize kernels. Genes 14(12):2204

Hang J et al (2024) Comparative transcriptomic and proteomic analyses of two salt-tolerant alfalfa (*Medicago sativa* L.) genotypes: investigation of the mechanisms underlying tolerance to salt. Front Plant Sci 15:1442963

Heringer AS et al (2018) Insights from proteomic studies into plant somatic embryogenesis. Proteomics 18(5–6):e1700265

Hochholdinger F et al (2018) Proteomics of maize root development. Front Plant Sci 9:143

Holleman C et al (2020) The impact of climate variability and extremes on agriculture and food security – an analysis of the evidence and case studies. FAO Agricultural Development Economics Technical Study 1(4). Food and Agriculture Organization of the United Nations, Agricultural Development Economics Division

Holtorf H et al (2002) Plant functional genomics. Naturwissenschaften 89(6):235–249

Hossain Z, Komatsu S (2013) Contribution of proteomic studies towards understanding plant heavy metal stress response. Front Plant Sci 3:310

Hsiao Y et al (2024) Analysis and visualization of quantitative proteomics data using fragpipe-analyst. J Proteome Res 23(10):4303–4315

Hu A et al (2016) Technical advances in proteomics: new developments in data-independent acquisition. F1000Res 5(419):419

Isaacson T et al (2006) Sample extraction techniques for enhanced proteomic analysis of plant tissues. Nat Protoc 1(2):769–774

Kamran M et al (2020) An overview of hazardous impacts of soil salinity in crops, tolerance mechanisms, and amelioration through selenium supplementation. Int J Mol Sci 21(1):148

Kanehisa M et al (2016) KEGG: new perspectives on genomes, pathways, diseases and drugs. Nucleic Acids Res 45(D1):D353–D361

Karmakar S et al (2024) A detailed proteomics and metabolomics landscape sheds light on the mechanistic insights into the resistance response of transgenic pigeon pea against wilt stress. Plant Cell Environ. https://doi.org/10.1111/pce.15261

Karunamoorthy J et al (2025) Proteomic analysis of the Papaya-Fusarium equiseti interaction: understanding mode of infection and plant response at the molecular level. Physiol Mol Plant Pathol 136:102583

Kausar R et al (2022) Crop proteomics under abiotic stress: from data to insights. Plants 11(21):2877

Keerthana U et al (2025) Comparative proteomic analysis of resistant and susceptible aromatic rice landraces in response to blast pathogen, *Magnaporthe oryzae*. Physiol Mol Plant Pathol 137:102629

Kolberg L et al (2023) g:Profiler—interoperable web service for functional enrichment analysis and gene identifier mapping (2023 update). Nucleic Acids Res 51(W1):W207–W212

Komatsu S et al (2021) Proteomic and biochemical analyses of the mechanism of tolerance in mutant soybean responding to flooding stress. Int J Mol Sci 22(16):9046

Kruaweangmol P et al (2025) Metabolomic and proteomic changes in leaves of rubber seedlings infected by *Phytophthora palmivora*. Tree Physiol 45(2):tpaf010

Lee GH et al (2023a) Dataset on post-translational modifications proteome analysis of MSP1-overexpressing rice leaf proteins. Data Brief 50:109573

Lee GH et al (2023b) Analysis of post-translational modification dynamics unveiled novel insights into rice responses to MSP1. J Proteome 287:104970

Lemcke R et al (2024) Integrative transcript to proteome analysis of barley during *Ramularia collo-cygni* leaf spot development identified several proteins that are related to fungal recognition and infection responses. Front Plant Sci 15:1367271

Li P et al (2020) Translocation of drought-responsive proteins from the chloroplasts. Cells 9(1):259

Li R et al (2023) Integrated proteomic analysis reveals interactions between phosphorylation and ubiquitination in rose response to *Botrytis infection*. Hortic Res 11(1):uhad238

Li J et al (2024) Egg-associated secretions from the brown planthopper (*Nilaparvata lugens*) activate rice immune responses. Insect Sci 31(4):1135–1149

Li L et al (2025a) Conservative mechanism through various rapeseed (*Brassica napus* L.) varieties respond to heavy metal (Cadmium, Lead, Arsenic) stress. Front Plant Sci 15:1521075
Li X et al (2025b) Evaluation of salt-tolerant germplasms and identification of salt tolerance-related proteins in upland cotton at the seedling stage. Int J Mol Sci 26(5):1982
Lin M-H et al (2022) Benchmarking differential expression, imputation and quantification methods for proteomics data. Brief Bioinform 23(3):bbac138
Lippman ZB, Zamir D (2007) Heterosis: revisiting the magic. Trends Genet 23(2):60–66
Liu Y et al (2017) Proteomics analysis identified a DRT protein involved in arsenic resistance in *Populus*. Plant Cell Rep 36(12):1855–1869
Liu J et al (2018) Proteomic analysis of kiwifruit in response to the postharvest pathogen, *Botrytis cinerea*. Front Plant Sci 9:158
Liu Y et al (2019) Proteomics: a powerful tool to study plant responses to biotic stress. Plant Methods 15(1):135
Liu J et al (2020) Exploring the molecular basis of heterosis for plant breeding. J Integr Plant Biol 62(3):287–298
Liu D et al (2021) Quantitative proteomics and relative enzymatic activities reveal different mechanisms in two peanut cultivars (arachis hypogaea l.) under waterlogging conditions. Front Plant Sci 12:716114
Liu M et al (2023) Candidate regulators of drought stress in tomato revealed by comparative transcriptomic and proteomic analyses. Front Plant Sci 14:1282718
Love MI et al (2014) Moderated estimation of fold change and dispersion for RNA-seq data with DESeq2. Genome Biol 15(12):550
Lu J et al (2023) Combined transcriptome and proteome analysis revealed the molecular regulation mechanisms of zinc homeostasis and antioxidant machinery in tobacco in response to different zinc supplies. Plant Physiol Biochem 202:107919
Lyon D et al (2016) Drought and recovery: independently regulated processes highlighting the importance of protein turnover dynamics and translational regulation in medicago truncatula*. Mol Cell Proteomics 15(6):1921–1937
Ma Z et al (2025) Phosphoproteomic analysis of maize seedlings provides insights into the mechanisms of heat-stress tolerance. Int J Mol Sci 26(6):2439
Manghwar H et al (2024) Waterlogging stress in plants: unraveling the mechanisms and impacts on growth, development, and productivity. Environ Exp Bot 224:105824
Marcon C et al (2013) Heterosis-associated proteome analyses of maize (Zea mays L.) seminal roots by quantitative label-free LC–MS. J Proteome 93:295–302
Massoud MB et al (2022) Calcium and citrate protect *Pisum sativum* roots against copper toxicity by regulating the cellular redox status. J Soil Sci Plant Nutr 22(1):345–358
Meng W et al (2024) Analysis of differentially expressed proteins and related metabolic pathways in response to lead stress in the leaves of *Pogonatherum crinitum*. Ecotoxicol Environ Saf 288:117438
Mergner J, Kuster B (2022) Plant proteome dynamics. Annu Rev Plant Biol 73:67–92
Mikulášek K et al (2021) SP3 protocol for proteomic plant sample preparation prior lc-ms/ms. Front Plant Sci 12:635550
Mittra PK et al (2023) Proteome insights of citric acid-mediated cadmium toxicity tolerance in *Brassica napus* L. Environ Sci Pollut Res 30(54):115461–115479
Mohayeji M et al (2014) Heterosis profile of sunflower leaves: a label free proteomics approach. J Proteome 99:101–110
Ni M-w et al (2017) Modified filter-aided sample preparation (FASP) method increases peptide and protein identifications for shotgun proteomics. Rapid Commun Mass Spectrom 31(2):171–178
Nikonorova N et al (2021) The *Arabidopsis* root tip (phospho)proteomes at growth-promoting versus growth-repressing conditions reveal novel root growth regulators. Cells 10(7):1665
Olson BJSC (2016) Assays for determination of protein concentration. Curr Protoc Pharmacol 73(1):A.3A.1–A.3A.32

Ouyang Z et al (2024) Ceramide-1-phosphate enhances defense responses against Sclerotinia sclerotiorum in *Brassica napus*. Plant Physiol 197(2):kiae649
Oveland E et al (2015) Viewing the proteome: how to visualize proteomics data? Proteomics 15(8):1341–1355
Pan L et al (2021) Comparative proteomic analysis of the fungal pathogen *Neoscytalidium dimidiatum* infection in the pitaya. Hortic Environ Biotechnol 62(4):649–659
Papavasileiou A et al (2020) Proteomic analysis upon peach fruit infection with *Monilinia fructicola* and *M. laxa* identify responses contributing to brown rot resistance. Sci Rep 10(1):7807
Passamani LZ et al (2017) Salt stress induces changes in the proteomic profile of micropropagated sugarcane shoots. PLoS One 12(4):e0176076
Peralta-Videa JR et al (2009) The biochemistry of environmental heavy metal uptake by plants: implications for the food chain. Int J Biochem Cell Biol 41(8):1665–1677
Perveen A et al (2025) Integrative physiological, biochemical, and proteomic analysis of the leaves of two cotton genotypes under heat stress. PLoS One 20(1):e0316630
Picotti P (2015) Phosphoproteomics takes it easy. Nat Biotechnol 33(9):929–930
Piersma SR et al (2013) Whole gel processing procedure for GeLC-MS/MS based proteomics. Proteome Sci 11(1):17
Pinto VB et al (2021) Deciphering the major metabolic pathways associated with aluminum tolerance in popcorn roots using label-free quantitative proteomics. Planta 254(6):132
Prinsi B et al (2018) Root proteomic and metabolic analyses reveal specific responses to drought stress in differently tolerant grapevine rootstocks. BMC Plant Biol 18(1):126
Qing D et al (2022) ITRAQ-based quantitative proteomic analysis of japonica rice seedling during cold stress. Breed Sci 72(2):150–168
Qing D et al (2024) TMT-based quantitative proteomics analysis of defense responses induced by the Bph3 gene following brown planthopper infection in rice. BMC Plant Biol 24(1):1092
Qu D et al (2023) A 4d proteome investigation of the potential mechanisms of sa in triggering resistance in kiwifruit to *Pseudomonas syringae* pv. Actinidiae. Int J Mol Sci 24(24):17448
Ramsey JS et al (2020) Longitudinal transcriptomic, proteomic, and metabolomic analysis of citrus limon response to graft inoculation by *Candidatus liberibacter* asiaticus. J Proteome Res 19(6):2247–2263
Reis-de-Oliveira G et al (2024) OmicScope unravels systems-level insights from quantitative proteomics data. Nat Commun 15(1):6510
Ren W et al (2022) iTRAQ-based quantitative proteomic analysis provides insight into the drought-stress response in maize seedlings. Sci Rep 12(1):9520
Ren J et al (2024) TaCRT3 is a positive regulator of resistance to *Blumeria graminis* f. sp. tritici in wheat. Phytopathology 114(3):641–652
Rockenbach MF et al (2018) Differentially abundant proteins associated with heterosis in the primary roots of popcorn. PLoS One 13(5):e0197114
Rodrigues SP et al (2024) Juvenile-related tolerance to papaya sticky disease (PSD): proteomic, ultrastructural, and physiological events. Plant Cell Rep 43(11):269
Roychowdhury R et al (2023) Multi-omics pipeline and omics-integration approach to decipher plant's abiotic stress tolerance responses. Genes 14(6):1281
Ruffatto K et al (2024) Unravelling soybean responses to early and late *Tetranychus urticae* (Acari: Tetranychidae) infestation. Plant Biol 26(7):1223–1239
Salvato F et al (2019) Label-free quantitative proteomics of enriched nuclei from sugarcane (*Saccharum* ssp) stems in response to drought stress. Proteomics 19(14):1900004
Schachtman DP, Goodger JQD (2008) Chemical root to shoot signaling under drought. Trends Plant Sci 13(6):281–287
Schessner JP et al (2022) A practical guide to interpreting and generating bottom-up proteomics data visualizations. Proteomics 22(8):2100103
Schmitt KFM et al (2024) Decoding the effects of drought stress on popcorn (*Zea mays* var. everta) flowering combining proteomics and physiological analysis. Plant Physiol Biochem 208:108444

Schubert OT et al (2017) Quantitative proteomics: challenges and opportunities in basic and applied research. Nat Protoc 12(7):1289–1294

Schwacke R et al (2019) MapMan4: a refined protein classification and annotation framework applicable to multi-omics data analysis. Mol Plant 12(6):879–892

Seleiman MF et al (2021) Drought stress impacts on plants and different approaches to alleviate its adverse effects. Plants 10(2):259

Shah P et al (2012) Proteomic analysis of ripening tomato fruit infected by *Botrytis cinerea*. J Proteome Res 11(4):2178–2192

Shannon P et al (2003) Cytoscape: a software environment for integrated models of biomolecular interaction networks. Genome Res 13(11):2498–2504

Sharma P et al (2005) The molecular biology of the low-temperature response in plants. BioEssays 27(10):1048–1059

Shen Z et al (2024) Integrated transmission electron microscopy and proteomic analyses reveal the cytoarchitectural response to cucumber mosaic virus infection in tobacco. Int J Biol Macromol 262(Pt 2):130100

Shi F et al (2024) Unveiling molecular mechanisms of pepper resistance to *Phytophthora capsici* through grafting using iTRAQ-based proteomic analysis. Sci Rep 14(1):4789

Song X et al (2007) Wheat (L.) root proteome and differentially expressed root proteins between hybrid and parents. Proteomics 7(19):3538–3557

Song X et al (2025) Proteomic and physiological analyses reveal the mechanisms through which melatonin ameliorates heat stress-induced photoinhibition in *Nicotiana tabacum*. Plant Physiol Biochem 221:109665

Stavi I et al (2021) Soil salinity and sodicity in drylands: a review of causes, effects, monitoring, and restoration measures. Front Environ Sci 9:712831

Szklarczyk D et al (2022) The STRING database in 2023: protein–protein association networks and functional enrichment analyses for any sequenced genome of interest. Nucleic Acids Res 51(D1):D638–D646

Tarolli P et al (2024) Soil salinization in agriculture: mitigation and adaptation strategies combining nature-based solutions and bioengineering. iScience 27(2):108830

Taylor CF et al (2007) The minimum information about a proteomics experiment (MIAPE). Nat Biotechnol 25(8):887–893

Terzi H, Yıldız M (2021) Proteomic analysis reveals the role of exogenous cysteine in alleviating chromium stress in maize seedlings. Ecotoxicol Environ Saf 209:111784

Theologidis I et al (2024) Proteome profiling of *Cucurbita pepo* phyllosphere after infection by podosphaera xanthii and application of *Reynoutria sachalinensis* extract. Appl Sci 14(21):10061

Tyanova S, Cox J (2018) Perseus: a bioinformatics platform for integrative analysis of proteomics data in cancer research. In: von Stechow L (ed) Cancer systems biology: methods and protocols. Springer New York, New York, pp 133–148

Tyanova S et al (2016a) The MaxQuant computational platform for mass spectrometry-based shotgun proteomics. Nat Protoc 11(12):2301–2319

Tyanova S et al (2016b) The Perseus computational platform for comprehensive analysis of (prote) omics data. Nat Methods 13(9):731–740

Tzortzakis N (2019) Physiological and proteomic approaches to address the active role of *Botrytis cinerea* inoculation in tomato postharvest ripening. Microorganisms 7(12):681

Uchimiya H, Takahashi N (1973) Kinetics of heterosis in growth of the leaf blade in *Zea mays* L. Ann Bot 37(2):247–252

Vale EM et al (2016) Comparative proteomic analysis of the heterosis phenomenon in papaya roots. Sci Hortic 209:178–186

Van de Graaf MW et al (2024) Benchmarking of protein interaction databases for integration with manually reconstructed signalling network models. J Physiol 602(18):4529–4542

Varnavides G et al (2022) In search of a universal method: a comparative survey of bottom-up proteomics sample preparation methods. J Proteome Res 21(10):2397–2411

Wang W et al (2004) Role of plant heat-shock proteins and molecular chaperones in the abiotic stress response. Trends Plant Sci 9(5):244–252
Wang W et al (2008) Proteomic profiling of rice embryos from a hybrid rice cultivar and its parental lines. Proteomics 8(22):4808–4821
Wang D et al (2021a) Comparative proteomic analysis reveals that the Heterosis of two maize hybrids is related to enhancement of stress response and photosynthesis respectively. BMC Plant Biol 21(1):34
Wang X et al (2021b) Comparative proteomic analysis of grapevine rootstock in response to waterlogging stress. Front Plant Sci 12:749184
Wang W et al (2023) Label-free quantitative proteomics reveal the mechanisms of young wheat (*Triticum aestivum* L.) ears' response to spring freezing. Int J Mol Sci 24(21):15892
Wang K et al (2024a) The proteome of *Penicillium expansum* during infection of postharvest apple is revealed using label-free and parallel reaction monitoring(PRM) techniques. J Proteome 298:105142
Wang M et al (2024b) Transcriptomic–proteomic analysis revealed the regulatory mechanism of peanut in response to *Fusarium oxysporum*. Int J Mol Sci 25(1):619
Wang K et al (2025) VOZ-dependent priming of salicylic acid-dependent defense against Rhizopus stolonifer by β-aminobutyric acid requires the TCP protein TCP2 in peach fruit. Plant J 121(1):e17176
Wijeweera S et al (2025) Spatial and development responses in the wheat leaf highlight the loss of chloroplast protein homeostasis during salt stress. J Proteome 316:105438
Wu L et al (2023) Comparative proteomic analysis of roots from a wild eggplant species solanum sisymbriifolium in defense response to *Verticillium dahliae* inoculation. Genes 14(6):1247
Xavier LR et al (2023) Time-dependent proteomic signatures associated with embryogenic callus induction in *Carica papaya* L. Plants 12(22):3891
Xie X et al (2025) Proteomics analysis revealed the activation and suppression of different host defense components challenged with mango leaf spot pathogen *Alternaria alternata*. BMC Plant Biol 25(1):227
Xing J et al (2016) Proteomic patterns associated with heterosis. Biochim Biophys Acta Proteins Proteomics 1864(8):908–915
Xu M et al (2022) Proteomic analysis of apple response to *Penicillium expansum* infection based on label-free and parallel reaction monitoring techniques. J Fungi 8(12):1273
Yan S et al (2022) Recent advances in proteomics and metabolomics in plants. Mol Hortic 2(1):17
Yang R et al (2022) iTRAQ proteomic analysis of wheat (*Triticum aestivum* L.) genotypes differing in waterlogging tolerance. Front Plant Sci 13:890083
Yang X et al (2023) iTRAQ-based quantitative proteomics unveils protein dynamics in the root of *Solanum melongena* L. under waterlogging stress conditions. Life 13(6):1399
Yang Y et al (2024) Comparative proteomic and metabolomic analyses reveal stress responses of hemp to salinity. Plant Cell Rep 43(6):154
You Z et al (2025) Developing a new ethylene glycol/H2O pretreatment system to achieve efficient enzymatic hydrolysis of sugarcane bagasse cellulose and recover highly active lignin: countercurrent extraction. Sep Purif Technol 354:128564
Yu L et al (2025) Transcriptomic, metabonomic and proteomic analyses reveal that terpenoids and flavonoids are required for *Pinus koraiensis* early defence against *Bursaphelenchus xylophilus* infection. BMC Plant Biol 25(1):185
Zaccaron AZ et al (2024) Transcriptome analysis of two isolates of the tomato pathogen *Cladosporium fulvum*, uncovers genome-wide patterns of alternative splicing during a host infection cycle. PLoS Pathog 20(12):e1012791
Zahedi SM et al (2025) Plant adaptation to drought stress: the role of anatomical and morphological characteristics in maintaining the water status. J Soil Sci Plant Nutr 25(1):409–427
Zahid MA et al (2024) Enhanced stress resilience in potato by deletion of Parakletos. Nat Commun 15(1):5224

Zhang C et al (2012) Comparative proteomic study reveals dynamic proteome changes between superhybrid rice LYP9 and its parents at different developmental stages. J Plant Physiol 169(4):387–398

Zhang Y et al (2013) Protein analysis by shotgun/bottom-up proteomics. Chem Rev 113(4):2343–2394

Zhang F et al (2020) Data-independent acquisition mass spectrometry-based proteomics and software tools: a glimpse in 2020. Proteomics 20(17–18):1900276

Zhang B-Q et al (2022a) Cold-induced physiological and biochemical alternations and proteomic insight into the response of *Saccharum spontaneum* to low temperature. Int J Mol Sci 23(22):14244

Zhang H et al (2022b) The potential role of plasma membrane proteins in response to Zn stress in rice roots based on iTRAQ and PRM under low Cd condition. J Hazard Mater 429:128324

Zhang J et al (2022c) Ascorbate peroxidase 1 confers resistance to southern corn leaf blight in maize. J Integr Plant Biol 64(6):1196–1211

Zhang Y et al (2023) Plants' response to abiotic stress: mechanisms and strategies. Int J Mol Sci 24(13):10915

Zhang H et al (2024) Physiological and proteomic analysis of various priming on rice seed under chilling stress. Plants 13(17):2430

Zhang B et al (2025a) RBB1 negatively regulates rice disease resistance by modulating protein glycosylation. J Integr Plant Biol 67(2):391–407

Zhang N et al (2025b) Integration of multi-omics data accelerates molecular analysis of common wheat traits. Nat Commun 16(1):2200

Zhao Y et al (2023) Stimulus-responsive proteins involved in multi-process regulation of storage substance accumulation during rice grain filling under elevated temperature. BMC Plant Biol 23(1):547

Zhao Y et al (2025a) Fine control of growth and thermotolerance in the plant response to heat stress. J Integr Agric 24(2):409–428

Zhao Y et al (2025b) Wheat yellow mosaic virus P1 INHIBITS ROS accumulation to facilitate viral infection. Int J Mol Sci 26(4):1455

Zheng Y et al (2016) iTAK: a program for genome-wide prediction and classification of plant transcription factors, transcriptional regulators, and protein kinases. Mol Plant 9(12):1667–1670

Zhou Y et al (2019) Metascape provides a biologist-oriented resource for the analysis of systems-level datasets. Nat Commun 10(1):1523

Zhou Y et al (2022) Proteomic investigation of molecular mechanisms in response to PEG-induced drought stress in soybean roots. Plants 11(9):1173

Zhu S et al (2024) Comprehensive physiology and proteomics analysis revealed the resistance mechanism of rice (*Oryza sativa* L) to cadmium stress. Ecotoxicol Environ Saf 278:116413

Plant Leaf Proteomics and Drought Stress Tolerance

Christin Bündig

Abstract Abiotic stress factors such as drought, heat, and soil salinity pose serious challenges to plant growth, especially in the face of increasing climate change. In Europe, major droughts and heatwaves have had a negative impact on agriculture, viticulture, and forestry. In addition, multiple stress factors—such as nutrient deficiencies, flooding, frost, and pollution—are also contributing to declining crop yields and threatening global food security. While breeding has traditionally focused on resistance to biotic stress and yield, abiotic stress tolerance has become increasingly important, with genetic studies identifying the most important traits. Proteomics evolved as a valuable tool in stress research, providing physiological insights into plant adaptation through changes in protein composition. Despite its potential, proteomics still faces challenges such as high costs and data complexity, but advances in omics technologies will further improve understanding of tolerance mechanisms and support breeding efforts for resilient plants. The chapter highlights previous results of proteomic studies from the last decade to identify major leaf proteins involved in drought tolerance and adaptation, as well as linkage to heat and salt stress responses.

Keywords Proteome · Drought · Stress response · Changes · Plants · Leaf Tolerance

Abbreviations

2-DE	Two-dimensional gel electrophoresis
ABA	Abscisic acid
Ala	Alanine
ALDHs	Aldehyde dehydrogenase

C. Bündig (✉)
Leibniz Universität Hannover, Institute of Plant Genetics, Section Reproduction and Development, Hannover, Germany
e-mail: buendig@baum.uni-hannover.de

L. M. Vaschetto (ed.), *Omics Technologies in Plant Science*, Interdisciplinary Biotechnological Advances, https://doi.org/10.1007/978-981-92-0533-2_7

APXs	Ascorbate peroxidases
Arg/Lys	Arginine/lysin
ASR	Abscisic acid-, stress-, and ripening-induced protein
ATP	Adenosine triphosphate
BPs	Ribulose-1,5-bisphosphat-carboxylase/-oxygenase binding proteins
CA	Carbonic anhydrase
Clp	Caseinolytic protease
CO_2	Carbon dioxide
CTA	Catalase
Cys	Cysteine
DAPs	Differential abundant proteins
DIGE	Difference gel electrophoresis
DNA	Deoxyribonucleic acid
e.g.	Example given
ER	Endoplasmic reticulum
ERD	Early responsive to dehydration
FC	Field capacity
FtsH	Filamentation temperature-sensitive H
GAPA	Glyceraldehyde-3-phosphate dehydrogenase A
GAPB	Glyceraldehyde-3-phosphate dehydrogenase B
GAPC	Glyceraldehyde-3-phosphate dehydrogenase C
GAPDH	Glyceraldehyde-3-phosphate dehydrogenase
Glu	Glutamic acid
Gly	Glycine
GRPs	Glycine-rich RNA-binding proteins
GST	Glutathione S-transferase
H_2O_2	Hydrogen peroxide
HCO_3^-	Hydrogen carbonate
HSPs	Heat shock proteins
ICSs	Intramolecular crowding sensors
iTRAQ	Isobaric tags for relative and absolute quantitation
LC-MS/MS	Liquid chromatography-mass spectrometry/mass spectrometry
LEA	Late embryogenesis abundant proteins
LHC	Light-harvesting complex
LHCB	Chlorophyll a-b binding proteins
MDA	Malondialdehyde
NADPH	Nicotinamide adenine dinucleotide phosphate
NH_3	Ammonia
NIPs	Nodulin 26-like intrinsic proteins
P5CS	1-Pyrroline-5-carboxylate synthase
PIPs	Membrane intrinsic proteins
PMs	Plasma membranes
POD	Peroxidase
RCA	Ribulose-1,5-bisphosphat-carboxylase/-oxygenase activase
RNA	Ribonucleic acid

ROS	Reactive oxygen species
RuBisCO	Ribulose-1,5-bisphosphat-carboxylase/-oxygenase
RWC	Relative water content
sHSPs	Small heat shock proteins
SIPs	Small basic intrinsic proteins
SOD	Superoxide dismutase
Thr	Threonine
TIPs	Tonoplast intrinsic proteins
Trp	Tryptophane
XIPs	X intrinsic proteins

1 Introduction

Abiotic stressors are environmental factors that are unfavorable for plant growth (Minhas et al. 2017). As manmade climate change is getting more severe, these unfavorable conditions are increasing in frequency and severity. The greatest constraints that threaten agricultural output, ecosystems, and their services are drought and high temperatures/heat (for review see Arif et al. 2025), as well as the salinity of soils (Krasensky and Jonak 2012). Especially the severe summer drought periods in 2003, 2015, and 2018 together with an extreme heat wave negatively impacted Europe's agricultural, vinicultural, and silvicultural sectors (Fink et al. 2004; Ionita et al. 2017; Kornhuber et al. 2019). However, other factors such as a lack of specific nutrients, late frost events, high light intensity, high concentrations of ozone, input of heavy metals in soils or flooding of dry or compacted soils also have been reported to increase and lead to drastic yield losses (Tan et al. 2005; Ainsworth 2017; Yang et al. 2019; Mohammad et al. 2020; Zhou et al. 2020; Guo et al. 2024). Together with an increase in population, the decline of arable land is a serious threat to food security worldwide.

During the last two decades, the primary foci of breeding were shifted from biotic stress resistance together with an increase in yield, to also include breeding for an increase in abiotic stress tolerance for many different plants. However, this has mainly been done by generating and analyzing information on quantitative trait loci (QTLs) and genome-wide association studies (GWASs). Omics data, such as transcriptomics and metabolomics, have been used to complement genomic information on stress responses, which together gave rise to further functional context into stress tolerance adaptation (for review see Gupta et al. 2020).

During the last three decades, plant proteomics has come into focus in stress tolerance research. Wilkins et al. (1996) first defined the proteome to be "the total protein complement of a genome." However, as a proteome is highly dynamic, Lottspeich (1999) later defined the proteome to be "the quantitative analysis of proteins present in an organism at a certain time and under certain conditions." This definition also more accurately describes that a proteome results from protein biosynthesis, degradation, and modification, such as phosphorylation at a given time point. As the protein composition depicts the actual physiological state of the cell, tissue, or organ at a certain time point during stress, this approach is the most promising to study changes that arise from unfavorable environmental conditions. Proteomics bridges the gap between transcriptomics and metabolomics, strongly enriching insights from genomics. Only through the link of all omics levels, it will be possible to unravel the complexity of biological stress adaptation (Yang et al. 2014b; Satrio et al. 2024; for reviews see Kosová et al. 2011; Ghosh and Xu 2014).

Among other structural and cellular adaptations, stressed plants are able to adapt to unfavorable environmental conditions due to alterations in protein composition. However, even though studying the proteome is a powerful tool to elevate stress tolerance research, it is still in its early stages because methods are usually limited to a few samples, as they are still costly and generate large amounts of data. Therefore, most studies are limited to control and stressed proteome comparisons of one or two genotypes. Nevertheless, as technology and especially omics approaches are evolving, using proteomics to further gain insights into adaptation processes of different plant species and genotypes has come more into focus over the last decade. For this purpose, different proteomic tools, such as 2-DE, DIGE, iTRAQ, or LC–MS/MS, have been used for comparative proteomics studies to detect differentially abundant proteins (DAPs) in response to diverse abiotic stressors in several plant species (for an overview on techniques, see Thikekar et al. 2023).

However, as abiotic stress responses are based on complex traits, understanding of the composition and function of specific proteins within major pathways during specific phases of stress is important to generate a detailed picture of tolerance traits and subsequently link these to breeding tolerant crops. Therefore, this chapter focuses on recent proteomic findings of the last decade on the major stressor, drought stress, and possible overlaps with heat and salt stress responses.

2 Drought and Osmotic Stress Tolerance

Drought stress is considered to be the most challenging abiotic stress for plants. It not only limits water availability but also limits, for example, nutrient uptake as this is linked to movement of water as well as cell growth prospects as turgor pressure is reduced, which leads to cell shrinking (Hussain et al. 2008; Bista et al. 2018). Many studies on crop stress responses at the morphological level have shown that plants´ primary responses comprise stomatal closure, which causes a decrease in CO_2 uptake, leading to a decline in cell growth through, e.g., a decrease in nutrient uptake and ultimately limiting shoot biomass build-up through these responses (Li and Liu 2022). However, plants cope with drought stress by countermeasures in terms of protein composition at specific time points. Proteomic studies have given insights into the complex responses to drought stress. In addition, many studies have been conducted using osmotic stress experiments, as this is part of drought stress.

Proteomic studies have been performed on different tissues and organs as well as organelles. Multiple studies on the leaf proteome for different plants have been performed (for review see Wang et al. 2016b). Fewer reports have been published for root proteomes (for review, see Ghosh and Xu 2014). This is mainly due to difficulties in harvesting root material compared to aboveground material. However, root proteome studies have increased in the last decade. The differences in timing of sample collection are also a critical factor in proteomic studies. These range from hours, a few days after drought or osmotic onset, to weeks. This influences the protein content of the plant and is an important parameter to keep in mind when interpreting and comparing studies. However, it is evident that different species respond very specifically in terms of leaf protein abundance depending on the severity of drought stress applied (Wu et al. 2016). Nevertheless, certain proteins were found in different studies on leaf proteome, leading to a first glance at the mechanisms underlying tolerance under drought, salt, and heat stress.

2.1 *Drought Stress Sensors and Receptors*

Receptors for drought stress perception have been elusive for many years (Wang et al. 2016b). Early reports indicated that drought-receptive photoreceptors like phytochromes might be part of the highly sought-after drought receptors, as these were reported to be involved in drought tolerance suppression in *Arabidopsis thaliana* (Boggs et al. 2010). However, recent studies have led to a multitude of different proposed osmosensors by intramolecular crowding sensors (ICSs; mechanosensors) that either sense osmolarity triggers and membrane tension mainly as ion channels (Hamilton et al. 2015; Meneses-Reyes et al. 2024) or through sensing of osmolarity changes by liquid–liquid phase separation through intercellular proteins (Dorone et al. 2021; Wang et al. 2022, 2024). However, the complete sensing mechanisms are still not fully understood, and proteins are elusive in proteomic studies.

2.2 *Photosynthesis-Related Proteins*

Under drought stress, photosynthesis is one of the first metabolic pathways that is altered. Due to fast stomatal closure and the related reduction in CO_2 uptake, regulation in photosynthetic activity has been shown to take place (Pirasteh-Anosheh et al. 2016). Likewise, analyses, which are specified in detail below, of the leaf proteome of different species, also displayed major alterations in proteins associated with photosynthesis. These include proteins of the light-harvesting complex, such as Ribulose-1,5-Bisphosphate Carboxylase/Oxygenase (RuBisCO) activase (RCA) (Ogden et al. 2020; Yahoueian et al. 2021), RuBisCO large subunit-binding proteins (BPs) (Cheng et al. 2016; Wu et al. 2016; Zhang et al. 2016), and chlorophyll a-b binding proteins (Zhao et al. 2016; Michaletti et al. 2018; Zenda et al. 2018; Vessal et al. 2020; Yahoueian et al. 2021).

In proteomic studies, RCA was found to be more abundant in tolerant cultivars of sorghum (*Sorghum bicolor*) and soybean (*Glycine max*) under drought (Ogden et al. 2020; Yahoueian et al. 2021). RCA regulates the activity of RuBisCO through the dissociation of inhibitory sugar phosphates from the active site of RuBisCO in an ATP-dependent manner (Spreitzer and Salvucci 2002). Studies showed that RCA comprises two isoforms, a redox-insensitive ß-isoform and a longer redox-sensitive α-isoform (Zhang and Portis 1999). Under drought stress, more tolerant cultivars of sorghum showed higher abundance of the α-isoform of RCA, indicating that this might display a more effective system for redox homeostasis under drought stress conditions (Ogden et al. 2020). A proteomic study on seedlings of rapeseed (*Brassica napus*) showed an increase under drought stress for RCA (Koh et al. 2015). However, contrasting reports indicate no changes in RCA abundance after drought stress in two divergently responding wheat parental lines (Nemati et al. 2019). In osmotic stress experiments, RCA was identified to be less abundant in drought-sensitive potato (*Solanum tuberosum*) genotypes (Bündig et al. 2016). In soybean, RCA was also reported to be more abundant in a slow-wilting, high-yielding cultivar under heat stress (Katam et al. 2020). Conversely, in a heat-tolerant spinach inbred line, RCA was less abundant under heat stress (Zhao et al. 2018). This was also reported under salt stress for saltwater cress (*Eutrema salsugineum*), where RCA was less abundant after stress (Goussi et al. 2021). RCA might therefore be an interesting candidate to further study in various species under drought stress at different time points.

RuBisCO is a photosynthetic enzyme that plays crucial roles in biological processes. RuBisCO BPs were among the ones to be identified as downregulated under drought stress in sensitive and tolerant genotypes of wheat (*Triticum aestivum*) (Cheng et al. 2016). This was described as one major factor for the reduction of photosynthesis besides stomatal closure (Galmés et al. 2013). However, in a study on maize (*Zea mays*), only a tolerant cultivar was shown to reduce BP subunit beta (Zhang et al. 2016). In contrast to previous results, a RuBisCO subunit in chickpea (*Cicer arietinum*) was identified to be increased in abundance in a tolerant cultivar (Vessal et al. 2020). Data suggest that the large subunit binding proteins play a versatile role in drought stress responses in different plant species.

Proteins of the light-harvesting complex (LHC) and associated proteins were identified as DAPs in multiple studies and show complex abundance. Chlorophyll a-b binding proteins (LHCBs) were among the most widely identified DAPs throughout proteomic studies under drought stress. LHCBs were shown to be downregulated in more tolerant cultivars of maize (Zenda et al. 2018; Li et al. 2021a, b), wheat (Michaletti et al. 2018; Moloi et al. 2024), chickpea (Vessal et al. 2020), and mutant lines of maize (Zhao et al. 2016). In soybean, an LHCB of the LHC decreased in the sensitive as well as in the tolerant cultivar after drought, however, to a different extent, as the sensitive cultivar showed a higher reduction (Yahoueian et al. 2021). While the reduction was rather high in the tolerant genotype, the sensitive genotype decreased the content even further. However, in the same study, a chlorophyll a-b binding protein was identified, which showed higher abundance in both genotypes. LHCB was also reported to be decreased after salt stress in saltwater cress (Goussi et al. 2021). As a decrease in LHCB was widely reported in proteomic studies under different stressors, this protein might be interesting to check in a time series in very early stages of plant stress for diverse stressors.

Several studies also reported glyceraldehyde-3-phosphate dehydrogenase (GAPDH) to be increased after drought stress (Cheng et al. 2016; Chintakovid et al. 2017; Yahoueian et al. 2021). GAPDH reversibly converts glycerate-3-phosphate to glyceraldehyde-3-phosphate and is prone to oxidative modifications (Hildebrandt et al. 2015). Glycerate-3-phosphate was identified to be an electron acceptor that receives electrons from NADPH and therefore protects photosystem II from oxidative damage through excessive H_2O_2 production (Takahashi and Murata 2006; Hwang et al. 2009). There are different GAPDHs in plants, e.g., GAPDH A (GAPA) and GAPDH B (GAPB), which are primarily found in chloroplasts, while GAPDH C (GAPC) is located within the cytoplasm (Zaffagnini et al. 2013). GAPDH B was identified to be increased under drought in wheat (Wang et al. 2020b). GAPDH was also reported to be increased after stress application in tolerant cultivars of rice (*Oryza sativa*) and wheat at early time points up to 24 h after stress onset (Cheng et al. 2016; Chintakovid et al. 2017). Two cytosolic GAPDHs, as well as a chloroplastic GAPDH, were reported to be increased in a rather sensitive genotype of potato after osmotic stress in vitro (Bündig et al. 2016). On the contrary, Cheng et al. (2016) also reported a decrease in abundance for GAPDH A, which decreased in a steady pattern throughout the experiment, even under rewatering conditions. However, a GAPDH was found to be increased only for a sensitive cultivar in soybean in a greenhouse pot experiment (Yahoueian et al. 2021). Moloi et al. (2024) reported GAPDH abundance to be decreased in both the sensitive and tolerant cultivars of wheat after mild and severe drought stress. This indicates that the period in which sample collection is done is of utmost importance when screening for stress tolerance on a proteomic level. Plants will respond differently when screening early time points up to 24 h compared to long-term progressing drought or osmotic stress. However, as different GAPDHs are localized in different compartments, as identified in *A. thaliana*, maize, and wheat, analyzing the different GAPDHs further might help to shed light on specific functions in different cellular compartments under drought. Nevertheless, due to differential patterns of abundance, GAPDH is an interesting candidate to further analyze in time series for different species to identify if cultivars that are more tolerant display an increase in abundance in very early time points of drought stress onset. As GAPDH was also reported to be increased after heat stress in soybean (Katam et al. 2020), this might indicate that this protein might be of ubiquitous importance to regulate general abiotic stress responses to different stressors.

2.3 Energy Cycle

The energy cycle of plants was also found to be altered when stress occurs. Proteomic studies identified ATP synthase alpha and beta subunits, as well as carbonic anhydrases, to be of importance when dealing with unfavorable environmental conditions.

The ATP synthase alpha and beta subunits, as well as a precursor, were identified to be upregulated under stress in chickpea and in wheat up to 48 h after drought onset

(Cheng et al. 2016; Vessal et al. 2020). In contrast, Michaletti et al. (2018) reported a down-regulation of ATP synthase beta subunit as well as a chloroplastic ATP synthase delta chain in a tolerant wheat cultivar under drought stress. In a sensitive maize cultivar, ATP synthase beta subunit 1 was more abundant under stress (Zhang et al. 2016). In a drought-tolerant wheat parental line, ATP synthase CF1 alpha chain was more abundant after drought stress; however, other identified ATP synthases were not altered in either the sensitive or the tolerant parental lines (Nemati et al. 2019). An abundance of ATP synthase proteins was also reported for a heat-tolerant spinach (*Spinacia oleracea*) line in leaves under heat stress, where different ATP-synthases were either up- or downregulated during heat stress (Zhao et al. 2018).

Carbonic anhydrase (CA) catalyzes the interconversion of CO_2 and HCO_3^- and is found in all living organisms (Rudenko et al. 2020; for review see Polishchuk 2021). CA was reported as more abundant under drought stress in a tolerant chickpea cultivar (Vessal et al. 2020) as well as in bermudagrass (Fang et al. 2022). That CAs might confer drought tolerance was also reported by Wang et al. (2016a). Especially, βCA1 was reported to be a drought-responsive phosphoprotein, which is active in the drought stress response of rapeseed (Wang et al. 2016a). However, Nemati et al. (2019) reported a lower abundance of CA under drought stress in two divergently responding wheat parental lines. CA was lower abundant also under heat stress in wheat (Lu et al. 2017). Even though divergent reports on the regulation of the CA exist, it is evident that CAs play a role in abiotic stress adaptation of plants.

2.4 ROS Detoxification and Signaling

Drought and osmotic stress induce reactive oxygen species (ROSs), which are byproducts of aerobic metabolism. Plants respond by inducing either enzymatic or non-enzymatic scavengers to reduce ROS-induced oxidative damage on important structures such as RNA, DNA, or membranes. Major proteins of the enzymatic response comprise peroxidases, superoxide dismutase, catalases, glyoxylases, oxidoreductases, thioredoxins, and glutathione transferases. These help to maintain a redox equilibrium during stress by detoxifying ROS. High levels of ROS are harmful to the cell. However, a basal level is required for accurate ROS or redox signaling (Mittler 2017), giving rise to the need for delicate balances of production and scavenging.

Ascorbate peroxidases (APXs) are enzymes that utilize the reduction of H_2O_2 using ascorbate as an electron donor. It functions as a central enzyme in the ascorbate–glutathione cycle, which increases the kinetics of H_2O_2 reduction (for review see Li 2023). In soybean, Yahoueian et al. (2021) identified APX to be more abundant in tolerant as well as susceptible cultivars under drought stress. In a study on potatoes, APX was identified to be increased after drought stress (Wellpott et al. 2023). APX and different peroxidases were also identified as DAPs by Cheng et al. (2016). They showed an increase over time and a decrease after rewatering. However, a peroxidase 24 precursor did not show alterations during the experiment. An increase was also detectable in maize under drought stress, where the tolerant cultivar showed a higher abundance of APX, whereas the sensitive cultivar displayed a lower abundance of a cytosolic isoform (Zhang et al. 2016).

Glutathione S-transferase (GST) is another enzyme functioning in the reduction of ROS. GST was induced under drought stress in a high-yield producing maize hybrid (Zhao et al. 2016). In chickpea, one GST was more abundant only in the tolerant genotype under stress (Vessal et al. 2020). A time series by Cheng et al. (2016) showed that different GSTs were all more abundant and reached a peak after 24 h of drought stress in a tolerant wheat cultivar. It is evident that proteins from the ascorbate-glutathione cycle are of major importance to set a ROS equilibrium and reduce oxidative stress.

More tolerant cultivars of maize also displayed a higher abundance of different peroxidases (POD) under drought stress (Li et al. 2021b). However, reports on POD in *Pandanus amaryllifolius*, a plant used as a spice, were identified to be decreased after drought onset, but increased after rewatering (Amnan et al. 2022). Additionally, in proteomic studies analyzing heat stress responses, POD4 was increased in *Clematis florida* (Jiang et al. 2020).

Another important ROS scavenging enzyme is superoxide dismutase (SOD). SODs transform superoxide to H_2O_2. Under drought stress, SODs were identified to be more abundant in chickpea (Vessal et al. 2020) and sorghum (Ogden et al. 2020). A manganese SOD was reported to be induced under drought stress in wheat, reaching a peak at 24 h of drought, but declined after 48 h and rewatering (Cheng et al. 2016). In a tolerant maize variety, SOD was identified to be increased during drought stress. This was also observed on the RNA level, indicating that a de novo synthesis might also take place (Li et al. 2021b).

Catalase (CAT) detoxifies H_2O_2 to H_2O. CAT was identified to be more abundant under drought stress in a tolerant and a sensitive wheat parental line (Nemati et al. 2019). However, Amnan et al. (2022) reported an increase in catalase isoenzymes A and B when comparing control and recovery phase plants, but not during drought stress. Under salt stress, CAT was reduced compared to control conditions in *Eutrema salsugineum* (Goussi et al. 2021).

However, POD, SOD, and CAT were also reported to be affected by other stressors in proteomic studies such as heat and salt stress as well as double stress studies for heat and drought, therefore this might show a general response to oxidative stress as part of the stress response (Zhao et al. 2016; Katam et al. 2020; Li et al. 2020; Zhu et al. 2020). The changes in protein content of these scavenging enzymes display a highly dynamic alteration to abiotic stresses, leading to their usage as biomarkers to monitor plant stress responses.

2.5 Chaperones

Chaperones are among the most widespread DAPs identified in plants under multiple abiotic stresses. Heat shock proteins (HSPs) are classified according to their approximate molecular weight (MW) into five distinct classes: HSP100, HSP90, HSP70, HSP60, and small HSPs (sHSPs; MW 15–42 kDa) (Vierling 1997). They were first described under heat stress; however, they are now known to function as chaperones under diverse suboptimal environmental conditions.

Among the widely identified DAPs, HSP70, as well as sHSPs, were reported in proteomic studies under abiotic stress.

Within the class of HSP70, there are different families depending on localization: cytosol, endoplasmic reticulum (ER), plastids, and mitochondria (for review, see Sung et al. 2001). HSP70s were identified to be increased under drought stress compared to control conditions in many studies (Vessal et al. 2020; Amnan et al. 2022; Liu et al. 2023; Wellpott et al. 2023). However, Cheng et al. (2016) reported a lower abundance of HSP70 in a tolerant wheat cultivar. No changes in two wheat parental lines were reported for HSP70–1 and HSP70–2 under drought stress (Nemati et al. 2019). In a more drought-tolerant potato genotype, HSP70 was identified as more abundant after drought stress (Wellpott et al. 2023). However, in an in vitro study on osmotic stress, HSP70–2-like and a HSP70 were identified to be more abundant under stress in a rather drought-sensitive potato genotype (Bündig et al. 2016). HSP70s were also identified to be induced in a slow-wilting and high-yielding soybean cultivar under heat stress (Katam et al. 2020), as well as for soybean seedlings in a previous study on heat stress (Ahsan et al. 2010). In a heat memory study on Rhododendron (*Rhododendron hainanense*), proteomic responses included an HSP70, which was increased after heat stress (Wang et al. 2020a). Jiang et al. (2020) also identified a HSP70 to be increased under heat stress in *Clematis florida* in a heat-tolerant line compared to a heat-sensitive line. The results indicate that HSP70 family members are essential when dealing with different abiotic stressors.

sHSPs were also identified as DAPs in a range of different species, ranging from rice, sorghum, maize, potato, and pandan (*Pandanus amaryllifolius*) (Wu et al. 2016; Zhao et al. 2016; Ogden et al. 2020; Amnan et al. 2022; Wellpott et al. 2023). However, while sHSP17 was reported to increase under drought stress in sorghum in a tolerant genotype, in rice, HSP17.9 increased in tolerant and sensitive cultivars, while HSP18.6 only increased after extreme drought (Wu et al. 2016; Ogden et al. 2020). In a high-yielding maize hybrid, small HSPs (16.9, 17.2, 17.4, and 17.5) were reported to be less abundant after drought stress; however, they increased after heat and double stress (Zhao et al. 2016). Li et al. (2021b) reported an increase of a 17.4 kDa HSP in a sensitive cultivar of maize. Heat and salt stress also induce sHSPs in soybean (Katam et al. 2020). Wellpott et al. (2023) also reported an increase of sHSPs in a more tolerant genotype of potato. Therefore, sHSPs seem to be of major importance under drought and osmotic stress and might be relevant as chaperones under unfavorable conditions.

Other chaperones were also among those that were identified as DAPs during abiotic stress. Chaperone protein ClpD1 (Caseinolytic proteases 1; designated to ERD1 = early response to drought 1) under extreme drought being higher abundant in tolerant rice cultivar (Wu et al. 2016) was shown to induce salt and osmotic stress tolerance in transgenic *A. thaliana* plants (Mishra et al. 2016). In a heat stress memory study on Rhododendron, chaperone ClpB1 was increased in abundance after heat stress (Wang et al. 2020a).

Chaperones, HSP70s, and sHSPs were identified under several stressors in a vast and diverse set of species through proteomic studies. Therefore, it might be interesting to study the regulation of these as well as their mechanistic function and their target proteins or molecules to maintain protein function and rectify misfolding of proteins under different stressors. This might lead to a better understanding of tolerance mechanisms in plants.

2.6 Protein Metabolism

Proteasomes are essential for protein degradation and amino acid flux. These multi-subunit, multicatalytic proteases are of great importance during stress, when many proteins are incorrectly folded or get dehydrated and lose their function.

Filamentation temperature-sensitive H (FtsH) is an ATP-dependent zinc metalloprotease with ATPase as well as proteolysis activity and acts like a molecular chaperone. Several FtsHs have been reported to be differentially abundant during stress. These include FtsH6-protease, which was reduced after drought and heat stress, but increased during double stress (Zhao et al. 2016). Also, ATP-dependent FtsH2 was identified to be reduced in a sensitive chickpea cultivar (Vessal et al. 2020). In seedlings of *Brassica napus*, FtsH8 was increased after 10 days of drought stress (Koh et al. 2015). In bermudagrass (*Cynodon dactylon*), FtsH6 and FtsH2 were reduced, while FtsH1 did not show alteration upon drought stress (Fang et al. 2022).

Another group of proteins readily identified in different plant species under drought stress are ribosomal proteins. Ribosomes are complexes essential for translation and protein biosynthesis. In a rather sensitive alfalfa (*Medicago sativa*) genotype, more ribosomal proteins were reduced in comparison to a tolerant genotype, which the authors stated to indicate that the sensitive genotype displayed a stronger reduction in protein biosynthesis (Quan and Liu 2024). In a sensitive maize cultivar, a putative ribosomal protein S12 was downregulated after drought stress (Zhang et al. 2016). No alteration was identified in two wheat parental lines for a 60S ribosomal protein when comparing two genotypes (Nemati et al. 2019). In a rather drought-tolerant wheat cultivar, a 30S ribosomal protein and the proteasome alpha subunit protein were induced up to 24 h after drought induction, after which the abundance declined. However, the 20S proteasome beta 7 unit was induced up to 48 h and declined after rehydration of the plants (Cheng et al. 2016). These findings indicate that the alteration of protein biosynthesis might be an effective response by plants to deal with environmental stressors. However, data also suggest that the regulation is highly specific depending on the subunit of the proteasome and the time of stress.

Delta 1-pyrroline-5-carboxylate synthase (P5CS) is a critical enzyme in the production of proline and ornithine under stress and was identified in several studies. This protein was, for example, identified to be more abundant under drought stress as well as double stress but reduced under heat stress in a high-yielding maize cultivar (Zhao et al. 2016). Proline, an essential amino acid for protein assembly, was reported to be of great importance as it not only serves as an osmoprotectant during stress but also plays a role in ROS scavenging, as it was identified to have metal chelator properties and might therefore act as a molecular chaperone. However, it was also proposed that an accumulation of proline during stress has a signaling function as it alters specific gene expressions during the recovery phase from different stressors (for review on proline, see Ghosh et al. 2022). P5CS was also reported to be more abundant under e.g., heat stress in a heat-tolerant chickpea cultivar (Parankusam et al. 2017). Amino acids are an important factor under stress in terms of signaling as well as metabolic adaptation (for review, see Heinemann and Hildebrandt 2021). Therefore, an understanding of anabolism and catabolism of proteins as well as amino acids will be an important focus when studying abiotic stress tolerance in the future, as this might be an essential part of stress responses of plants.

2.7 Abscisic Acid Responsive Proteins

Proteomic studies identified several DAPs described as abscisic acid (ABA) responsive. ABA is an important hormone, which is essential for, e.g., drought stress responses. It was reported to be involved in many functions from stomatal closure to specific ABA-dependent signal transduction and responses (for review see Chen et al. 2020).

Among the ABA-responsive proteins are late embryogenesis abundant (LEA) proteins, which were identified to be essential during drought, heat, and cold stress. LEAs are classified according to their sequence similarity, such as a lack or low amount of Cys and Trp residues, high hydrophilicity, as well as an excess of Ala, Arg/Lys, Gly, Glu, and Thr content (Dure et al. 1989; Banerjee and Roychoudhury 2016; for review see Battaglia et al. 2008). These include 7 groups, where group 1–4 are the primary described LEAs from cotton, group 5 includes atypical and hydrophobic LEAs, group 6 proteins with sequence similarities to LEA18, and group 7 proteins with similarities to abscisic acid-, stress-, and ripening-induced 1 (ASR1). A LEA protein was identified to be increased in abundance after drought as well as salt stress in bermudagrass (Fang et al. 2022). Two LEA proteins of group 3 were found to be increased compared to control conditions under drought, heat, and double stress in maize leaves (Zhao et al. 2016).

Dehydrins belong to LEA group 2 and were first and foremost described to function as chaperones. However, recent studies have also shown that these proteins also regulate stress-responsive genes and were recently suggested to play a role in histone modification. Further, they were reported to function as chelators for, e.g., metals and as cryoprotectant, but also safeguard plasma membranes (PMs) when binding to PMs, as they were described to lead to higher stability (for review see Tiwari and Chakrabarty 2021). In a proteomic study, a dehydrin was found to be increased under drought stress in tomato (*Solanum lycopersicum*) (Liu et al. 2023) but also in two parental wheat lines divergently responding to drought (Nemati et al. 2019). Dehydrins ERD10 and ERD14-like were increased in abundance under drought stress in seedlings of *Brassica napus* as early as 3 and 7 days after drought onset, respectively (Koh et al. 2015). Therefore, LEA proteins, including dehydrins, are of utmost interest to further study drought and dehydration stress.

ASR proteins are transcription factors that have been reported to be involved in abiotic stress responses to divergent stressors. They have been reported in tomato for the first time; however, until now, homologs have been identified in several other species, including mono and dicots. ASRs have multiple functions, such as the involvement in fruit development and ripening or stress responses (Dominguez et al. 2021; for review, see Zhang et al. 2024). However, it is noteworthy that no orthologues were identified in *A. thaliana* (González and Iusem 2014; Zhang et al. 2024). ASR4 increased in tomato under drought (Liu et al. 2023). ASR4 was also reported to be increased in a rather drought-sensitive potato genotype in an in vitro experiment under osmotic stress (Bündig et al. 2016). In addition, when ZmASR3 was introduced to *A. thaliana*, a decrease in malondialdehyde (MDA) content and an increase in relative water content (RWC), as well as proline, were detected in mutant lines, which correlated with increased drought stress tolerance in these lines (Liang et al. 2019). Lower MDA and ROS levels, as well as an increase in proline and activity of ROS detoxification enzymes, were also shown in a rice cultivar over-expressing ASR6 (Srivastava et al. 2022). ASR5 was also reported to be more abundant after salt stress in a tolerant rice cultivar, whereas a sensitive cultivar displayed a lower abundance (Frukh et al. 2020). Therefore, ASR proteins are ideal candidates to be considered in further studies on stress tolerance in plants.

As water is an essential molecule for all living beings, special membrane pores have been identified, which facilitate the transport of small molecules such as water, as well as glycerol, but also volatiles such as NH_3 and CO_2. They are classified into five distinct subfamilies, including membrane intrinsic proteins (PIPs), tonoplast intrinsic proteins (TIPs), nodulin 26-like intrinsic proteins (NIPs), small basic intrinsic proteins (SIPs), and X intrinsic proteins (XIPs) (for review see Kapilan et al. 2018). In addition, they are an essential part of a plant's water status, and their functions in abiotic stress responses have been the focus for the last few decades (for review see Afzal et al. 2016). Aquaporin PIP 2–6 was identified under drought stress to be more abundant in a high-yielding maize hybrid and was not regulated under heat but increased under double stress (Zhao et al. 2016).

Aldehyde dehydrogenases (ALDHs) oxidize aldehydes, which are produced in many different forms under stress. They are essential to minimize ROS damage, for instance, to membranes, DNA, and damage to mitochondria (Tola et al. 2021). Many different ALDHs were identified during the last decades to have an important role in drought and desiccation tolerance. Plants contain 14 ALDHs protein families, of which the families ALDH11, ALDH12, ALDH19, ALDH21, ALDH22, ALDH23, and ALDH24 are only found in plants (Brocker et al. 2013). An increase of ALDH 7A1 was reported when comparing control to rewatered plants on a proteomic level in pandan (Amnan et al. 2022). In other proteomic studies on two parental wheat lines divergently responding to drought stress, no alteration under stress was identified for ALDH (Nemati et al. 2019). ALDH was also increased under drought stress in potato in a rather drought-sensitive genotype (Wellpott et al. 2023).

Many studies report 14-3-3 proteins to be involved in signal transduction during abiotic stress. 14-3-3 proteins have various functions, e.g., post-transcriptional and post-translational modifications of receptors, ion pumps, channels, and protein kinases (Cotelle and Leonhardt 2015; Zhao et al. 2021; Jiang et al. 2023) as well as hormone signaling (for review see Camoni et al. 2018). Koh et al. (2015) showed an increase in 14-3-3 gf14 psi, lambda, and nu; 14-3-3 gf14 omega displayed alternate responses within *Brassica napus*. In a tolerant maize cultivar, 14-3-3 g and f proteins were more abundant after drought stress, while a sensitive cultivar showed a higher abundance of g and b proteins under drought stress (Zhang et al. 2016). A high abundance of three 14-3-3 proteins was also reported by Zhao et al. (2018) for spinach under heat stress. As these proteins are essential for modifications of various proteins and, therefore, regulation of cell function, a time series of different species to study specific 14-3-3 proteins under different environmental conditions might lead to further insights into regulatory mechanisms regarding stress tolerance.

2.8 RNA Processing and Metabolism

Fast changes through the utilization of the spliceosome have been reported, leading to a change in the proteome. Alternative splicing produces multiple transcripts and is an important response to different environmental conditions (Laloum et al. 2018).

Glycine-rich RNA-binding proteins (GRPs) were identified in many proteomic studies to be differentially abundant under drought stress. In molecular studies, it was shown that GRP7 was highly affected by alternative splicing (Du et al. 2022) and GRP1 was proposed to be a novel splicing factor (Gao et al. 2019). Under drought stress, GRP2 was less abundant in a proteomic study on maize (Zhao et al. 2016). However, in rice, a GRP was identified to be more abundant under mild drought stress (field capacity (FC) 70%) while it decreased under moderate drought stress (FC of 50%) (Habibpourmehraban et al. 2022). Increase of expression of GRP2 and GRP7 was also reported to be beneficial under drought stress for rice, both in vegetative and in generative growth stage, leading to an increased grain yield (Yang et al. 2014a). Also, under heat stress in a soybean cultivar, a GRP was identified to be more abundant under heat stress (Katam et al. 2020). Kim et al. (2008) showed that GRP7 was induced under cold stress in *A. thaliana*; however, it had a negative impact on drought and salt stress responses as it was involved in the opening and closing of stomata. The grp7 knockout and knock-down mutants had a higher seedling fresh weight under control as well as osmotic and salt stress conditions compared to the wild type (Kim et al. 2008). Together, this leads to the hypothesis that protein components of the spliceosome and post-transcriptional modification apparatus might be an integral part of stress responses and correct adjustment to the severity of drought stress. Therefore, further analysis of spliceosome components under different stressors might be interesting to gain further insights into stress responses based on alterations in spliceosome function.

3 General and Specific Responses of Plants to Drought Stress

In the last decade, proteomic research on abiotic stress tolerance has gained helpful data to further understand the response of plants at a certain time point under stress. Together with metabolomics, transcriptomics, genomics, and further omic approaches, the overall knowledge displays a rather complex picture of protein abundances and pathways, which are dynamically altered during the onset and intensification as well as the release of abiotic stress. However, as drought stress responses are variable between species and even genotypes, a general stress response is hard to decipher. Nevertheless, proteins described are differentially abundant throughout many plant species under stress and are therefore displaying an overhead response mechanism when battling abiotic stress, even though the specific response might vary between species. However, it is noteworthy that it is still very difficult to compare abiotic stress experiments throughout the literature, as the intensity and duration of stress application, as well as the system (in vitro experiments, climate chamber, greenhouse, or field experiments), make a correlation of findings difficult.

Knowledge of specific proteins, which contribute to the adaptation to unfavorable conditions for plant growth and yield, is essential to understand which alterations in plants are favorable for developing strategies to breed for tolerance. An overview of recently published proteins, which show up under different stressors, is a first start to generate a picture of abiotic stress tolerance on the proteomic level (Fig. 1). However, until now, most studies have only included one stressor, are limited to a few time points, and only analyze one or very few stress levels. Nevertheless, recent advances display a need to understand responses also to a wide variety of multiple stressors. Usually, unfavorable conditions are coupled, such as drought and heat stress; double or multiple stressors should be applied, as plant responses to these might drastically vary. Another important part is stress memory (for a mini review on stress priming and proteomics, see Tanou et al. 2012). Priming of plants might be a way of imprinting specific favorable stress responses; whether this is also possible over long periods or even generations is poorly understood. Through the advances in subcellular proteomics (organelles) (Komatsu 2019) and single-cell proteomics (for review see Bennett et al. 2023), they are becoming more interesting as these help to understand processes better (Kausar et al. 2022). Further, modifications such as phosphorylation of complete proteomes should be shifted into focus to unravel the regulation of proteins and stress-induced signaling under drought (Chen et al. 2020; Zhang et al. 2023).

Nevertheless, when drought occurs, plants adapt by responding in a very defined way depending on the severity, longitude, etc., of the stress and plant species analyzed. While studying the proteome, it should be mentioned that a payoff with yield

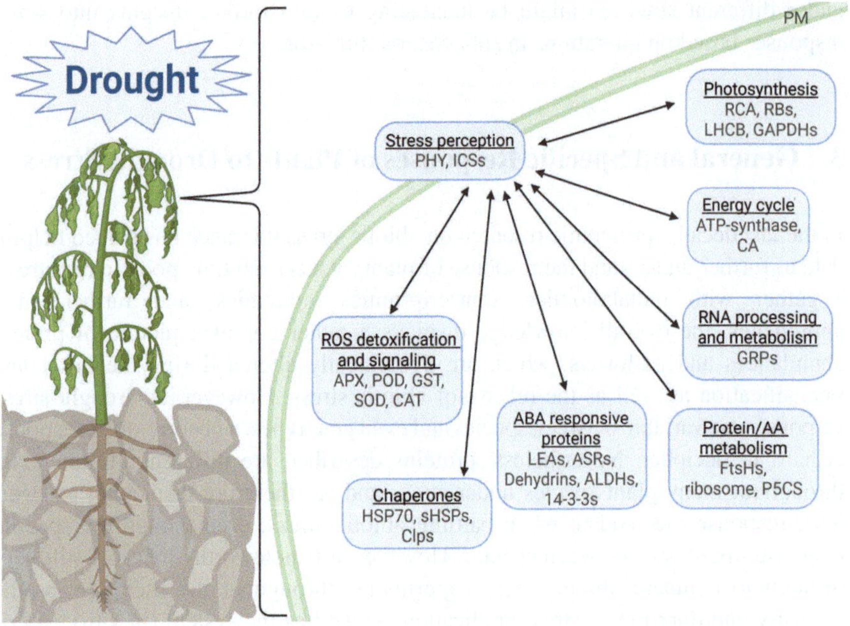

Fig. 1 Overview of DAPs of leaf proteome in multiple proteomic studies under drought stress. PHY = phytochrome, ICSs = intramolecular crowding sensors, RCA = RuBisCO activase, RBs = RuBisCO-binding proteins, LHCB = chlorophyll a-b binding proteins, GAPDHs = glyceraldehyde-3-phosphate dehydrogenases, CA = carbonic anhydrase, APX = ascorbate peroxidase, POD = peroxidase, GST = gluthatione S-transferase, SOD = superoxide dismutase, CAT = catalase, HSP70 = heat shock protein 70 family members, sHSPs = small heat shock proteins, Clp = caseinolytic proteases, FtsH = filamentation temperature-sensitive H, P5CS = 1-pyrroline-5-carboxylate synthase, LEAs = late embryogenic abundant protein, ASRs = abscisic acid- stress-, ripening induced proteins, ALDHs = aldehyde dehydrogenases, 14-3-3s = 14-3-3 proteins, GRPs = glycine-rich RNA-binding protein. ABA = Abscisic acid, AA = Amino acids. Note: ICSs are not present in studies. Proteins might interact; due to better readability, not all are linked. (Figure created in BioRender)

will most likely occur, as the production of proteins that are favorable under stress is often accompanied by a growth arrest due to a decrease in photosynthesis and other important processes. Therefore, it would be ideal to generate a complete picture by facilitating multi-omics studies. These studies might open a new field for proteomic comparisons to further unravel stress responses of plants.

4 Conclusion

Proteomic analyses are promising approaches to studying the current state of the plant under diverse environmental conditions. This is shifting more into focus as climate change progresses and unfavorable conditions arise with increased severity and frequency. Through time series, proteomics also holds the power to analyze the velocity of plants/genotypes to adapt over time to abiotic stress, as it shows the actual state of the protein makeup. Several proteins have been identified throughout several different studies, which might benefit the plant under abiotic stress, but might stand in a payoff situation with yield. These include proteins without homologs in *A. thaliana*, which makes it relevant to screen several plant species on a proteomic level under different stressors and not only rely on a few model plants. The study of the proteome offers a different approach compared to analyzing the state of the plant through the transcriptome or genome. While the transcriptome and genome reveal the plant's potential, factors such as alternative splicing or mRNA degradation under stress can change the protein composition. This, in turn, can alter the plant's response. Further, it will also be of great interest to check proteins for post-translational modifications, such as methylation, to determine the activity of the specific proteins identified through proteomic studies. This will further improve the picture of a plant's response under different stressors. In summary, proteomic studies are essential to generate a complete picture in combination with transcriptomic, metabolomics, epigenomics, etc., approaches, as well as studying post-translational modifications to further unravel specific plant responses.

Acknowledgments Language was refined with the assistance of an artificial intelligence (AI) tool, GPT-4.1, OpenAI, 2025.

Every section was carefully reviewed, revised, and approved, and where necessary, altered by the author to ensure that the text is appropriate concerning content and scientific correctness and maintains the integrity of the original content.

Conflict of Interest The author declares no conflict of interest.

References

Afzal Z, Howton TC, Sun Y, Mukhtar MS (2016) The roles of aquaporins in plant stress responses. J Dev Biol 4(1):9. https://doi.org/10.3390/jdb4010009

Ahsan N, Donnart T, Nouri M-Z, Komatsu S (2010) Tissue-specific defense and thermo-adaptive mechanisms of soybean seedlings under heat stress revealed by proteomic approach. J Proteome Res 9(8):S4189–S4204. https://doi.org/10.1021/pr100504j

Ainsworth EA (2017) Understanding and improving global crop response to ozone pollution. Plant J 90(5):S886–S897. https://doi.org/10.1111/tpj.13298

Amnan MAM, Aizat WM, Khaidizar FD, Tan BC (2022) Drought stress induces morpho-physiological and proteome changes of *Pandanus amaryllifolius*. Plants 11(2):221. https://doi.org/10.3390/plants11020221

Arif M, Fiaz S, Kandegama WMWW, Sheth S, Li L (2025) Evaluating the impacts of environmental stresses on agriculture in the context of climate resilience. Plant Mol Biol 115(4):S73. https://doi.org/10.1007/s11103-025-01598-2

Banerjee A, Roychoudhury A (2016) Group II late embryogenesis abundant (LEA) proteins: structural and functional aspects in plant abiotic stress. Plant Growth Regul 79(1):S1–S17. https://doi.org/10.1007/s10725-015-0113-3

Battaglia M, Olvera-Carrillo Y, Garciarrubio A, Campos F, Covarrubias AA (2008) The enigmatic LEA proteins and other hydrophilins. Plant Physiol 148(1):S6–S24. https://doi.org/10.1104/pp.108.120725

Bennett HM, Stephenson W, Rose CM, Darmanis S (2023) Single-cell proteomics enabled by next-generation sequencing or mass spectrometry. Nat Methods 20(3):S363–S374. https://doi.org/10.1038/s41592-023-01791-5

Bista DR, Heckathorn SA, Jayawardena DM, Mishra S, Boldt JK (2018) Effects of drought on nutrient uptake and the levels of nutrient-uptake proteins in roots of drought-sensitive and -tolerant grasses. Plants 7(2):28. https://doi.org/10.3390/plants7020028

Boggs JZ, Loewy K, Bibee K, Heschel MS (2010) Phytochromes influence stomatal conductance plasticity in *Arabidopsis thaliana*. Plant Growth Regul 60(2):S77–S81. https://doi.org/10.1007/s10725-009-9427-3

Brocker C, Vasiliou M, Carpenter S, Carpenter C, Zhang Y, Wang X et al (2013) Aldehyde dehydrogenase (ALDH) superfamily in plants: gene nomenclature and comparative genomics. Planta 237(1):S189–S210. https://doi.org/10.1007/s00425-012-1749-0

Bündig C, Jozefowicz AM, Mock H-P, Winkelmann T (2016) Proteomic analysis of two divergently responding potato genotypes (*Solanum tuberosum* L.) following osmotic stress treatment in vitro. J Proteome 143:S227–S241. https://doi.org/10.1016/j.jprot.2016.04.048

Camoni L, Visconti S, Aducci P, Marra M (2018) 14-3-3 Proteins in plant hormone signaling: doing several things at once. Front Plant Sci 9:297. Online verfügbar unter https://www.frontiersin.org/journals/plant-science/articles/10.3389/fpls.2018.00297

Chen K, Li G-J, Bressan RA, Song C-P, Zhu J-K, Zhao Y (2020) Abscisic acid dynamics, signaling, and functions in plants. J Integr Plant Biol 62(1):S25–S54. https://doi.org/10.1111/jipb.12899

Cheng L, Wang Y, He Q, Li H, Zhang X, Zhang F (2016) Comparative proteomics illustrates the complexity of drought resistance mechanisms in two wheat (*Triticum aestivum L.*) cultivars under dehydration and rehydration. BMC Plant Biol 16(1):S188. https://doi.org/10.1186/s12870-016-0871-8

Chintakovid N, Maipoka M, Phaonakrop N, Mickelbart MV, Roytrakul S, Chadchawan S (2017) Proteomic analysis of drought-responsive proteins in rice reveals photosynthesis-related adaptations to drought stress. Acta Physiol Plant 39(10):S240. https://doi.org/10.1007/s11738-017-2532-4

Cotelle V, Leonhardt N (2015) 14-3-3 proteins in guard cell signaling. Front Plant Sci 6:S1210. https://doi.org/10.3389/fpls.2015.01210

Dominguez PG, Conti G, Duffy T, Insani M, Alseekh S, Asurmendi S et al (2021) Multiomics analyses reveal the roles of the ASR1 transcription factor in tomato fruits. J Exp Bot 72(18):S6490–S6509. https://doi.org/10.1093/jxb/erab269

Dorone Y, Boeynaems S, Flores E, Jin B, Hateley S, Bossi F et al (2021) A prion-like protein regulator of seed germination undergoes hydration-dependent phase separation. Cell 184(16):4284–4298.e27. https://doi.org/10.1016/j.cell.2021.06.009

Du C, Bai H-y, Chen J-j, Wang J-h, Wang Z-f, Zhang Z-h (2022) Alternative splicing regulation of glycine-rich proteins via target of rapamycin-reactive oxygen species pathway in Arabidopsis seedlings upon glucose stress. Front Plant Sci 13:830140. https://doi.org/10.3389/fpls.2022.830140

Dure L, Crouch M, Harada J, Ho T-HD, Mundy J, Quatrano R et al (1989) Common amino acid sequence domains among the LEA proteins of higher plants. Plant Mol Biol 12(5):S475–S486. https://doi.org/10.1007/BF00036962

Fang Z, Liu C, Ye T, Huang C, Zhang Z, Wang Y, Chan Z (2022) Integrated transcriptome and proteome analyses provide insight into abiotic stress crosstalks in bermudagrass. Environ Exp Bot 199:S104864. https://doi.org/10.1016/j.envexpbot.2022.104864

Fink AH, Brücher T, Krüger A, Leckebusch GC, Pinto JG, Ulbrich U (2004) The 2003 European summer heatwaves and drought –synoptic diagnosis and impacts. Weather 59(8):S209–S216. https://doi.org/10.1256/wea.73.04

Frukh A, Siddiqi TO, Khan MIR, Ahmad A (2020) Modulation in growth, biochemical attributes and proteome profile of rice cultivars under salt stress. Plant Physiol Biochem 146:S55–S70. https://doi.org/10.1016/j.plaphy.2019.11.011

Galmés J, Aranjuelo I, Medrano H, Flexas J (2013) Variation in Rubisco content and activity under variable climatic factors. Photosynth Res 117(1–3):S73–S90. https://doi.org/10.1007/s11120-013-9861-y

Gao X, Yin C, Liu X, Peng J, Chen D, He D et al (2019) A glycine-rich protein MoGrp1 functions as a novel splicing factor to regulate fungal virulence and growth in *Magnaporthe oryzae*. Phytopathol Res 1(1):S2. https://doi.org/10.1186/s42483-018-0007-1

Ghosh D, Xu J (2014) Abiotic stress responses in plant roots: a proteomics perspective. Front Plant Sci 5:6. https://doi.org/10.3389/fpls.2014.00006

Ghosh UK, Islam MN, Siddiqui MN, Cao X, Khan MAR (2022) Proline, a multifaceted signalling molecule in plant responses to abiotic stress: understanding the physiological mechanisms. Plant Biol 24(2):227–239. https://doi.org/10.1111/plb.13363

González RM, Iusem ND (2014) Twenty years of research on Asr (ABA-stress-ripening) genes and proteins. Planta 239(5):941–949. https://doi.org/10.1007/s00425-014-2039-9

Goussi R, Manfredi M, Marengo E, Derbali W, Cantamessa S, Barbato R, Manaa A (2021) Thylakoid proteome variation of *Eutrema salsugineum* in response to drought and salinity combined stress. Biochim Biophys Acta Bioenerg 1862(12):S148482. https://doi.org/10.1016/j.bbabio.2021.148482

Guo W, Hangyu Dai, Junhao Qian, Jinglu Tan, Zhenyu Xu, Ya Guo (2024) An assessment of the relationship between spring frost indicators and global crop yield losses. Sci Total Environ 954:S176560. https://doi.org/10.1016/j.scitotenv.2024.176560

Gupta S, Mishra SK, Misra S, Pandey V, Agrawal L, Nautiyal CS, Chauhan PS (2020) Revealing the complexity of protein abundance in chickpea root under drought-stress using a comparative proteomics approach. Plant Physiol Biochem 151:S88–S102. https://doi.org/10.1016/j.plaphy.2020.03.005

Habibpourmehraban F, Atwell BJ, Haynes PA (2022) Unique and shared proteome responses of Rice plants (*Oryza sativa*) to individual abiotic stresses. Int J Mol Sci 23(24):15552. https://doi.org/10.3390/ijms232415552

Hamilton ES, Jensen GS, Maksaev G, Katims A, Sherp AM, Haswell ES (2015) Mechanosensitive channel MSL8 regulates osmotic forces during pollen hydration and germination. Science 350(6259):S438–S441. https://doi.org/10.1126/science.aac6014

Heinemann B, Hildebrandt TM (2021) The role of amino acid metabolism in signaling and metabolic adaptation to stress-induced energy deficiency in plants. J Exp Bot 72(13):S4634–S4645. https://doi.org/10.1093/jxb/erab182

Hildebrandt T, Knuesting J, Berndt C, Morgan B, Scheibe R (2015) Cytosolic thiol switches regulating basic cellular functions: GAPDH as an information hub? Biol Chem 396(5):S523–S537. https://doi.org/10.1515/hsz-2014-0295

Hussain M, Malik MA, Farooq M, Ashraf MY, Cheema MA (2008) Improving drought tolerance by exogenous application of glycinebetaine and salicylic acid in sunflower. J Agron Crop Sci 194(3):S193–S199. https://doi.org/10.1111/j.1439-037X.2008.00305.x

Hwang NR, Yim S-H, Kim YM, Jeong J, Song EJ, Lee Y et al (2009) Oxidative modifications of glyceraldehyde-3-phosphate dehydrogenase play a key role in its multiple cellular functions. Biochem J 423(2):S253–S264. https://doi.org/10.1042/BJ20090854

Ionita M, Tallaksen LM, Kingston DG, Stagge JH, Laaha G, van Lanen HAJ et al (2017) The European 2015 drought from a climatological perspective. Hydrol Earth Syst Sci 21(3):1397–1419. https://doi.org/10.5194/hess-21-1397-2017

Jiang C, Bi Y, Mo J, Zhang R, Qu M, Feng S, Essemine J (2020) Proteome and transcriptome reveal the involvement of heat shock proteins and antioxidant system in thermotolerance of *Clematis Florida*. Sci Rep 10(1):S8883. https://doi.org/10.1038/s41598-020-65699-2

Jiang W, He J, Babla M, Wu T, Tong T, Riaz A et al (2023) Molecular evolution and interaction of 14-3-3 proteins with H+-ATPases in plant abiotic stresses. J Exp Bot 75(3):S689–S707. https://doi.org/10.1093/jxb/erad414

Kapilan R, Vaziri M, Zwiazek JJ (2018) Regulation of aquaporins in plants under stress. Biol Res 51(1):S4. https://doi.org/10.1186/s40659-018-0152-0

Katam R, Shokri S, Murthy N, Singh SK, Suravajhala P, Khan MN et al (2020) Proteomics, physiological, and biochemical analysis of cross tolerance mechanisms in response to heat and water stresses in soybean. PLoS One 15(6):e0233905. https://doi.org/10.1371/journal.pone.0233905

Kausar R, Wang X, Komatsu S (2022) Crop proteomics under abiotic stress: from data to insights. Plants 11(21):2877. https://doi.org/10.3390/plants11212877

Kim JS, Jung HJ, Lee HJ, Kim KA, Goh C-H, Woo Y et al (2008) Glycine-rich RNA-binding protein7 affects abiotic stress responses by regulating stomata opening and closing in *Arabidopsis thaliana*. Plant J 55(3):S455–S466. https://doi.org/10.1111/j.1365-313X.2008.03518.x

Koh J, Chen G, Yoo M-J, Zhu N, Dufresne D, Erickson JE et al (2015) Comparative proteomic analysis of *Brassica napus* in response to drought stress. J Proteome Res 14(8):S3068–S3081. https://doi.org/10.1021/pr501323d

Komatsu S (2019) Plant proteomic research 2.0: trends and perspectives. Int J Mol Sci 20(10):2495. https://doi.org/10.3390/ijms20102495

Kornhuber K, Osprey S, Coumou D, Petri S, Petoukhov V, Rahmstorf S, Gray L (2019) Extreme weather events in early summer 2018 connected by a recurrent hemispheric wave-7 pattern. Environ Res Lett 14(5):S54002. https://doi.org/10.1088/1748-9326/ab13bf

Kosová K, Vítámvás P, Prášil IT, Renaut J (2011) Plant proteome changes under abiotic stress—contribution of proteomics studies to understanding plant stress response. J Proteome 74(8):S1301–S1322. https://doi.org/10.1016/j.jprot.2011.02.006

Krasensky J, Jonak C (2012) Drought, salt, and temperature stress-induced metabolic rearrangements and regulatory networks. J Exp Bot 63(4):S1593–S1608. https://doi.org/10.1093/jxb/err460

Laloum T, Martín G, Duque P (2018) Alternative splicing control of abiotic stress responses. Trends Plant Sci 23(2):S140–S150. https://doi.org/10.1016/j.tplants.2017.09.019

Li S (2023) Novel insight into functions of ascorbate peroxidase in higher plants: more than a simple antioxidant enzyme. Redox Biol 64:102789. https://doi.org/10.1016/j.redox.2023.102789

Li S, Liu F (2022) Chapter 2 – Crop exposure to drought stress under elevated CO_2: responses in physiological, biochemical, and molecular levels. In: Liu F, Li X, Hogy P, Jiang D, Brestic M, Liu B (eds) Sustainable crop productivity and quality under climate change. Academic Press, pp S21–S42. Online verfügbar unter https://www.sciencedirect.com/science/article/pii/B9780323854498000129

Li J, Essemine J, Shang C, Zhang H, Zhu X, Yu J et al (2020) Combined proteomics and metabolism analysis unravels prominent roles of antioxidant system in the prevention of alfalfa (*Medicago sativa* L.) against salt stress. Int J Mol Sci 21(3):909. https://doi.org/10.3390/ijms21030909

Li H-J, Wang Y-F, Zhao C-F, Yang M, Wang G-X, Zhang R-H (2021a) The quantitative proteomic analysis provides insight into the effects of drought stress in maize. Photosynthetica 59(1):S1–S11. https://doi.org/10.32615/ps.2020.078

Li H, Yang M, Zhao C, Wang Y, Zhang R (2021b) Physiological and proteomic analyses revealed the response mechanisms of two different drought-resistant maize varieties. BMC Plant Biol 21(1):S513. https://doi.org/10.1186/s12870-021-03295-w

Liang Y, Jiang Y, Du M, Li B, Chen L, Chen M et al (2019) ZmASR3 from the maize ASR gene family positively regulates drought tolerance in transgenic Arabidopsis. Int J Mol Sci 20(9):2278. https://doi.org/10.3390/ijms20092278

Liu M, Zhao G, Huang X, Pan T, Chen W, Qu M et al (2023) Candidate regulators of drought stress in tomato revealed by comparative transcriptomic and proteomic analyses. Front Plant Sci 14:1282718. https://doi.org/10.3389/fpls.2023.1282718

Lottspeich F (1999) Proteome analysis: a pathway to the functional analysis of proteins. Angew Chem Int Ed Engl 38(17):S2476–S2492. https://doi.org/10.1002/(sici)1521-3773(19990903)38:17<2476::aid-anie2476>3.0.co;2-s

Lu Y, Li R, Wang R, Wang X, Zheng W, Sun Q et al (2017) Comparative proteomic analysis of flag leaves reveals new insight into wheat heat adaptation. Front Plant Sci 8:1086. https://doi.org/10.3389/fpls.2017.01086

Meneses-Reyes GI, Rodriguez-Bustos DL, Cuevas-Velazquez CL (2024) Macromolecular crowding sensing during osmotic stress in plants. Trends Biochem Sci 49(6):S480–S493. https://doi.org/10.1016/j.tibs.2024.02.002

Michaletti A, Naghavi MR, Toorchi M, Zolla L, Rinalducci S (2018) Metabolomics and proteomics reveal drought-stress responses of leaf tissues from spring-wheat. Sci Rep 8(1):S5710. https://doi.org/10.1038/s41598-018-24012-y

Minhas PS, Rane J, Pasala RK (2017) Abiotic stresses in agriculture: an overview. In: Minhas PS, Rane J, Pasala RK (eds) Abiotic stress management for resilient agriculture. Springer, Singapore, pp S3–S8

Mishra RC, Richa, Grover A (2016) Constitutive over-expression of rice ClpD1 protein enhances tolerance to salt and desiccation stresses in transgenic *Arabidopsis* plants. Plant Sci 250:S69–S78. https://doi.org/10.1016/j.plantsci.2016.06.004

Mittler R (2017) ROS are good. Trends Plant Sci 22(1):S11–S19. https://doi.org/10.1016/j.tplants.2016.08.002

Mohammad F, Saquib Q, Alatar AA, Al-Khedhairy AA (2020) Cellular and molecular phytotoxicity of heavy metals. Springer International Publishing, Cham

Moloi SJ, Alqarni AO, Brown AP, Goche T, Shargie NG, Moloi MJ et al (2024) Comparative physiological, bio-chemical, and leaf proteome responses of contrasting wheat varieties to drought stress. Plants 13(19):2797. https://doi.org/10.3390/plants13192797

Nemati M, Piro A, Norouzi M, Vahed MM, Nisticò DM, Mazzuca S (2019) Comparative physiological and leaf proteomic analyses revealed the tolerant and sensitive traits to drought stress in two wheat parental lines and their F6 progenies. Environ Exp Bot 158:S223–S237. https://doi.org/10.1016/j.envexpbot.2018.10.024

Ogden AJ, Abdali S, Engbrecht KM, Zhou M, Handakumbura PP (2020) Distinct Preflowering drought tolerance strategies of *Sorghum bicolor* genotype RTx430 revealed by subcellular protein profiling. Int J Mol Sci 21(24):9706. https://doi.org/10.3390/ijms21249706

Parankusam S, Bhatnagar-Mathur P, Sharma KK (2017) Heat responsive proteome changes reveal molecular mechanisms underlying heat tolerance in chickpea. Environ Exp Bot 141:S132–S144. https://doi.org/10.1016/j.envexpbot.2017.07.007

Pirasteh-Anosheh H, Saed-Moucheshi A, Pakniyat H, Pessarakli M (2016) Stomatal responses to drought stress. In: Water stress and crop plants. Wiley, pp S24–S40

Polishchuk OV (2021) Stress-related changes in the expression and activity of plant carbonic anhydrases. Planta 253(2):S58. https://doi.org/10.1007/s00425-020-03553-5

Quan W, Liu X (2024) Tandem mass tag (TMT)-based quantitative proteomics analysis reveals the different responses of contrasting alfalfa varieties to drought stress. BMC Genomics 25(1):S806. https://doi.org/10.1186/s12864-024-10702-7

Rudenko, Natalia N, Borisova-Mubarakshina MM, Ignatova LK, Fedorchuk TP, Nadeeva-Zhurikova EM, Ivanov BN (2020) Role of plant carbonic anhydrases under stress conditions. 16. In: Hossain A (ed) Plant stress physiology. IntechOpen, Rijeka

Satrio RD, Fendiyanto MH, Miftahudin M (2024) Tools and techniques used at global scale through genomics, transcriptomics, proteomics, and metabolomics to investigate plant stress responses at the molecular level. In: Shahid M, Gaur R (eds) Molecular dynamics of plant stress and its management. Springer Nature, Singapore, pp S555–S607

Spreitzer RJ, Salvucci ME (2002) RUBISCO: structure, regulatory interactions, and possibilities for a better enzyme. Annu Rev Plant Biol 53:S449–S475. https://doi.org/10.1146/annurev.arplant.53.100301.135233

Srivastava D, Verma G, Chawda K, Chauhan AS, Pande V, Chakrabarty D (2022) Overexpression of Asr6, abscisic acid stress-ripening protein, enhances drought tolerance and modulates gene expression in rice (*Oryza sativa* L.). Environ Exp Bot 202:S105005. https://doi.org/10.1016/j.envexpbot.2022.105005

Sung D-Y, Kaplan F, Guy CL (2001) Plant Hsp70 molecular chaperones: protein structure, gene family, expression and function. Physiol Plant 113(4):S443–S451. https://doi.org/10.1034/j.1399-3054.2001.1130402.x

Takahashi S, Murata N (2006) Glycerate-3-phosphate, produced by CO_2 fixation in the Calvin cycle, is critical for the synthesis of the D1 protein of photosystem II. Biochim Biophys Acta 1757(3):S198–S205. https://doi.org/10.1016/j.bbabio.2006.02.002

Tan ZX, Lal R, Wiebe KD (2005) Global soil nutrient depletion and yield reduction. J Sustain Agric 26(1):S123–S146. https://doi.org/10.1300/J064v26n01_10

Tanou G, Fotopoulos V, Molassiotis A (2012) Priming against environmental challenges and proteomics in plants: update and agricultural perspectives. Front Plant Sci 3:216. https://doi.org/10.3389/fpls.2012.00216

Thikekar AK, Rathod VS, Panchal VP, Raut SA, Raut RS, Jain KS (2023) A review on-analytical tools in proteomics. J Proteins Proteomics 14(3):S201–S221. https://doi.org/10.1007/s42485-023-00108-6

Tiwari P, Chakrabarty D (2021) Dehydrin in the past four decades: from chaperones to transcription co-regulators in regulating abiotic stress response. Curr Res Biotechnol 3:S249–S259. https://doi.org/10.1016/j.crbiot.2021.07.005

Tola AJ, Jaballi A, Germain H, Missihoun TD (2021) Recent development on plant aldehyde dehydrogenase enzymes and their functions in plant development and stress signaling. Genes 12(1):51. https://doi.org/10.3390/genes12010051

Vessal S, Arefian M, Siddique KHM (2020) Proteomic responses to progressive dehydration stress in leaves of chickpea seedlings. BMC Genomics 21(1):S523. https://doi.org/10.1186/s12864-020-06930-2

Vierling E (1997) The small heat shock proteins in plants are members of an ancient family of heat induced proteins. Acta Physiol Plant 19(4):S539–S547. https://doi.org/10.1007/s11738-997-0051-4

Wang L, Jin X, Li Q, Wang X, Li Z, Wu X (2016a) Comparative proteomics reveals that phosphorylation of β carbonic anhydrase 1 might be important for adaptation to drought stress in *Brassica napus*. Sci Rep 6(1):S39024. https://doi.org/10.1038/srep39024

Wang X, Cai X, Xu C, Wang Q, Dai S (2016b) Drought-responsive mechanisms in plant leaves revealed by proteomics. Int J Mol Sci 17(10):1706. https://doi.org/10.3390/ijms17101706

Wang X, Li Z, Liu B, Zhou H, Elmongy MS, Xia Y (2020a) Combined proteome and transcriptome analysis of heat-primed azalea reveals new insights into plant heat acclimation memory. Front Plant Sci 11:1278. https://doi.org/10.3389/fpls.2020.01278

Wang Y, Li X, Liu N, Wei S, Wang J, Qin F, Suo B (2020b) The iTRAQ-based chloroplast proteomic analysis of *Triticum aestivum* L. leaves subjected to drought stress and 5-aminolevulinic acid alleviation reveals several proteins involved in the protection of photosynthesis. BMC Plant Biol 20(1):S96. https://doi.org/10.1186/s12870-020-2297-6

Wang B, Zhang H, Huai J, Peng F, Wu J, Lin R, Fang X (2022) Condensation of SEUSS promotes hyperosmotic stress tolerance in *Arabidopsis*. Nat Chem Biol 18(12):S1361–S1369. https://doi.org/10.1038/s41589-022-01196-z

Wang Y, Li S, Mokbel M, May AI, Liang Z, Zeng Y et al (2024) Biomolecular condensates mediate bending and scission of endosome membranes. Nature 634(8036):S1204–S1210. https://doi.org/10.1038/s41586-024-07990-0

Wellpott K, Jozefowicz AM, Meise P, Schum A, Seddig S, Mock H-P, Winkelmann T, Bündig C (2023) Combined nitrogen and drought stress leads to overlapping and unique proteomic responses in potato. Planta 257(3):S58. https://doi.org/10.1007/s00425-023-04085-4

Wilkins MR, Sanchez J-C, Gooley AA, Appel RD, Humphery-Smith I, Hochstrasser DF, Williams KL (1996) Progress with proteome projects: why all proteins expressed by a genome should be identified and how to do it. Biotechnol Genet Eng Rev 13(1):S19–S50. https://doi.org/10.1080/02648725.1996.10647923

Wu Y, Mirzaei M, Pascovici D, Chick JM, Atwell BJ, Haynes PA (2016) Quantitative proteomic analysis of two different rice varieties reveals that drought tolerance is correlated with reduced

abundance of photosynthetic machinery and increased abundance of ClpD1 protease. J Proteome 143:S73–S82. https://doi.org/10.1016/j.jprot.2016.05.014
Yahoueian SH, Bihamta MR, Babaei HR, Bazargani MM (2021) Proteomic analysis of drought stress response mechanism in soybean (*Glycine max* L.) leaves. Food Sci Nutr 9(4):S2010–S2020. https://doi.org/10.1002/fsn3.2168
Yang DH, Kwak KJ, Kim MK, Park SJ, Yang K-Y, Kang H (2014a) Expression of Arabidopsis glycine-rich RNA-binding protein AtGRP2 or AtGRP7 improves grain yield of rice (*Oryza sativa*) under drought stress conditions. Plant Sci 214:S106–S112. https://doi.org/10.1016/j.plantsci.2013.10.006
Yang D, Du X, Yang Z, Liang Z, Guo Z, Liu Y (2014b) Transcriptomics, proteomics, and metabolomics to reveal mechanisms underlying plant secondary metabolism. Eng Life Sci 14(5):S456–S466. https://doi.org/10.1002/elsc.201300075
Yang B, Tang J, Yu Z, Khare T, Srivastav A, Datir S, Kumar V (2019) Light stress responses and prospects for engineering light stress tolerance in crop plants. J Plant Growth Regul 38(4):S1489–S1506. https://doi.org/10.1007/s00344-019-09951-8
Zaffagnini M, Fermani S, Costa A, Lemaire SD, Trost P (2013) Plant cytoplasmic GAPDH: redox post-translational modifications and moonlighting properties. Front Plant Sci 4:450. https://doi.org/10.3389/fpls.2013.00450
Zenda T, Liu S, Wang X, Jin H, Liu G, Duan H (2018) Comparative proteomic and physiological analyses of two divergent maize inbred lines provide more insights into drought-stress tolerance mechanisms. Int J Mol Sci 19(10):3225. https://doi.org/10.3390/ijms19103225
Zhang N, Portis AR (1999) Mechanism of light regulation of Rubisco: a specific role for the larger Rubisco activase isoform involving reductive activation by thioredoxin-f. Proc Natl Acad Sci 96(16):S9438–S9443. https://doi.org/10.1073/pnas.96.16.9438
Zhang H, Ni Z, Chen Q, Guo Z, Gao W, Su X, Qu Y (2016) Proteomic responses of drought-tolerant and drought-sensitive cotton varieties to drought stress. Mol Genet Genomics 291(3):S1293–S1303. https://doi.org/10.1007/s00438-016-1188-x
Zhang WJ, Zhou Y, Zhang Y, Su YH, Xu T (2023) Protein phosphorylation: a molecular switch in plant signaling. Cell Rep 42(7):S112729. https://doi.org/10.1016/j.celrep.2023.112729
Zhang Y, Wang M, Kitashov AV, Yang L (2024) Development history, structure, and function of ASR (Abscisic Acid-Stress-Ripening) transcription factor. Int J Mol Sci 25(19):10283. https://doi.org/10.3390/ijms251910283
Zhao Y, Wang Y, Yang H, Wang W, Wu J, Hu X (2016) Quantitative proteomic analyses identify ABA-related proteins and signal pathways in maize leaves under drought conditions. Front Plant Sci 7:1827. https://doi.org/10.3389/fpls.2016.01827
Zhao Q, Chen W, Bian J, Xie H, Li Y, Xu C et al (2018) Proteomics and phosphoproteomics of heat stress-responsive mechanisms in spinach. Front Plant Sci 9:800. https://doi.org/10.3389/fpls.2018.00800
Zhao X, Li F, Li K (2021) The 14-3-3 proteins: regulators of plant metabolism and stress responses. Plant Biol 23(4):S531–S539. https://doi.org/10.1111/plb.13268
Zhou W, Chen F, Meng Y, Chandrasekaran U, Luo X, Yang W, Shu K (2020) Plant waterlogging/flooding stress responses: from seed germination to maturation. Plant Physiol Biochem 148:S228–S236. https://doi.org/10.1016/j.plaphy.2020.01.020
Zhu Y, Jia X, Wu Y, Hu Y, Cheng L, Zhao T et al (2020) Quantitative proteomic analysis of *Malus halliana* exposed to salt-alkali mixed stress reveals alterations in energy metabolism and stress regulation. Plant Growth Regul 90(2):S205–S222. https://doi.org/10.1007/s10725-019-00563-6

Post-translational Modifications: Dynamic Regulators of Protein Function to Unlock Genetic Potential of Crop Plants

Udit Nandan Mishra, Debanjana Saha, Ankita Mohanty, Bulbul Boblina, Bishal Binaya Bhatta, and Mahipal Singh Kesawat

Abstract Post-translational modifications (PTMs) have become pivotal regulators of protein functionality, serving as dynamic molecular switches that translate genetic information into adaptive plant phenotypes. Through precise and reversible alterations in protein activity, stability, subcellular localization, and interaction networks, PTMs (i.e., phosphorylation, ubiquitination, acetylation, glycosylation, and redox-mediated modifications) coordinate a wide spectrum of biological processes, such as growth and development, stress resilience, nutrient-use efficiency, and immune regulation. This multilayered regulatory capacity confers exceptional plasticity to plant systems, enabling rapid, finely tuned responses to ever-changing environmental cues. Recent advances in proteomics, high-resolution mass spectrometry, and computational biology have substantially deepened our understanding of PTM landscapes in crop plants, uncovering extensive crosstalk between PTMs and other regulatory layers of gene expression. The integration of PTM-centric proteomics with genomics, transcriptomics, and metabolomics has enabled the identification of key regulatory nodes that bridge genotype and phenotype. These insights are increasingly being translated into practical applications, including precision breeding, genome editing, and metabolic engineering, to improve crop resilience and productivity while maintaining yield stability. This chapter

U. N. Mishra · A. Mohanty
Department of Crop Physiology and Biochemistry, Faculty of Agriculture, Sri Sri University, Cuttack, Odisha, India

D. Saha · M. S. Kesawat (✉)
Department of Genetics and Plant Breeding, Faculty of Agriculture, Sri Sri University, Cuttack, Odisha, India
e-mail: mahibiotech@snu.ac.kr

B. Boblina
Department of Plant Pathology, Faculty of Agriculture, Sri Sri University, Cuttack, Odisha, India

B. B. Bhatta
Department of Crop Physiology, Institute of Agricultural Sciences (IAS), Siksha 'O' Anusandhan, Bhubaneswar, Odisha, India

L. M. Vaschetto (ed.), *Omics Technologies in Plant Science*, Interdisciplinary Biotechnological Advances, https://doi.org/10.1007/978-981-92-0533-2_8

integrates current insights into the mechanistic roles of PTMs in controlling plant stress responses, developmental processes, and nutrient acquisition. It further emphasizes emerging multi-omics and genome-editing strategies that strategically target PTM-regulated pathways to develop climate-resilient, high-performing crop varieties. Therefore, the chapter highlights the transformative potential of PTM-informed approaches in driving sustainable agricultural innovation and strengthening global food security.

Keywords Post-translational modifications (PTMs) · Protein regulation · Crop improvement · Sustainable agriculture · Climate resilience

Abbreviations

ABA	Abscisic acid
BAK1	BRI-associated kinase 1
BRs	Brassinosteroids
CDPK	Calcium-dependent protein kinase
COR	Cold responsive
db	Database
HM	Heavy metal
HOS1	High expression of osmotically responsive genes 1
HSR	Heat shock response
IAA	Indole-3-acetic acid
MAPKs	Mitogen-activated protein kinases
NAS	Nicotianamine synthase
NO	Nitric oxide
PPI	Protein-protein interaction
PTI	PAMP-triggered immunity
PTM	Post translational modification
RLK	Receptor-like kinase
ROS	Reactive oxygen species
SLAC	Slow anion channel
SNF1	Sucrose non-fermenting 1
SnRKs	SNF1 related protein kinases
SOS	Salt overly sensitive
SSR	SUMO stress response
TOR	Target of rapamycin
UPS	Ubiquitin proteasome system

1 Introduction

Plant complexity arises not only from the genetic blueprint but also from dynamic regulatory processes that operate on proteins after their synthesis. The transient and reversible nature of post-translational modifications (PTMs) enables precise modulation of protein stability, subcellular localization, and protein–protein interactions (Lee et al. 2023; Millar et al. 2019). Through this highly specific regulation of protein function, PTMs generate extensive proteomic diversity and constitute an additional, distinct regulatory layer within the plant regulatory hierarchy. This layer serves as a crucial link between genotype and phenotype, equipping plants with the capacity for finely regulated molecular responses to changing environmental conditions (Li et al. 2025b; Yang 2022). To date, more than 400 distinct types of PTMs have been identified in plants, with phosphorylation, ubiquitination, acetylation, methylation, glycosylation, lipidation, and redox-mediated modifications representing some of the most prevalent and functionally significant classes (Chung et al. 2025). These modifications are mediated by specialized enzyme systems, including kinases, phosphatases, acetyltransferases, and ubiquitin ligases, which impart high substrate specificity and reversibility. The coordinated interplay of these enzymes establishes complex, context-dependent regulatory networks that fine-tune cellular processes with exceptional precision (Lee et al. 2023). Phosphorylation is the most extensively studied PTM in plants and operates as a reversible molecular switch that regulates protein conformation, enzymatic activity, and interaction capacity. This modification involves the covalent attachment of phosphate groups to serine, threonine, or tyrosine residues, enabling rapid and dynamic control of target protein function (Niño-González and Duque 2025). Mitogen-activated protein kinase (MAPK) and SNF1-related kinase (SnRK)–mediated signaling cascades are central to the regulation of abiotic stress responses, hormone signaling pathways, and key developmental transitions in plants (Jamsheer et al. 2021). In parallel, ubiquitination regulates protein turnover and quality control via the ubiquitin–proteasome system, enabling the timely degradation or stabilization of key regulatory proteins (Sharma et al. 2016). Additionally, PTMs, such as acetylation, methylation, and glycosylation, modulate chromatin architecture, transcriptional regulation, membrane trafficking, and receptor-mediated signaling, thereby broadening the functional scope of PTM-driven regulatory control (Reumann and Parasyri 2025). PTMs exert profound structural effects on proteins by introducing charged moieties, altering hydrophobic properties, or generating novel interaction surfaces. These modifications can trigger conformational rearrangements that activate or inactivate enzymes, expose or conceal degradation signals, and regulate the assembly or disassembly of protein complexes (Li et al. 2025a; Reumann and Parasyri 2025). For instance, lysine acetylation can neutralize positive charges on target residues, thereby modulating protein–DNA interactions and protein oligomerization, as demonstrated in both peroxisomal and nuclear proteins. Moreover, glycosylation and lipidation modulate protein folding, trafficking, and membrane association, thereby influencing essential processes including nutrient uptake, hormone perception, and pathogen recognition (Tian et al. 2025; Yuan et al. 2024). Beyond their individual roles, PTMs often engage in extensive crosstalk, adding layers of regulatory complexity. Sequential or combinatorial modifications such as phosphorylation-dependent ubiquitination or the interplay between acetylation and other acylations allow plants to integrate diverse signals into coordinated physiological responses. PTMs-mediated regulation of crop traits is summarized in Table 1. This integrated regulation is especially crucial under stress conditions, where rapid proteome remodeling can determine survival and productivity.

Table 1 PTMs-mediated regulation of crop traits

Crop/plant	PTM type	Protein target	Alteration in structure/function/interaction	References
Rice	Phosphorylation	XA21 receptor kinase	Induces conformational change, activating immune signaling and protein–protein interactions in defense pathways	Yang et al. (2013)
Wheat	Ubiquitination	Rust susceptibility proteins	Tags proteins for proteasomal degradation, modulating defense response	Li et al. (2025a)
Maize	Acetylation	ZmDREB2A transcription factor	Stabilizes drought-responsive TF by blocking ubiquitination, enhancing stress tolerance	Zhao et al. (2014)
Barley	SUMOylation	SUMO proteases	Enhances fungal resistance by stabilizing stress-related transcription factors	Sharma et al. (2021a, b)
Soybean	Phosphorylation	BZL2 protein	Modulates brassinosteroid signaling, affecting protein function	Song et al. (2019)
Tomato	Tyrosine phosphorylation	Protein tyrosine phosphatases	Regulates drought stress responses via signaling cascades	Liu et al. (2023)
Chestnut	Phosphorylation	DSP4 phosphatase	Mediates cold acclimation through kinase pathway modulation	Berrocal-Lobo et al. (2011)

(continued)

Table 1 (continued)

Crop/plant	PTM type	Protein target	Alteration in structure/ function/interaction	References
Arabidopsis	Deamidation	Aquaporins (PIP2)	Alters water channel activity and protein interaction, coupled with phosphorylation status	Di Pietro et al. (2013)
Arabidopsis	IsoAsp formation/repair	PRH75 RNA helicase	IsoAsp formation inactivates the protein; PIMT1 repair restores structure and translation function	Nayak et al. (2013)
Soybean	Phosphorylation	Nuclear phosphoproteins (e.g., zinc finger, Rrp5)	Regulates nuclear translocation and early flooding stress response	Yin and Komatsu (2015)
Rice	Glycosylation	β1,2-Xylosyltransferase, α1,3-fucosyltransferase	Disruption affects root aerenchyma formation and gravitropic responses, altering protein localization and interaction	Takano et al. (2015)
Maize	Histone acetylation	Histone H3 (K9)	Increases cell wall protein expression and plasma membrane proton pump activity under salt stress	Li et al. (2014)
Brassica napus	Histone methylation	Histones	Alters epigenetic status, regulating gene expression, and stress memory	Fang et al. (2017)
African/ Asian Rice	Phosphorylation	Tubulin α	Promotes microtubule depolymerization under salt/heat stress, altering cytoskeletal structure	Muleya et al. (2022)
Soybean	Phosphorylation	Tubulin α	Modifies cytoskeletal dynamics under cadmium stress	Gzyl et al. (2015)
Palm tree	Glycosylation	Glycoproteins	Stepwise editing of sugar chains affects seed development and protein function	Hashiguchi and Komatsu (2017)
Barley	Phosphorylation	CPK2 kinase	Modulates calcium-mediated drought response pathways, altering protein–protein interactions	Aliniaeifard et al. (2020)
Arabidopsis	Citrullination	Various proteins	Alters protein charge and function, affecting protein–protein interactions	Marondedze et al. (2021)

Recent advances in high-throughput proteomics, including mass spectrometry–based acetylome and phosphoproteome profiling, have greatly expanded our understanding of PTM landscapes in plants. These approaches have facilitated the identification of thousands of PTM sites, the mapping of dynamic PTM changes in response to environmental cues, and the elucidation of PTM-mediated regulatory networks. Integrated resources, such as dbPTM, now offer comprehensive platforms for exploring PTM sites, functional annotations, and regulatory circuits across diverse plant species (Chung et al. 2025). The study of PTMs extends beyond basic science, offering significant potential for crop improvement and sustainable agriculture. By targeting PTM pathways through conventional breeding, genetic engineering, or genome editing, it is possible to enhance stress tolerance, optimize growth and development, and improve nutrient-use efficiency, thereby opening new avenues for sustainable food production and security. For example, engineering PTM sites in key regulatory proteins have been demonstrated to confer increased resistance to drought, salinity, and pathogens in both model and crop species. This chapter provides a comprehensive review of the major classes of PTMs and their mechanistic roles in regulating stress tolerance, growth, development, and nutrient-use efficiency in crop plants. It further examines how PTM-focused research, when integrated with multi-omics approaches and genome-editing technologies, can guide next-generation breeding strategies for the development of climate-resilient, high-yielding crops.

2 Role of PTM in Enhancing Stress Tolerance and Adaptive Capacity of Crops

Over 400 distinct PTMs have been identified, highlighting the extraordinary diversity of protein regulatory mechanisms in plants (Chung et al. 2025). Among the numerous PTMs identified, phosphorylation, acetylation, methylation, ubiquitination, sumoylation, and nitrosylation are the most widely studied and functionally significant (Ali et al. 2019; Li et al. 2022). An overview of the roles of PTMs in regulating protein dynamics, cellular signaling, and regulatory networks in crop plants is presented in Table 2. In this section, we focus on the roles of PTMs in mediating plant responses to both abiotic and biotic stresses.

2.1 PTMs in Drought Stress Response and Mitigation

Drought is among the most severe and persistent environmental stresses, significantly constraining plant growth and productivity. To withstand water deficit, plants deploy a range of physiological and molecular adaptations, with stomatal regulation serving as a key protective mechanism. Stomatal closure is primarily governed by the phytohormone abscisic acid (ABA) and the signaling molecule nitric oxide

Table 2 Role of PTMs in protein dynamics, cellular signaling, and regulatory networks in crop plants

PTM type	Effect on protein networks	Role in cellular signaling	Impact on protein turnover	Regulatory significance	References
Phosphorylation	Modulates assembly/disassembly of protein complexes (e.g., kinases, phosphatases, metabolic enzymes) Controls oligomeric state, subcellular localization, and protein–protein interaction specificity Enables crosstalk with other PTMs (e.g., phosphorylation-ubiquitination cascade in immune receptors) Alters metabolic enzyme activity across primary metabolism (Calvin cycle, glycolysis, C4 shuttle)	Acts as a reversible molecular switch in signaling cascades (e.g., MAPK, hormone, and stress pathways) Regulates the distribution of light energy between photosystems (state transitions) and PSII repair in chloroplasts Integrates environmental signals (light, stress, hormones) for rapid response Controls activation/inactivation of transcription factors and metabolic regulators	Marks proteins for activation or degradation (e.g., phosphorylation-dependent ubiquitination) Enables dynamic, reversible regulation of protein half-life and activity Regulates turnover of photosynthetic and stress-responsive proteins	Central to stress adaptation (drought, salt, cold, pathogen attack) Fine-tunes growth, development, and metabolic flux Key for rapid adjustment of physiological processes in response to environmental cues Targeted by pathogens to suppress plant immunity Used in breeding for stress-resilient and high-yielding crops	Li et al. (2025a)

(continued)

Table 2 (continued)

PTM type	Effect on protein networks	Role in cellular signaling	Impact on protein turnover	Regulatory significance	References
Ubiquitination	Directs proteins to the 26S proteasome, shaping network composition and abundance of regulatory proteins Modifies interaction networks by removing or stabilizing key nodes (e.g., E3 ligases targeting TFs) Interacts with SUMOylation and phosphorylation for layered regulation	Terminates or modulates signaling by degrading key regulators (e.g., negative regulators of defense) Controls the duration and intensity of hormone and immune signaling Involved in cell cycle, hormone, and defense signaling	Targets misfolded, damaged, or regulatory proteins for degradation Controls the abundance of short-lived signaling proteins and metabolic enzymes Maintains proteome quality and homeostasis	Maintains protein quality, removes damaged proteins, and fine-tunes stress and defense pathways Essential for resetting signaling networks after stress or developmental cues Exploited by pathogens to suppress plant defense Manipulation can improve disease resistance and stress tolerance in crops	Friso and van Wijk (2015)
SUMOylation	Stabilizes protein complexes and modifies protein–protein interaction specificity Alters nuclear-cytoplasmic shuttling and localization of transcription factors Crosstalk with ubiquitination for fine-tuning protein fate	Regulates activity, localization, and stability of signaling proteins, especially in stress and immune responses Modulates transcription factor activity and stress signaling pathways	Protects proteins from ubiquitin-mediated degradation Extends the half-life of stress-responsive and regulatory proteins	Enhance resistance to biotic and abiotic stresses by stabilizing key regulators Modulation can improve crop resilience to environmental challenges	Gough and Sadanandom (2021)

Acetylation/ methylation (histone and non-histone)	Alters chromatin structure, affecting accessibility of large gene/protein networks Modifies non-histone proteins, influencing metabolic enzyme activity and protein interactions	Modulates transcriptional responses to environmental and developmental signals (epigenetic regulation) Regulates gene expression in response to stress, development, and memory formation	Epigenetic "memory" can influence long-term protein expression patterns Alters protein stability and interaction landscape	Enables rapid and reversible gene expression changes for adaptation and stress memory (e.g., salt/heat stress, vernalization) Potential target for improving crop plasticity and adaptation	Grabsztunowicz et al. (2017)
Glycosylation	Influences protein folding, stability, and trafficking Affects cell wall and membrane protein networks, impacting development and defense	Modulates receptor function and cell surface signal transduction Involved in pathogen recognition and immune signaling	Increase protein stability and half-life Protects against proteolytic degradation	Supports cell wall integrity, pathogen recognition, and developmental processes Manipulation can enhance disease resistance and developmental traits in crops	Muleya et al. (2022)
Redox-based PTMs such as S-nitrosylation, disulfide bonds, and glutathionylation	Modifies enzyme activity and protein-protein interactions, especially in chloroplasts and mitochondria Alters oligomeric state and assembly of metabolic complexes	Integrates redox and environmental signals into metabolic and stress signaling pathways Regulates photosynthetic efficiency and stress adaptation	Controls reversible activation/inactivation of enzymes (e.g., Calvin cycle enzymes) Fine-tunes protein turnover in response to oxidative stress	Essential for rapid adaptation to oxidative stress, drought conditions, and pathogen attacks Serves as a target for enhancing photosynthetic efficiency and strengthening stress resilience in crops	Martí-Guillén et al. (2022)

(continued)

Table 2 (continued)

PTM type	Effect on protein networks	Role in cellular signaling	Impact on protein turnover	Regulatory significance	References
Lipidation includes N-myristoylation and S-acylation	Anchors proteins to membranes, organizing signaling complexes and spatially restricting protein networks	Facilitates spatial regulation of signaling pathways (e.g., hormone, stress, immune signaling)	Modulates protein half-life, intracellular trafficking, and subcellular localization	Critical for membrane-associated signaling, stress adaptation, and pathogen response Manipulation can improve signal specificity and stress tolerance in crops	Fhu and Ali (2021), Chamberlain and Shipston (2015)
Other PTMs such as deamidation, citrullination, and carbonylation	Alters protein charge, conformation, and interaction specificity Capable of regulating protein activity, which in turn shapes the topology and behavior of complex interaction networks	Modifies signal transduction and stress responses, often under extreme or aging conditions	Functions as a cue for repairing damaged proteins or directing them for degradation, exemplified by isoAsp modification and its restoration	Influences plant longevity, aging processes, and responses to various stress	Muleya et al. (2022)

(NO), which act synergistically to modulate guard cell activity and minimize water loss through transpiration (Tuteja and Sopory 2008). At the molecular level, ABA-induced stomatal closure is mediated by the activation of slow anion channels (SLACs) in guard cells, a process tightly controlled by multiple protein kinases, including stress-activated protein kinases (SAPKs), SNF1-related kinases (SnRKs), and calcium-dependent protein kinases (CPKs). Functional studies further highlight the involvement of transcriptional regulators in this pathway; for example, silencing the tomato transcription factor SlWRKY1 enhances stomatal closure, thereby reducing stomatal conductance and transpiration under drought conditions (Birhanu 2020). Beyond transcriptional regulation, PTMs are increasingly recognized as key modulators of drought adaptation. Protein phosphorylation serves as a rapid and central event in ABA signaling, fine-tuning guard cell responses. Simultaneously, the ubiquitin–proteasome system (UPS) controls the stability of essential signaling components, thereby influencing drought resilience. Notably, E3 ubiquitin ligases have been demonstrated to enhance dehydration tolerance through ABA-dependent pathways (Kong et al. 2015). In soybean, the GmPUB E3 ligase negatively regulates drought tolerance through its effects on ABA-mediated signaling and osmotic stress response pathways (Wang et al. 2016). Collectively, these findings establish PTMs as dynamic regulators that integrate hormonal and environmental signals to coordinate adaptive drought responses in plants.

2.2 Regulatory Functions of PTMs Under Salinity Stress Conditions

Similar to drought, salinity stress imposes both ionic and oxidative challenges, frequently leading to accelerated senescence and increased plant mortality (Munns 2002). To cope with salinity stress, plants must eliminate excess Na^+ and maintain K^+ homeostasis, a process regulated by PTMs such as phosphorylation, ubiquitination, and nitrosylation. These PTMs play crucial roles in the Salt Overly Sensitive (SOS) pathway, where SOS1, SOS2, and SOS3 proteins mediate signal perception and confer salinity tolerance (Mahajan et al. 2008). Upon salt exposure, SOS3 detects calcium signals triggered by Na^+ stress and directly interacts with, and activates, SOS2. The activated SOS2 kinase then phosphorylates the Na^+/H^+ antiporter SOS1, H^+-ATPases, and cation exchangers, promoting the extrusion of excess Na^+ and restoring ionic homeostasis (Assaha et al. 2017; Chen et al. 2021). In addition to phosphorylation, ubiquitination serves as a key regulatory mechanism in salinity adaptation. For example, overexpression of CaAIP1 in Capsicum enhances tolerance to both drought and salinity via ubiquitin-mediated protein modulation, underscoring the significance of post-translational control independent of transcriptional regulation (Park et al. 2016). Similarly, S-nitrosylation acts as a crucial modulator of antioxidant defenses under salt stress. In tomato, overexpression of SlMDHR elevates S-nitrosylated SlMDHR levels, enhances antioxidant enzyme activity, and reduces ROS accumulation, highlighting the protective role of nitrosylation in maintaining cellular redox homeostasis during salinity stress (Qi et al. 2020). Thus, these findings suggest that PTMs integrate ionic, redox, and hormonal signaling pathways to regulate plant adaptation to salinity stress.

2.3 Regulatory Role of PTMs Under Elevated Temperature Conditions

Climate change–driven increases in the frequency and intensity of heat waves represent a major threat to global agricultural productivity. Crop species such as wheat are especially vulnerable when heat stress coincides with sensitive developmental stages, leading to substantial yield losses (Sun et al. 2019; Tigchelaar et al. 2018; Zhao et al. 2017). To cope with elevated temperatures, plants activate multilayered adaptive mechanisms, among which PTMs serve as central regulators of heat stress perception and response. Among these, ubiquitination has emerged as a key determinant of thermotolerance. In *Arabidopsis thaliana*, overexpression of AtPUB48, a U-box–type E3 ubiquitin ligase, markedly enhances heat tolerance during both vegetative growth and seed germination, underscoring the critical role of ubiquitin-mediated protein turnover in heat stress adaptation (Peng et al. 2019). On the other hand, SUMOylation plays a prominent role in heat stress signaling. Elevated temperatures trigger widespread conjugation of SUMO1/2 to target proteins in *Arabidopsis*, accompanied by increased expression of heat shock proteins such as HSP70, indicating tight functional integration between SUMOylation and the heat shock regulon (Kurepa et al. 2003).

Protein phosphorylation further integrates heat stress signaling with cellular homeostasis. Cyclin-dependent kinases (CDKs) phosphorylate the heat shock transcription factor HSF1, thereby coordinating heat stress responses with cell cycle progression and growth regulation. In addition to these rapid signaling events, chromatin-based PTMs introduce an additional layer of control in heat adaptation. Heat stress promotes nucleosome eviction and enhances histone H3 lysine 56 acetylation (H3K56ac) at heat-responsive loci, including HSFA2 and HSA32, thereby facilitating rapid transcriptional activation (Weng et al. 2014). Moreover, acquired thermotolerance following prior heat exposure is associated with elevated levels of histone H3 lysine 4 di- and tri-methylation (H3K4me2/3) at stress-inducible genes such as APX2 and HSP22.0, underscoring the role of histone methylation in establishing transcriptional memory under heat stress. Collectively, these findings demonstrate that PTMs function across signaling, proteostasis, and chromatin-remodeling pathways to enable plants to perceive, respond to, and retain memory of heat stress, thereby enhancing resilience under increasingly frequent and intense temperature regimes.

2.4 *Significance of PTMs in Cold Stress Resilience in Plants*

Both chilling (0–5 °C) and freezing (<0 °C) stresses impose severe constraints on crop growth and productivity by compromising membrane integrity and inhibiting enzymatic activity. To mitigate low-temperature damage, plants activate a tightly coordinated network of physiological, transcriptional, and post-translational regulatory mechanisms. Genome-wide expression analyses in *Arabidopsis thaliana* reveal that nearly 10% of the transcriptome is differentially regulated during cold stress, underscoring the extensive molecular reprogramming that underpins cold acclimation (Kreps et al. 2002; Bajwa et al. 2014). Cold stress signaling in plants is centrally governed by the ICE–CBF–COR regulatory module. Within this pathway, the basic helix–loop–helix transcription factor INDUCER OF CBF EXPRESSION 1 (ICE1) functions as a master upstream regulator, activating the CBF/DREB transcriptional cascade that drives low-temperature–responsive gene expression (Tang et al. 2020). Within this cascade, CBF1/DREB1B and CBF3/DREB1A act as positive regulators of cold-responsive (COR) genes, whereas CBF2/DREB1C functions as a negative regulator by repressing the expression of CBF1 and CBF3. This antagonistic regulation fine-tunes both the magnitude and duration of downstream COR gene activation (Novillo et al. 2004; Novillo et al. 2007).

PTMs add an additional regulatory layer that fine-tunes the intensity and duration of cold stress signaling. For example, SUMOylation of ICE1 by the SUMO E3 ligase SIZ1 stabilizes ICE1 and enhances CBF3 induction, thereby amplifying downstream expression of COR genes during cold acclimation. Early cold perception is further coordinated through phosphorylation-dependent activation of multiple protein kinase families, including receptor-like kinases (RLKs), calcium-dependent protein kinases (CDPKs), and mitogen-activated protein kinases (MAPKs), which collectively mediate signal perception and transduction in response to low-temperature exposure (Kamal et al. 2020). In contrast to these positive regulatory mechanisms, ubiquitin-mediated protein turnover can negatively impact cold tolerance. Overexpression of the RING-finger E3 ubiquitin ligase HIGH EXPRESSION OF OSMOTICALLY RESPONSIVE GENES 1 (HOS1) reduces freezing tolerance in Arabidopsis by promoting the degradation of key cold-responsive regulators, highlighting the antagonistic role of selective proteolysis in cold stress adaptation (Dong et al. 2006).

2.5 *Role of PTMs Under Heavy Metal Stress Conditions*

Among the numerous abiotic stresses, heavy metal (HM) toxicity poses a significant threat to plant growth, development, and overall productivity (Wani et al. 2016; Khan et al. 2025). HMs such as cadmium (Cd), lead (Pb), arsenic (As), and mercury (Hg) disrupt cellular homeostasis by interfering with key physiological and biochemical processes, including photosynthesis, membrane stability, and phytohormone regulation (Hafeez et al. 2023; Aslam et al. 2023; Ashraf et al. 2021; Bali and Sidhu 2021). Plants counteract these harmful effects through PTMs, which serve as dynamic modulators of protein function under stress conditions (Naik et al. 2023). For example, the phosphorylation of transcription factors belonging to the bZIP and WRKY families enhances heavy metal tolerance by modulating the MAPK signaling cascade (Hashiguchi and Komatsu 2017; Niekerk et al. 2024). Simultaneously, ubiquitination regulates antioxidant defenses and the function of metal transporters, such as NRAMP and ZIP family proteins, thereby mitigating heavy metal toxicity (Fang et al. 2023). Glycosylation further enhances heavy metal tolerance by strengthening cell wall integrity, fine-tuning immune responses, and coordinating adaptive mechanisms to abiotic stress (Kosov'a et al. 2021). Together, these PTM-mediated mechanisms constitute an integrated network that allows plants to detect, respond to, and adapt under HM stress.

2.6 Mechanistic Insights into PTM-Mediated Regulation Under Nutrient Stress Conditions

Plant cells detect and respond to nutrient deficiencies through dynamic PTMs, which orchestrate metabolic and physiological adaptations. Among these, O-linked N-acetylglucosamine (O-GlcNAc) modification has emerged as a key regulatory mechanism, often acting in concert with, or parallel to, protein phosphorylation in nutrient sensing and signal transduction. Nutrient limitation triggers extensive molecular reprogramming, predominantly mediated by PTM-dependent protein–protein interactions (PPIs) that regulate nutrient uptake, transport, and metabolic adjustments. For example, nitrogen deficiency modulates the GS-GOGAT cycle and the TOR kinase complex (TOR–RAPTOR–LST8), thereby coordinating nitrogen assimilation, remobilization, and overall nitrogen use efficiency (Wang et al. 2025). TOR-dependent signaling links cellular energy status with amino acid biosynthesis, allowing plants to maintain metabolic balance under nitrogen-limited conditions (Zhang et al. 2023). Similarly, phosphorus deficiency modulates the interactions between SPX domain-containing proteins and PHT transporters, thereby promoting phosphate uptake and sustaining phosphate homeostasis (Yang et al. 2021). Under potassium deficiency, the HAK5–CIPK23 module is activated, promoting high-affinity potassium uptake and enabling long-distance transport (Wang 2019; Xian et al. 2023). During iron deficiency, interactions between NAS1 and YELLOW STRIPE-LIKE (YSL) transporters enhance iron chelation and facilitate root-to-shoot translocation (Zenda et al. 2021; Shen et al. 2023). Taken together, these examples illustrate how PTM-mediated regulation links nutrient sensing with metabolic and developmental plasticity, allowing plants to dynamically adjust to changing nutrient availability.

2.7 *Influence of PTMs in Modulating Plant Immunity to Biotic Challenges*

Biotic stress restricts plant growth and development, driving adaptive changes in molecular mechanisms. These responses can be explored through genomic, transcriptomic, and proteomic approaches, which uncover the dynamic regulation of cellular machinery. For example, proteomic analyses of potatoes infected with *Phytophthora infestans* reveal differential expressions of defense-related proteins (Ali et al. 2014; Muthamilarasan and Prasad 2013). Beyond transcriptional changes, PTMs of functional proteins play a central role in plant resistance against insects and pathogens. In *Arabidopsis thaliana*, recognition of the pathogen-associated molecular pattern (PAMP) flg22 by BRI1-associated kinase 1 (BAK1) initiates a phosphorylation cascade that activates PAMP-triggered immunity (PTI) (Xiao et al. 2019). Conversely, MAPK phosphatase 1 dephosphorylates MAPKs, thereby dampening kinase-mediated immune responses. Ubiquitination of the FLS2 receptor further fine-tunes intracellular signaling, acting in concert with phosphorylation during biotic stress. In contrast, certain pathogen effectors exploit PTMs to suppress plant immunity; for instance, N-glycosylation followed by C-terminal proteolysis of the GPP protein from the MgGPP effector inhibits host defenses while promoting parasitism. Therefore, these studies highlight the pivotal role of PTMs in regulating plant–pathogen interactions, balancing immune activation with suppression.

3 PTM-Mediated Regulation of Growth, Development, and Nutrient Efficiency in Plants

3.1 *Hormonal Regulation and Signal Transduction*

Indole-3-acetic acid (IAA), the primary natural auxin, is transported throughout mature plants via phloem bulk flow from source tissues such as young leaves and flowers. PTMs play a critical role in auxin signaling by regulating protein stability, subcellular localization, and interaction networks, thereby modulating the transcriptional activity of key regulatory factors. Phosphorylation and dephosphorylation of serine, threonine, or tyrosine residues act as molecular switches that activate or repress auxin-responsive proteins. Ubiquitination of Aux/IAA proteins represents a central regulatory node, and further PTMs of Aux/IAAs are likely to be identified. The functional significance of Aux/IAA phosphorylation in photomorphogenesis was first demonstrated by the *shy2-1D* gain-of-function mutant, whereas recombinant *Oryza sativa* phytochrome A could bypass phytochrome requirements for specific developmental responses (Colon-Carmona et al. 2000). Light receptors PHYB and CRY1 further integrate environmental signals by enhancing TIR1/AFB-mediated Aux/IAA degradation via activation of CK2 kinase (Xu et al. 2018).

MAPK cascades intersect with the auxin signaling pathway to fine-tune plant growth responses. Specifically, MPK6 regulates auxin biosynthesis and lateral root development, with upstream MAPKKK–MAPKK–MPK modules transmitting signals that modulate Aux/IAA gene expression (Zhang et al. 2025). The MAPKKK NPK1 suppresses early auxin-responsive transcription, whereas the MEKK1–MPK4 module selectively regulates specific subsets of Aux/IAA genes. MPK3 and MPK6 interact with BZR1-mediated brassinosteroid (BR) signaling to regulate hypocotyl elongation and phosphorylate IAA8 and IAA9 during stress responses, thereby maintaining auxin homeostasis (Wang et al. 2025). Receptor-like kinases, including members of the TMK family, modulate auxin signaling by phosphorylating plasma membrane H^+-ATPases to promote cell wall acidification and elongation, and by forming an extracellular auxin perception complex with ABP1 (Xu et al. 2017). Several auxin response factors (ARFs) are regulated via phosphorylation: ARF7 and ARF19 are targeted by MPK6 to control lateral root development, while ARF2 is phosphorylated by BIN2, integrating auxin and brassinosteroid signaling (Tan et al. 2025). Additional PTMs, such as acylation and myristoylation, regulate the membrane association and protein–protein interactions of ARF regulators and auxin transporters. Histone PTMs influence chromatin accessibility, facilitating transcriptional reprogramming of numerous auxin-responsive genes. Consequently, epigenetic mechanisms provide an additional layer of regulation over auxin-mediated growth and stress responses.

The phosphorylation activity of diversc kinases plays a central role in regulating ABA signaling in response to stress, particularly in mediating ABA-induced responses under drought conditions. Multiple studies have demonstrated that ABA receptors are subject to phosphorylation. Specifically, PYL receptors are phosphorylated by TOR kinase, cytosolic ABA receptor kinase 1 (CARK1), C-terminally encoded peptide receptor 2 (CEPR2), and *Arabidopsis* Early flowering 1 (EL1)-like casein kinase (AEL) (Li et al. 2019). Phosphorylation acts as a positive regulator of ABA signaling. Under abiotic stress, ABA accumulation activates SnRK2 kinases, including SnRK2.2, SnRK2.3, and SnRK2.6. Evidence from triple mutants (snrk2.2/2.3/2.6) indicates that these kinases are essential for ABA perception and signaling, highlighting their positive regulatory role in mediating stress-induced ABA responses (Soma et al. 2021). E3 ligases have been shown to indirectly support SnRK2 stability. Specifically, AtPP2-B11, a component of the SKP1/Cullin/F-box complex, directly interacts with SnRK2.3, as demonstrated using cell-free degradation assays (Cheng et al. 2017). In addition to SnRK2s, PYL1 and PYL4 members of the PYR/PYL/RCAR family of ABA receptors play a critical role in ABA signal transduction. These receptors are central to ABA-mediated regulation of plant responses to abiotic stresses such as drought and salinity, as well as seed dormancy and stomatal closure. Upon ABA binding, PYL1 and PYL4 undergo conformational changes that enable the ABA–PYL complex to inhibit PP2C phosphatases, thereby releasing SnRK2 kinases to activate downstream targets (Weiner et al. 2010). SnRK2 also phosphorylates plasma membrane proteins. For example, the guard cell kinase SnRK2.6/OST1 targets ion channels such as SLAC1, KAT1, and NADPH oxidase. Recent studies have identified additional kinases, including the abiotic stress-responsive Raf-like kinase (ARK), casein kinase 2 (CK2), ABA, and BR-insensitive 2 (BIN2), that regulate SnRK2 stability and activity (Saruhashi et al. 2015).

BR signaling heavily relies on phosphorylation to regulate downstream responses. BR binding to the extracellular domain of its receptor initiates a phosphorylation cascade that ultimately activates the phosphatase BRI1 SUPPRESSOR 1 (BSU1). A critical regulatory step involves the phosphorylation status of the transcription factors BZR1 and BES1. In their dephosphorylated state, BZR1 and BES1 accumulate in the nucleus to activate transcription of BR-responsive genes, whereas phosphorylated BZR1 and BES1 are exported to the cytoplasm, preventing their function as transcriptional activators (Xiong et al. 2022). Phosphorylated BZR1 and BES1 interact with 14-3-3 proteins, which further promote their cytoplasmic sequestration. The GSK3-like kinase BIN2 functions as a key negative regulator of BR signaling by phosphorylating BZR1 and BES1, thereby inhibiting their DNA-binding activity, reducing transcriptional output, and promoting their degradation. Through BIN2-mediated repression, BR signaling remains inactive in the absence of brassinosteroids. Therefore, these PTM-mediated mechanisms coordinate auxin, ABA, and BR signaling pathways, allowing plants to dynamically modulate growth, development, and stress responses.

3.2 PTM-Mediated Control of Flowering Time and Seed Development

Plant growth and development, including seed germination, seedling establishment, flowering, fruiting, and seed maturation, entail complex physiological processes accompanied by extensive morphological and molecular changes. Phosphorylation occurs extensively throughout these developmental stages. In maize, 31,595 phosphorylation sites were identified across 6227 proteins from 23 tissues during vegetative and reproductive growth, underscoring the pivotal role of PTMs in regulating plant development (Li et al. 2025a). The production of FLOWERING LOCUS T (FT), a florigen protein that senses changes in photoperiod during early vegetative growth to prevent premature flowering, initiates age-dependent molecular regulatory pathways in plants. In response to floral induction signals, meristematic tissues transition from vegetative to floral meristems. Floral integrators, including FT, florigen, and SUPPRESSOR OF OVEREXPRESSION OF CONSTANS 1 (SOC1), drive this transformation. In *Rosa chinensis*, the long noncoding RNA lncWD83 binds to PLANTU-BOX PROTEIN 11 (PUB11), promoting flowering by inhibiting the ubiquitination and accelerating the degradation of the floral repressor MYC2 LIKE (RcMYC2L) (Yeqing et al. 2023). Other key regulators of flowering initiation, including FLOWERING LOCUS C (FLC) and CONSTANS (CO), suppress responses to environmental cues until floral integrators are activated (Meir et al. 2021). Long days (LDs) induce the production of FT in leaves, which then translocates to the shoot apical meristem. The transition to the reproductive phase is mediated by FT's bZIP partner, FLOWERING LOCUS D (FD), which activates downstream floral integrators such as SOC1, FRUITFUL (FUL), and APETALA1 (AP1) (Notaguchi et al. 2008).

Gene expressions in response to environmental cues, such as temperature fluctuations, is largely governed by epigenetic regulation. In *Arabidopsis*, the histone deacetylase NRPD1 and acetyltransferase 5 have been found to be upregulated in the DNA-binding bromodomain, suggesting their potential roles in modulating gene expression during cold stress. In *Arabidopsis*, FLC is a key regulator of the transition from vegetative growth to reproductive development. Vernalization, or prolonged exposure to low temperatures, represses FLC expression. This process is mediated by epigenetic modifications at the FLC locus, where MADS-box transcriptional repressors play a crucial role in maintaining the vegetative state and controlling vernalization-induced flowering (Deng et al. 2015). Reprogramming during early embryogenesis, zygote formation, and subsequent development is essential to ensure that each generation requires cold exposure for timely flowering. Repressive chromatin modifications established during development are maintained throughout growth but are reset during gametogenesis.

Seed germination, the critical transition from dormancy to active seedling growth, requires extensive reprogramming of multiple physiological and molecular systems. Across species, germination involves metabolic changes, transcriptional activation, morphological adjustments, and the restoration of cellular structures. Increasing evidence indicates that PTMs are also crucial for regulating seed germination. Recent studies investigating the roles of PTMs have provided important insights into the molecular mechanisms underlying this process. Protein phosphorylation and dephosphorylation act as critical molecular switches regulating protein activity during seed germination. ABI5, a central component of the ABA signaling pathway, is a well-established target whose phosphorylation status directly influences germination. In the presence of ABA, ABI5 is phosphorylated and stabilized by the protein kinase SnRK2, thereby inhibiting seed germination (Piskurewicz et al. 2008). The *Arabidopsis* RING-ANK family protein KEEP ON GOING (KEG) contains a kinase domain with phosphorylation activity and interacts with ABI5 to negatively regulate ABA signaling (Liu and Stone 2010). Thus, these findings demonstrate that the ABA-mediated signaling pathway, which governs seed germination and seedling establishment, precisely regulates the phosphorylation and dephosphorylation of ABI5. Recent research further revealed that the complex formed by *Arabidopsis* Shewanella-like protein phosphatase 2 (AtSLP2) and the mitochondrial oxidoreductase import and assembly protein 40 (AtMIA40) negatively regulates GA-dependent processes during seed germination. The phosphatase activity of AtSLP2 relies on its interaction with AtMIA40 within the mitochondrial intermembrane space (Carrie et al. 2010). The transcription factors SPATULA (SPA) and PHYTOCHROME INTERACTING FACTOR3-LIKE5 (PIL5) suppress GA biosynthesis during seed dormancy, thereby inhibiting germination. Elevated levels of H3K4me3, H3K36me3, and H3K9Ac at their regulatory regions sustain high SPA and PIL5 expression throughout dormancy. Additionally, GA-mediated processes during seed germination are negatively regulated by AtMIA40 (Luján-Soto et al.

2021). PIL5 promotes ABA accumulation and suppresses GA biosynthesis by positively regulating the SOMNUS (SOM) protein, which induces seed dormancy. SOM activity inhibits histone arginine demethylases, including JUMONJI domain-containing proteins JMJ20 and JMJ22, thereby reinforcing the dormancy state (Torres et al. 2019). Together with ABA, DELAY OF GERMINATION-1 (DOG1) promotes seed dormancy throughout seed maturation. DOG1 expression peaks 14–16 days after pollination and subsequently declines during seed ripening and imbibition. The helicase-family chromatin remodeler ENHANCED PHOTOMORPHOGENIC 1/PICKLE (EPP1/PKL) physically interacts with PHYTOCHROME INTERACTING FACTOR 3 (PIF3) and BRASSINAZOLE RESISTANT 1 (BZR1) to promote hypocotyl elongation and facilitate seed germination by suppressing H3K27 trimethylation at cell elongation loci (Zhang et al. 2015). Together, the evidence suggests that PTMs such as phosphorylation, ubiquitination, and histone modifications operate as interconnected regulatory modules that integrate environmental cues with hormonal and epigenetic signaling, thereby orchestrating precise developmental transitions from germination to flowering.

3.3 Mechanisms for Enhanced Nutrient Acquisition Efficiency

PTMs play a central role in regulating nutrient uptake in plants by modulating the activity, localization, and stability of transporters, sensors, and signaling proteins in response to nutrient availability. AMT activity is tightly regulated by phosphorylation. In *Arabidopsis thaliana*, AtAMT1;1 is phosphorylated at Thr460, the first identified AMT phosphorylation site. Elevated ammonium levels trigger this phosphorylation, rapidly inactivating the transporter to prevent ammonium toxicity (Wu et al. 2019). However, the precise mechanisms by which protein kinases and phosphatases regulate AMT phosphorylation and dephosphorylation remain unclear. Another serine/threonine/tyrosine protein kinase, OsACTPK1, regulates the phosphorylation of T452 in OsAMT1.2 (analogous to T460 in AtAMT1.1) in response to elevated ammonium levels. Notably, the transcriptional regulation of OsACTPK1 is critical for controlling AMT1.2 activity. Under ammonium-deficient conditions, OsACTPK1 expression is low, leading to dephosphorylation and activation of OsAMT1.2 to enhance ammonium uptake. Conversely, when ammonium is abundant, OsACTPK1 transcript levels increase rapidly, and accumulation of OsACTPK1 protein promotes phosphorylation of OsAMT1.2, inactivating the transporter and preventing ammonium toxicity (Hao et al. 2025). AtAMT1;3 contains at least four phosphorylation sites, such as T464, S480, S487, and T494, with T464 corresponding to T460 in AtAMT1;1. Elevated ammonium triggers phosphorylation of T464, inactivating AtAMT1;3. Notably, T464 in AtAMT1;3 is not phosphorylated by AtCIPK23 or OsACTPK1, kinases that target AtAMT1;1 and AtAMT1;2, highlighting isoform-specific regulation of AMT activity (Xiangyu et al. 2019). During ammonium deficiency, AtABI2 and AtABI1 inhibit AtCIPK15, leading to activation of AtAMT1;3 and enhanced ammonium uptake. This intricate regulatory mechanism allows plant ammonium transporters to adjust activity in response to different nitrogen forms, optimizing ammonium acquisition (Loqué et al. 2006).

In addition to ammonium, plants acquire nitrogen in the form of nitrate. Two gene families, including the NPF (NRT1/PTR FAMILY, 53 members) and NRT2 (seven members), mediate nitrate uptake and transport. Post-translational regulatory mechanisms have been identified in two key members, AtNRT1.1 (AtNPF6.3) and AtNRT2.1, which encode dual- and high-affinity nitrate transporters, respectively. Remarkably, phosphorylation and dephosphorylation of Thr101 in AtNRT1.1 control the switch between high- and low-affinity nitrate transport in response to environmental nitrate availability (Wen et al. 2019). AtNRT1.1 functions as a high-affinity nitrate transporter when CIPK23 phosphorylates Thr101. Under low nitrate availability, an as-yet unidentified phosphatase dephosphorylates Thr101, converting AtNRT1.1 into a low-affinity nitrate transporter (Wang et al. 2020). High nitrate uptake via non-phosphorylated AtNRT1.1 is balanced with ammonium uptake by non-phosphorylated, active AMT1 transporters to maintain ionic homeostasis. Under moderate ammonium levels or high nitrogen demand, CIPK23 remains inactive and cytoplasmically sequestered. In contrast, when ammonium is abundant, or nitrogen demand is low, CIPK23 is activated and recruited to the plasma membrane, where it phosphorylates both AtNRT1.1 and AMT1 transporters to modulate their activity (Fang et al. 2021). Under these conditions, inactive AMT1 transporters and high-affinity AtNRT1.1 together satisfy the plant's nitrogen demand. CIPK23 likely plays a central role in coordinating nitrate and ammonium utilization by integrating multiple nitrogen signaling pathways.

Plants possess two phosphate uptake systems: a high-affinity system and a low-affinity system. Plasma membrane-localized PHT1 family phosphate transporters (PTs) mediate phosphate acquisition through both pathways. Rather than directly absorbing phosphate from the soil, PHT1;5 plays a critical role in mobilizing phosphate from source to sink tissues, facilitating root-to-shoot phosphate translocation (Nagrajan et al. 2011). The rice genome encodes 13 OsPHT1 proteins. Root phosphate uptake is thought to be mediated by 12 of these, as OsPHT1;7 is not expressed in roots. Functional analyses using single and double-gene deletion and knockdown lines indicate that OsPHT1.3 contributes significantly to phosphate absorption in rice roots (Chang et al. 2019). Under low phosphate conditions, OsPHT1.1, OsPHT1.2, OsPHT1.4, OsPHT1.6, OsPHT1.8, OsPHT1.9, and OsPHT1.10 contribute to phosphate uptake, whereas at moderate phosphate levels, OsPHT1.1, OsPHT1.4, OsPHT1.6, OsPHT1.8, OsPHT1.9, and OsPHT1.10 remain active. In *Arabidopsis*, AtPHT1;1 contains at least two phosphorylation sites, S514 and S520. Dephosphorylation of S514 facilitates trafficking of AtPHT1;1 from the endoplasmic reticulum to the plasma membrane under phosphate deficiency ($\leq$10 μM), while phosphorylation of S520 by CASEIN KINASE 2 (AtCK2) enhances phosphate uptake without altering its localization (Wang et al. 2020). In rice, OsPP95 (Clade F1 PP2C, PROTEIN PHOSPHATASE 95) dephosphorylates the S517 site in OsPHT1;8 (analogous to S514 in AtPHT1.1) under low phosphate conditions, facilitating its trafficking from the endoplasmic reticulum to the plasma membrane (Yang et al. 2020). Under high phosphate conditions, OsCK2 phosphorylates

S517 in OsPHT1;8, preventing its interaction with OsPHF1 (Phosphate Transporter Traffic Facilitator) and causing greater retention of OsPHT1;8 in the endoplasmic reticulum. Several transcription factors also regulate PHT1 gene expression, including ATE-INDUCIBLE, GARP-TYPE TRANSCRIPTIONAL REPRESSOR1.2 (AtNIGT1.2), AtWRKY45, OsWRKY10, OsWRKY21, and OsWRKY108. Among these, OsWRKY10 acts as a negative regulator of PHT1 expression, whereas AtPHR1, OsPHR2, AtNIGT1.2, AtWRKY45, OsWRKY21, and OsWRKY108 function as positive regulators. Together, these examples illustrate the intricate PTM-dependent regulatory networks that coordinate nutrient uptake and signaling in plants, enabling dynamic adaptation to fluctuating environmental nutrient availability.

4 Harnessing PTM Proteomics in Conjunction with Multi-Omics for Advanced Precision Breeding

Precision plant breeding can be greatly advanced by integrating PTM-focused proteomics with multi-omics approaches, providing a comprehensive understanding of protein-level regulatory mechanisms that underpin stress resilience, yield, and quality traits. Dynamic PTMs, including phosphorylation, acetylation, and ubiquitination, modulate protein activity and cellular signaling, enabling rapid adaptation to environmental fluctuations. By connecting genomic variation with transcriptomic and metabolomic profiles, PTM proteomics bridges the gap between genotype and phenotype, offering a systems-level perspective on complex agronomic traits. This integrative strategy not only elucidates the molecular networks governing plant performance but also identifies critical regulatory nodes that can be targeted to accelerate the development of climate-resilient, high-yielding crop varieties.

4.1 High-Throughput Approaches Driving Progress in Proteomics and PTM Profiling

Unlocking the genetic potential of crops to enhance stress tolerance, yield, and nutritional quality relies on proteomics and PTM analysis. Advances in high-throughput mass spectrometry (MS) and bioinformatics have provided unprecedented insights into plant molecular pathways. By enabling precise mapping of stress-responsive proteins and regulatory networks, these developments in proteomics and PTM profiling are transforming strategies for crop improvement (Hu et al. 2015). Recent advances in mass spectrometry, including Orbitrap technology and data-independent acquisition (DIA), enable large-scale profiling of low-abundance proteins and PTMs, such as phosphorylation, glycosylation, and ubiquitination, in crops like soybean and barley (Bobalova et al. 2023). These PTMs regulate critical processes, including pathogen defense and abiotic stress adaptation, such as phosphorylation-mediated flooding tolerance in soybean (Gough and Sadanandom 2021). The development of high-yielding cultivars has been accelerated by integrating proteomic data with genomic information and marker-assisted breeding, enabling rapid identification of quantitative trait loci (QTLs) and stress-resilient biomarkers. Targeted proteomics approaches, such as parallel reaction monitoring (PRM), enhance reproducibility in PTM validation, supporting the optimization of disease resistance and nutrient use efficiency. Thus, these technologies provide a robust framework for breeding climate-resilient crops by linking agronomic traits with underlying molecular mechanisms (Yali and Begna 2024).

4.2 Synergistic Interactions Between PTMs and Other Omics: A Focus on Metabolomic, Genomic, and Transcriptomic Convergence

Integrating PTMs with metabolomics, genomics, and transcriptomics has enabled a systems-level understanding of plant molecular networks, revealing how dynamic protein modifications regulate genetic potential under environmental stress. Transcriptomic analyses identify upstream regulatory genes, such as transcription factors and noncoding RNAs that orchestrate these PTM-driven pathways, while PTMs like phosphorylation and ubiquitination act as molecular switches, modulating protein activity and stability in response to abiotic stresses, including drought and salinity (Wang et al. 2024). This understanding is further strengthened by metabolomic profiling, which functionally validates metabolic adaptations by linking PTM-modified enzymes to alterations in primary and secondary metabolites, such as polyamines in drought-tolerant rice or anthocyanins in stress-adapted tomato (Yan et al. 2024). Genomics facilitates marker-assisted selection of crops with enhanced stress resilience by identifying genetic variations such as SNPs or QTLs associated with PTM-enriched regulatory hubs, thereby anchoring these molecular interactions. Cross-omics integration is exemplified by the correlation between transcriptomic upregulation of sugar metabolism genes, vacuolar metabolite accumulation, and phosphorylation of tonoplast transporters in cold-stressed barley, as revealed by phosphoproteomics. This convergence of multi-omics data accelerates the identification of pleiotropic genes and PTM-modified proteins that bridge genotype-to-phenotype gaps, providing valuable targets for breeding climate-resilient crops (Chao et al. 2024).

4.3 Genome Editing and Synthetic Biology Strategies: Targeting PTM-Related Genes Using CRISPR

CRISPR-mediated genome editing and synthetic biology are transforming crop development by enabling precise modification of genes involved in PTMs, which are key regulators of plant stress responses, growth, and productivity. CRISPR/Cas9 technology allows targeted editing of PTM-related genes such as those encoding kinases, phosphatases, or ubiquitin ligases to fine-tune protein activity and signaling pathways that govern desirable agronomic traits (Saini et al. 2023). Researchers can accelerate breeding programs and reduce regulatory constraints by introducing or eliminating specific alleles that enhance yield, abiotic stress tolerance, and disease resistance without incorporating foreign DNA. These advances are complemented by synthetic biology, which enables the design of synthetic gene circuits and regulatory elements to control PTM machinery, thereby allowing context-specific, precisely regulated expression of desirable traits in crops (Khan and Lister 2025). These approaches, demonstrated in model species such as *Nicotiana benthamiana* and *Arabidopsis thaliana*, are now being extended to major crops, offering a powerful toolkit to unlock genetic potential and develop high-yielding, climate-resilient cultivars.

5 Precision Breeding Strategies for Developing Climate-Resilient Crops

Among PTMs, phosphorylation, ubiquitination, and acetylation play pivotal roles in regulating gene expression, signal transduction, protein stability, and plant immune responses, enabling rapid adaptation to drought, heat, and pathogen stress (Friso and van Wijk 2015). Advances in high-throughput proteomics and genome editing tools, such as CRISPR/Cas9, are accelerating the development of crops with enhanced resilience and productivity under changing climatic conditions by enabling precise identification and targeted modification of PTM-related genes (Nerkar et al. 2022). By integrating PTM knowledge with genomic selection and marker-assisted breeding, researchers can more effectively select for traits associated with climate adaptation, including enhanced water-use efficiency, disease resistance, and metabolic flexibility. This strategy not only expands the diversity and functionality of crop proteomes but also establishes a robust framework for developing varieties capable of withstanding multiple, concurrent stressors, thereby contributing to global food security under climate change (Koshariya 2022).

6 Conclusion and Future Perspectives

PTMs form a crucial regulatory layer that controls protein function, stability, subcellular localization, and interaction networks in plants. Their reversible and dynamic nature allows plants to rapidly adjust cellular processes in response to environmental stresses, developmental cues, and nutrient fluctuations. Despite their pivotal role in regulating plant physiology, PTMs remain relatively underutilized in practical plant breeding and crop improvement strategies. This chapter highlights the multifaceted roles of PTMs in regulating stress tolerance, growth, immune responses, and nutrient-use efficiency across diverse plant systems. Recent advances in proteomics and multi-omics technologies have significantly deepened our understanding of PTM landscapes, uncovering complex regulatory networks that connect molecular signaling to phenotypic outcomes. These insights provide a robust framework for exploiting PTMs as precise molecular targets in crop improvement strategies. In addition, integrating PTM proteomics with transcriptomics, genomics, metabolomics, and ionomics provides a comprehensive framework for identifying key regulatory genes and molecular markers. This integrated multi-omics approach can be effectively utilized in precision breeding, marker-assisted selection, and the development of transgenic crops tailored to withstand the challenges posed by climate change. Ultimately, such strategies contribute to the development of climate-smart crops, thereby strengthening global food security and ensuring sufficient food supply for the growing population. Furthermore, researchers should prioritize the development of comprehensive PTM databases for diverse crop species, alongside the creation of cost-effective analytical tools for accurate PTM identification. Such advancements are essential to unravel the intricate regulatory mechanisms and complex crosstalk among various phytohormone pathways and PTMs under realistic field conditions. Moreover, translating laboratory discoveries into practical field applications requires strong interdisciplinary collaboration among agronomists, plant physiologists, plant biochemists, plant breeders, and molecular biologists. Therefore, these concerted efforts will accelerate the integration of PTM-based insights into crop improvement programs, eventually contributing to sustainable agriculture and global food security amid the challenges posed by a changing climate.

Conflict of Interest The authors declare no conflict of interest.

References

Ali A, Alexandersson E, Sandin M, Resjö S, Lenman M, Hedley P, Levander F, Andreasson E (2014) Quantitative proteomics and transcriptomics of potato in response to *Phytophthora infestans* in compatible and incompatible interactions. BMC Genomics 15:1–18

Ali F, Dar JS, Magray AR, Ganai BA, Chishti MZ (2019) Posttranslational modifications of proteins and their role in biological processes and associated diseases. In Protein Modificomics (pp. 1–35). Academic Press.

Aliniaeifard S, Shomali A, Seifikalhor M, Lastochkina O (2020) Calcium signaling in plants under drought. In: Salt and drought stress tolerance in plants: signaling networks and adaptive mechanisms. Springer, pp 259–298

Ashraf A, Bhardwaj S, Ishtiaq H, Devi YK, Kapoor D (2021) Lead uptake, toxicity and mitigation strategies in plants. Plant Archives 21(1):712–721

Aslam MM, Okal EJ, Waseem M (2023) Cadmium toxicity impacts plant growth and plant remediation strategies. J Plant Growth Regul 99(3):397–412. https://doi.org/10.1007/s10725-022-00917-7

Assaha DVM, Ueda A, Saneoka H, Al-Yahyai R, Yaish MW (2017) The role of Na^+ and K^+ transporters in salt stress adaptation in glycophytes. Front Physiol 8:509

Bajwa VS, Shukla MR, Sherif SM, Murch SJ, Saxena PK (2014) Role of melatonin in alleviating cold stress in *Arabidopsis thaliana*. J Pineal Res 56:238–245

Bali AS, Sidhu GPS (2021) Arsenic acquisition, toxicity and tolerance in plants—from physiology to remediation: a review. Chemosphere 283:131050. https://doi.org/10.1016/j.chemosphere.2021.131050

Berrocal-Lobo M, Ibanez C, Acebo P, Ramos A, Perez-Solis E, Collada C, Casado R, Aragoncillo C, Allona I (2011) Identification of a homolog of Arabidopsis DSP4 (SEX4) in chestnut: its induction and accumulation in stem amyloplasts during winter or in response to the cold. Plant Cell Environ 34(10):1693–1704

Birhanu MW (2020) Transgenic approaches of improving tomato (*Solanum lycopersicum*) to salt stress tolerance. J Biol Agric Healthc. https://doi.org/10.7176/JBAH/10-14-01

Bobalova J, Strouhalova D, Bobal P (2023) Common post-translational modifications (PTMs) of proteins: analysis by up-to-date analytical techniques with an emphasis on barley. J Agric Food Chem 71(41):14825–14837

Carrie C, Giraud E, Duncan O, Xu L, Wang Y, Huang S, Whelan J (2010) Conserved and novel functions for Arabidopsis thaliana MIA40 in assembly of proteins in mitochondria and peroxisomes. J Biol Chem 285(46):36138–36148

Chamberlain LH, Shipston MJ (2015) The physiology of protein S-acylation. Physiol Rev 95(2):341–376

Chang MX, Gu M, Xia YW, Dai XL, Dai CR, Zhang J, Xu GH (2019) OsPHT1; 3 mediates uptake, translocation, and remobilization of phosphate under extremely low phosphate regimes. Plant Physiol 179(2):656–670

Chao H, Zhang S, Hu Y, Ni Q, Xin S, Zhao L, Ivanisenko VA, Orlov YL, Chen M (2024) Integrating omics databases for enhanced crop breeding. J Integr Bioinform 20(4):20230012

Chen X, Ding Y, Yang Y, Song C, Wang B, Yang S, Guo Y, Gong Z (2021) Protein kinases in plant responses to drought, salt, and cold stress. J Integr Plant Biol 63:53–78

Cheng C, Wang Z, Ren Z, Zhi L, Yao B, Su C, Li X (2017) SCFAtPP2-B11 modulates ABA signaling by facilitating SnRK2. 3 degradation in Arabidopsis thaliana. PLoS Genet 13(8):e1006947

Chung C-R, Tang Y, Chiu Y-P, Li S, Hsieh W-K, Yao L, Chiang Y-C, Pang Y, Chen G-T, Chou K-C (2025) dbPTM 2025 update: comprehensive integration of PTMs and proteomic data for advanced insights into cancer research. Nucleic Acids Res 53(D1):D377–D386

Colón-Carmona A, Chen DL, Yeh KC, Abel S (2000) Aux/IAA proteins are phosphorylated by phytochrome in vitro. Plant Physiol 124(4):1728–1738

Deng W, Casao MC, Wang P, Sato K, Hayes PM, Finnegan EJ, Trevaskis B (2015) Direct links between the vernalization response and other key traits of cereal crops. Nat Commun 6(1):5882

Di Pietro M, Vialaret J, Li G-W, Hem S, Prado K, Rossignol M, Maurel C, Santoni V (2013) Coordinated post-translational responses of aquaporins to abiotic and nutritional stimuli in Arabidopsis roots. Mol Cell Proteomics 12(12):3886–3897

Dong C-H, Agarwal M, Zhang Y, Xie Q, Zhu J-K (2006) The negative regulator of plant cold responses, HOS1, is a RING E3 ligase that mediates the ubiquitination and degradation of ICE1. Proc Natl Acad Sci 103:8281–8286

Fang Y, Li J, Jiang J, Geng Y, Wang J, Wang Y (2017) Physiological and epigenetic analyses of *Brassica napus* seed germination in response to salt stress. Acta Physiol Plant 39(6):128

Fang P, Hu Y, Guo Q, Li L, Xu P (2023) Karrikins, redox and plant abiotic stress tolerance: a focal review. Plant Stress 9:100185

Fang XZ, Fang SQ, Ye ZQ, Liu D, Zhao KL, Jin CW (2021) NRT1. 1 dual-affinity nitrate transport/signalling and its roles in plant abiotic stress resistance. Front Plant Sci 12:715694

Fhu CW, Ali A (2021) Protein lipidation by palmitoylation and myristoylation in cancer. Front Cell Dev Biol 9:673647

Friso G, van Wijk KJ (2015) Posttranslational protein modifications in plant metabolism. Plant Physiol 169(3):1469–1487

Gough C, Sadanandom A (2021) Understanding and exploiting post-translational modifications for plant disease resistance. Biomolecules 11(8):1122

Grabsztunowicz M, Koskela MM, Mulo P (2017) Post-translational modifications in regulation of chloroplast function: recent advances. Front Plant Sci 8:240

Gzyl J, Chmielowska-Bąk J, Przymusiński R, Gwóźdź EA (2015) Cadmium affects microtubule organization and post-translational modifications of tubulin in seedlings of soybean (*Glycine max* L.). Front Plant Sci 6:937

Hafeez A, Rasheed R, Ashraf MA, Qureshi FF, Hussain I, Iqbal M (2023) Effect of heavy metals on growth, physiological and biochemical responses of plants. In: Plants and their interaction to environmental pollution. Elsevier, Amsterdam, pp 139–159

Hao DL, Zhou JY, Qu J, Zhai JY, Zhao Y, Li L, Zong JQ (2025) Genome-wide identification and functional characterization of the ammonium transporter family in centipedegrass suggest their roles in acidic soil adaptation. Plant Physiol Biochem 110287

Hashiguchi A, Komatsu S (2017) Posttranslational modifications and plant–environment interaction. Methods Enzymol 586:97–113

Hu J, Rampitsch C, Bykova NV (2015) Advances in plant proteomics toward improvement of crop productivity and stress resistance. Front Plant Sci 6:209

Jamsheer KM, Kumar M, Srivastava V (2021) SNF1-related protein kinase 1: the many-faced signaling hub regulating developmental plasticity in plants. J Exp Bot 72(17):6042–6065

Kamal MM, Ishikawa S, Takahashi F, Suzuki K, Kamo M, Umezawa T, Shinozaki K, Kawamura Y, Uemura M (2020) Large-scale phosphoproteomic study of *Arabidopsis* membrane proteins reveals early signaling events in response to cold. Int J Mol Sci 21:8631

Khan A, Lister R (2025) Synthetic gene circuits in plants: recent advances and challenges. Quant Plant Biol 6:e6

Khan M, Nizamani MM, Asif M, Kamran A, He G, Li X, Yang S, Xie X (2025) Comprehensive approaches to heavy metal bioremediation: integrating microbial insights and genetic innovations. J Environ Manag 374:123969

Kong L, Cheng J, Zhu Y, Ding Y, Meng J, Chen Z, Xie Q, Guo Y, Li J, Yang S (2015) Degradation of the ABA co-receptor ABI1 by PUB12/13 U-box E3 ligases. Nat Commun 6:8630

Koshariya AK (2022) Climate-resilient crops: breeding strategies for extreme weather conditions. Plant Sci Arch 7(2):1–3

Kosov'a K, Vítamvas P, Prasil IT, Klíma M, Renaut J (2021) Plant proteoforms under environmental stress: functional proteins arising from a single gene. Front Plant Sci 12:793113

Kreps JA, Wu Y, Chang H-S, Zhu T, Wang X, Harper JF, J.F. (2002) Transcriptome changes for *Arabidopsis* in response to salt, osmotic, and cold stress. Plant Physiol 130:2129–2141

Kurepa J, Walker JM, Smalle J, Gosink MM, Davis SJ, Durham TL, Sung D-Y, Vierstra RD (2003) The small ubiquitin-like modifier (SUMO) protein modification system in *Arabidopsis*: accumulation of SUMO1 and -2 conjugates is increased by stress. J Biol Chem 278(9):6862–6872

Lee JM, Hammarén HM, Savitski MM, Baek SH (2023) Control of protein stability by post-translational modifications. Nat Commun 14(1):201

Li H, Yan S, Zhao L, Tan J, Zhang Q, Gao F, Wang P, Hou H, Li L (2014) Histone acetylation associated up-regulation of the cell wall related genes is involved in salt stress induced maize root swelling. BMC Plant Biol 14(1):105

Li J, Jia Y, Dong R, Huang R, Liu P, Li X, Wang Z, Liu G, Chen Z (2019) Advances in the mechanisms of plant tolerance to manganese toxicity. Int J Mol Sci 20(20):5096

Li S, Li T, Zhang P, Wang X, Feng W, Zhang Y, Chen B, Liu Y, Zhan G, Hao C (2025a) The E3 ubiquitin ligase TaGW2 facilitates TaSnRK1γ and TaVPS24 degradation to enhance stripe rust susceptibility in wheat. Plant Biotechnol J 23(3):750–765

Li Z, Li S, Luo M, Jhong JH, Li W, Yao L, Lee TY (2022) dbPTM in 2022: an updated database for exploring regulatory networks and functional associations of protein post-translational modifications. Nucleic Acids Res 50(D1):D471–D479.

Li Z, Qin L, Wang D, Cui T, Liang Y, Zang S, Sun T, Zhao W, Wu Q, Que Y (2025b) Protein post-translational modification in plants: regulation and beyond. Hortic Plant J. https://doi.org/10.1016/j.hpj.2025.02.003

Liu H, Stone SL (2010) Abscisic acid increases Arabidopsis ABI5 transcription factor levels by promoting KEG E3 ligase self-ubiquitination and proteasomal degradation. Plant Cell 22(8):2630–2641.

Liu M, Zhao G, Huang X, Pan T, Chen W, Qu M, Ouyang B, Yu M, Shabala S (2023) Candidate regulators of drought stress in tomato revealed by comparative transcriptomic and proteomic analyses. Front Plant Sci 14:1282718

Loqué D, Yuan L, Kojima S, Gojon A, Wirth J, Gazzarrini S, Von Wirén N (2006) Additive contribution of AMT1; 1 and AMT1; 3 to high-affinity ammonium uptake across the plasma membrane of nitrogen-deficient Arabidopsis roots. Plant J 48(4):522–534

Luján-Soto E, Dinkova TD (2021) Time to wake up: Epigenetic and small-RNA-mediated regulation during seed germination. Plants 10(2):236.

Mahajan S, Pandey GK, Tuteja N (2008) Calcium-and salt-stress signaling in plants: shedding light on SOS pathway. Arch Biochem Biophys 471:146–158

Marondedze C, Elia G, Thomas L, Wong A, Gehring C (2021) Citrullination of proteins as a specific response mechanism in plants. Front Plant Sci 12:638392

Martí-Guillén JM, Pardo-Hernández M, Martínez-Lorente SE, Almagro L, Rivero RM (2022) Redox post-translational modifications and their interplay in plant abiotic stress tolerance. Front Plant Sci 13:1027730

Meir Z, Aviezer I, Chongloi GL, Ben-Kiki O, Bronstein R, Mukamel Z, Eshed Y (2021) Dissection of floral transition by single-meristem transcriptomes at high temporal resolution. Nat Plants 7(6):800–813.

Millar AH, Heazlewood JL, Giglione C, Holdsworth MJ, Bachmair A, Schulze WX (2019) The scope, functions, and dynamics of posttranslational protein modifications. Annu Rev Plant Biol 70(1):119–151

Muleya V, Lois LM, Chahtane H, Thomas L, Chiapello M, Marondedze C (2022) (De) activation (ir) reversibly or degradation: dynamics of post-translational protein modifications in plants. Life 12(2):324

Munns R (2002) Comparative physiology of salt and water stress. Plant Cell Environ 25:239–250

Muthamilarasan M, Prasad M (2013) Plant innate immunity: an updated insight into defense mechanism. J Biosci 38:433–449

Nagarajan VK, Jain A, Poling MD, Lewis AJ, Raghothama KG, Smith AP (2011) Arabidopsis Pht1; 5 mobilizes phosphate between source and sink organs and influences the interaction between phosphate homeostasis and ethylene signaling. Plant Physiol 156(3):1149–1163.

Naik B, Kumar V, Rizwanuddin S, Chauhan M, Choudhary M, Gupta AK, Kumar P, Kumar V, Saris PEJ, Rather MA, Bhuyan S, Neog PR, Mishra S, Rustagi S (2023) Genomics, proteomics, and metabolomics approaches to improve abiotic stress tolerance in tomato plant. Int J Mol Sci 24(3):3025

Nayak NR, Putnam AA, Addepalli B, Lowenson JD, Chen T, Jankowsky E, Perry SE, Dinkins RD, Limbach PA, Clarke SG (2013) An Arabidopsis ATP-dependent, DEAD-box RNA helicase loses activity upon IsoAsp formation but is restored by PROTEIN ISOASPARTYL METHYLTRANSFERASE. Plant Cell 25(7):2573–2586

Nerkar G, Devarumath S, Purankar M, Kumar A, Valarmathi R, Devarumath R, Appunu C (2022) Advances in crop breeding through precision genome editing. Front Genet 13:880195

Niekerk LA, Gokul A, Basson G, Badiwe M, Nkomo M, Klein A, Keyster M (2024) HMs stress and mitogen activated kinase transcription factors in plants: exploring heavy metal-ROS influences on plant signalling pathways. Plant Cell Environ 47(8):2793–2810

Niño-González M, Duque P (2025) Posttranslational regulation of plant membrane transporters. Plant J 121(3):e17262

Notaguchi M, Abe M, Kimura T, Daimon Y, Kobayashi T, Yamaguchi A, Araki T (2008) Long-distance, graft-transmissible action of Arabidopsis FLOWERING LOCUS T protein to promote flowering. Plant Cell Physiol 49(11):1645–1658.

Novillo F, Alonso JM, Ecker JR, Salinas J (2004) CBF2/DREB1C is a negative regulator of CBF1/DREB1B and CBF3/DREB1A expression and plays a central role in stress tolerance in *Arabidopsis*. Proc Natl Acad Sci 101:3985–3990

Novillo F, Medina J, Salinas J (2007) *Arabidopsis* CBF1 and CBF3 have a different function than CBF2 in cold acclimation and define different gene classes in the CBF regulon. Proc Natl Acad Sci 104:21002–21007

Park C, Lim CW, Lee SC (2016) The pepper RING-type E3 ligase, CaAIP1, functions as a positive regulator of drought and high salinity stress responses. Plant Cell Physiol 57:2202–2212

Peng L, Wan X, Huang K, Pei L, Xiong J, Li X, Wang J (2019) AtPUB48 E3 ligase plays a crucial role in the thermotolerance of *Arabidopsis*. Biochem Biophys Res Commun 509(1):281–286

Piskurewicz U, Jikumaru Y, Kinoshita N, Nambara E, Kamiya Y, Lopez-Molina L (2008) The gibberellic acid signaling repressor RGL2 inhibits Arabidopsis seed germination by stimulating abscisic acid synthesis and ABI5 activity. Plant Cell 20(10):2729–2745.

Qi Q, Yanyan D, Yuanlin L, Kunzhi L, Huini X, Xudong S (2020) Overexpression of SlMDHAR in transgenic tobacco increased salt stress tolerance involving S-nitrosylation regulation. Plant Sci 299:110609

Reumann S, Parasyri A (2025) Protein acetylation as a novel post-translational modification of plant peroxisomal proteins. J Exp Bot 76:3311–3323

Saini H, Thakur R, Gill R, Tyagi K, Goswami M (2023) CRISPR/Cas9-gene editing approaches in plant breeding. GM Crops Food 14(1):1–17

Saruhashi M, Kumar Ghosh T, Arai K, Ishizaki Y, Hagiwara K, Komatsu K, Takezawa D (2015) Plant Raf-like kinase integrates abscisic acid and hyperosmotic stress signaling upstream of SNF1-related protein kinase2. Proc Natl Acad Sci 112(46):E6388-E6396.

Sharma B, Joshi D, Yadav PK, Gupta AK, Bhatt TK (2016) Role of ubiquitin-mediated degradation system in plant biology. Front Plant Sci 7:806

Sharma M, Fuertes D, Perez-Gil J, Lois LM (2021a) SUMOylation in phytopathogen interactions: balancing invasion and resistance. Front Cell Dev Biol 9:703795

Sharma S, Prasad A, Sharma N, Prasad M (2021b) Role of ubiquitination enzymes in abiotic environmental interactions with plants. Int J Biol Macromol 181:494–507

Shen S, Zhan C, Yang C, Fernie AR, Luo J (2023) Metabolomics centered mining of plant metabolic diversity and function: past decade and future perspectives. Mol Plant 16(1):43–63

Soma F, Takahashi F, Yamaguchi-Shinozaki K, Shinozaki K (2021) Cellular phosphorylation signaling and gene expression in drought stress responses: ABA-dependent and ABA-independent regulatory systems. Plants 10(4):756.

Song L, Chen W, Wang B, Yao Q-M, Valliyodan B, Bai M-Y, Zhao M-Z, Ye H, Wang Z-Y, Nguyen HT (2019) GmBZL3 acts as a major BR signaling regulator through crosstalk with multiple pathways in *Glycine max*. BMC Plant Biol 19(1):86

Sun Q, Miao C, Hanel M, Borthwick AG, Duan Q, Ji D, Li H (2019) Global heat stress on health, wildfires, and agricultural crops under different levels of climate warming. Environ Int 128:125–136

Takano S, Matsuda S, Funabiki A, Furukawa J, Yamauchi T, Tokuji Y, Nakazono M, Shinohara Y, Takamure I, Kato K (2015) The rice RCN11 gene encodes β1, 2-xylosyltransferase and is required for plant responses to abiotic stresses and phytohormones. Plant Sci 236:75–88

Tan W, Yue X, Pan Y, Hu J, Huang R, Tan H, Yao Y (2025) BIN2-mediated phosphorylation of KAN1 integrates brassinosteroid and auxin signaling during poplar secondary growth. Plant Cell 37(10):koaf222.

Tang K, Zhao L, Ren Y, Yang S, Zhu J, Zhao C (2020) The transcription factor ICE1 functions in cold stress response by binding to the promoters of CBF and COR genes. J Integr Plant Biol 62:258–263

Tian M, Li X, Yu L, Qian J, Bai X, Yang J, Deng R, Lu C, Zhao H, Liu Y (2025) Glycosylation as an intricate post-translational modification process takes part in glycoproteins related immunity. Cell Commun Signal 23(1):214

Tigchelaar M, Battisti DS, Naylor RL, Ray DK (2018) Future warming increases probability of globally synchronized maize production shocks. Proc Natl Acad Sci USA 115(26):6644–6649

Torres O, Abad-Sojos S, Sánchez K, Carvalho RF, Carvalho SD (2019) Epigenetics of light signaling during plant development. In Epigenetics in plants of agronomic importance: fundamentals and applications: transcriptional regulation and chromatin remodelling in plants (pp. 223-246). Cham: Springer International Publishing

Tuteja N, Sopory SK (2008) Chemical signaling under abiotic stress environment in plants. Plant Signal Behav 3:525–536

Wang F, Deng M, Chen J, He Q, Jia X, Guo H, Mao C (2020) CASEIN KINASE2-dependent phosphorylation of PHOSPHATE2 fine-tunes phosphate homeostasis in rice. Plant Physiol 183(1):250–262

Wang X (2019) Protein and proteome atlas for plants under stresses: new highlights and ways for integrated omics in post-genomics era. Int J Mol Sci 20(20):5222

Wang N, Liu Y, Cong Y, Wang T, Zhong X, Yang S, Li Y, Gai J (2016) Genome-wide identification of soybean U-box E3 ubiquitin ligases and roles of GmPUB8 in negative regulation of drought stress response in *Arabidopsis*. Plant Cell Physiol 57:1189–1209

Wang Z, Xie C, Wu Y, Liu H, Zhang X, Du H, Li X, Zhang C (2024) Integrated metabolomics and transcriptomics analyses reveal the regulatory mechanisms of anthocyanin and carotenoid accumulation in the Peel of *Coffea arabica*. Int J Mol Sci 25(19):10754

Wang L, Ju C, Han C, Yu Z, Bai M-Y, Wang C (2025) The interaction of nutrient uptake with biotic and abiotic stresses in plants. J Integr Plant Biol 67(3):455–487

Wang W, Hu B, Li A, Chu C (2020) NRT1. 1s in plants: functions beyond nitrate transport. J Exp Bot 71(15):4373–4379

Wani SH, Kumar V, Shriram V, Sahd SK (2016) Phytohormones and their metabolic engineering for abiotic stress tolerance in crop plants. Crop J 4:162–176

Weiner JJ, Peterson FC, Volkman BF, Cutler SR (2010) Structural and functional insights into core ABA signaling. Curr Opin Plant Biol 13(5):495–502

Wen B, Li C, Fu X, Li D, Li L, Chen X, Gao D (2019) Effects of nitrate deficiency on nitrate assimilation and chlorophyll synthesis of detached apple leaves. Plant Physiol Biochem 142:363–371

Weng M, Yang YUE, Feng H, Pan Z, SHEN WH, Zhu YAN, Dong A (2014) Histone chaperone ASF1 is involved in gene transcription activation in response to heat stress in A rabidopsis thaliana. Plant Cell Physiol 37(9):2128–2138.

Wu X, Liu T, Zhang Y, Duan F, Neuhäuser B, Ludewig U, Yuan L (2019) Ammonium and nitrate regulate NH4+ uptake activity of Arabidopsis ammonium transporter AtAMT1; 3 via phosphorylation at multiple C-terminal sites. J Exp Bot 70(18):4919–4930.

Xian L, Tian J, Long Y, Ma H, Tian M, Liu X, Yin G, Wang L (2023) Metabolomics and transcriptomics analyses provide new insights into the nutritional quality during the endosperm development of different ploidy rice. Front Plant Sci 14:1210134

Xiao C, Gao J, Zhang Y, Wang Z, Zhang D, Chen Q, Ye X, Xu Y, Yang G, Yan L (2019) Quantitative proteomics of potato leaves infected with *Phytophthora infestans* provides insights into coordinated and altered protein expression during early and late disease stages. Int J Mol Sci 20:136

Xiong M, Feng GN, Gao Q, Zhang CQ, Li QF, Liu QQ (2022) Brassinosteroid regulation in rice seed biology. Seed Biology 1(1):1–9

Xu F, He S, Zhang J, Mao Z, Wang W, Li T, Yang HQ (2018) Photoactivated CRY1 and phyB interact directly with AUX/IAA proteins to inhibit auxin signaling in Arabidopsis. Mol Plant 11(4):523–541

Xu YX, Xiao MZ, Liu Y, Fu JL, He Y, Jiang DA (2017) The small auxin-up RNA OsSAUR45 affects auxin synthesis and transport in rice. Plant Mol Biol 94(1):97–107

Yali W, Begna T (2024) Clinical endocrinology and metabolism. Metabolism 3:6

Yan Q, Zhang G, Zhang X, Huang L (2024) A review of transcriptomics and metabolomics in plant quality and environmental response: from bibliometric analysis to science mapping and future trends. Meta 14(5):272

Yang P (2022) Proteomics of post-translational modifications in plants. Front Plant Sci 13:894314

Yang B, Ruan R, Cantu D, Wang X, Ji W, Ronald PC, Dubcovsky J (2013) A comparative approach expands the protein–protein interaction node of the immune receptor XA21 in wheat and rice. Genome 56(6):315–326

Yang Y, Saand MA, Huang L, Abdelaal WB, Zhang J, Wu Y, Li J, Sirohi MH, Wang F (2021) Applications of multi-omics technologies for crop improvement. Front Plant Sci 12:563953

Yang Z, Yang J, Wang Y, Wang F, Mao W, He Q, Mao C (2020) PROTEIN PHOSPHATASE95 regulates phosphate homeostasis by affecting phosphate transporter trafficking in rice. Plant Cell 32(3):740–757.

Yeqing C, Jun L, Weinan W, Chunguo F, Guozhen Y, Jingjing S, Changquan W (2023) Rose long noncoding RNA lncWD83 promotes flowering by modulating ubiquitination of the floral repressor RcMYC2L. Plant Physiol 193(4):2573–2591

Yin X, Komatsu S (2015) Quantitative proteomics of nuclear phosphoproteins in the root tip of soybean during the initial stages of flooding stress. J Proteome 119:183–195

Yuan Y, Li P, Li J, Zhao Q, Chang Y, He X (2024) Protein lipidation in health and disease: molecular basis, physiological function and pathological implication. Signal Transduct Target Ther 9(1):60

Zenda T, Liu S, Dong A, Li J, Wang Y, Liu X, Wang N, Duan H (2021) Omics-facilitated crop improvement for climate resilience and superior nutritive value. Front Plant Sci 12:774994

Zhang P, Tang Y, Liu Y, Liu J, Wang Q, Wang H, Li H, Li L, Qin P (2023) Metabolic characteristics of self-pollinated wheat seed under red and blue light during early development. Planta 258(3):63

Zhang RX, Liu Y, Xiao W, Huang H, Zhao Y, Han Y (2025) Diverse actions of auxin on fruit development and ripening. Hortic Adv 3(1):29

Zhang Y, Liu Z, Chen Y, He JX, Bi Y (2015) PHYTOCHROME-INTERACTING FACTOR 5 (PIF5) positively regulates dark-induced senescence and chlorophyll degradation in Arabidopsis. Plant Sci 237:57–68

Zhao L, Wang P, Yan S, Gao F, Li H, Hou H, Zhang Q, Tan J, Li L (2014) Promoter-associated histone acetylation is involved in the osmotic stress-induced transcriptional regulation of the maize ZmDREB2A gene. Physiol Plant 151(4):459–467

Zhao C, Liu B, Piao S, Wang X, Lobell DB, Huang Y, Huang M, Yao Y, Bassu S, Ciais P (2017) Temperature increase reduces global yields of major crops in four independent estimates. Proc Natl Acad Sci USA 114(35):9326–9331

Integrating Omics Data for Phenomic Prediction

Tiziana M. Sirangelo and Natasha Damiana Spadafora

Abstract A widespread diffusion of genotyping is still a challenge in certain breeding programs, primarily due to prohibitive costs and the lack of appropriate genotyping resources for some plant species. Consequently, there is a growing interest in alternative approaches that can provide more easily accessible means of phenotype prediction. Phenomic prediction (PP) captures epigenetic, environmental, and dynamic influences that cannot be fully explained by genomic data alone. Recently, PP methods have been employed to complement genomic prediction (GP) approaches, addressing a key limitation of linear genomic prediction models in predicting complex traits, namely, their inability to account for genotype-by-environment interactions. On the other hand, multi-omics technologies have led to novel strategies for PP and GP, and the integration of complementary omics layers, by providing a more comprehensive view of the molecular mechanisms underlying phenotypic variation, has emerged as a promising approach to enhance prediction accuracy. The use of integrative models, capable of combining heterogeneous variables, allows to weight each level according to its predictive power and to capture nonlinear interactions between levels. Datasets can be managed by machine learning and deep learning tools, which are becoming crucial to extract meaningful patterns from these vast and heterogeneous resources.

Keywords Phenomic prediction · Genomic selection · Omics · Multi-omics · Machine learning · Deep learning

---

T. M. Sirangelo
Division Biotechnologies, ENEA—Italian National Agency for New Technologies, Energy and Sustainable Economic Development, Casaccia Research Center, Rome, Italy
e-mail: tiziana.sirangelo@enea.it

N. D. Spadafora (✉)
Department of Environmental and Prevention Sciences, University of Ferrara, Ferrara, Italy
e-mail: damiana.spadafora@unife.it

L. M. Vaschetto (ed.), *Omics Technologies in Plant Science*, Interdisciplinary Biotechnological Advances, https://doi.org/10.1007/978-981-92-0533-2_9

Abbreviations

BLUP	Best Linear Unbiased Prediction
CNN	Convolutional Neural Network
DL	Deep learning
GBLUP	Genomic Best Linear Unbiased Prediction
GEBV	Genomic estimated breeding values
GP	Genomic prediction
GS	Genomic selection
KNNs	K-nearest neighbors
METs	Multi-environment trials
ML	Machine learning
MLR	Multiple linear regression
NGS	Next Generation Sequencing
NIRS	Near-infrared spectroscopy
PA	Prediction accuracy
PP	Phenomic prediction
QTL	Quantitative trait locus
RF	Random forest
SNPs	Single-nucleotide polymorphisms

1 Introduction

The main objective of plant breeding is to select accessions with desirable traits. For this purpose, breeding material must be phenotyped; however, this process can be time-consuming and expensive. Genomic selection (GS) has made it possible for breeders to make informed decisions about unphenotyped crops by predicting their performance based on genotypic information (Crossa et al. 2017). In fact, this breeding strategy uses a statistical and computational process that, by a large number of genetic markers, is able to predict genomic estimated breeding values (GEBV). This process, known as genomic prediction (GP), is based on prediction models that use a training dataset comprising both genotypic and phenotypic information. A widely adopted model for GP is the Genomic Best Linear Unbiased Prediction Model (GBLUP), in which all single-nucleotide polymorphisms (SNPs) markers are used to estimate a genomic relationship matrix (Crossa et al. 2017; Bernardo 2020, 2021). This matrix is then used to estimate the GEBVs, reducing data dimensionality by avoiding individual marker effects. The efficiency of genotyping technologies has noticeably increased over time, and the adoption of GP in crop breeding programs has consequently increased (Crossa et al. 2021).

However, the wide implementation of genotyping remains challenging in some breeding programs due to the prohibitive costs and the lack of suitable genotyping resources for certain plant species (Crossa et al. 2017). Consequently, there is a growing interest in alternative, more accessible methods for phenotype prediction. Phenomic Prediction (PP), first described by Rincent et al. (2018), has emerged as a promising alternative to GP. It employs high-dimensional phenotypic information, such as data derived from near-infrared spectroscopy (NIRS) (Aguate et al. 2017; Hayes et al. 2017), by incorporating it into statistical models analogous to GBLUP. This allows the prediction of traits across different environments and facilitates the selection of desirable lines, similarly to the way SNPs are used in GP. The relationship matrix based on NIRS is expected to have similarities with the genomic relationship matrix, because the reflectance spectrum is related to the biochemical composition of the investigated sample (Beer-Lambert law), which in turn is correlated with genetic and environmental factors (Brault et al. 2022). NIRS can be applied directly to intact seeds of many plant species, allowing non-destructive measurements suitable for PP (Parmley et al. 2019; Galán et al. 2021). Therefore, PP is characterized by several advantages, including its applicability to different samples with minimal preparation and rapid data processing, making it highly suitable for high-throughput phenotyping.

Rincent et al. (2018) showed that the prediction accuracy (PA) of yield in winter wheat using PP exceeded that of GS models. Numerous other studies confirmed the successful application of PP based on NIRS data in wheat (Krause et al. 2019; Cuevas et al. 2019; Robert et al. 2022). More recently, Winn et al. (2023) compared GP (using GEBVs generated by a GBLUP model) and PP (using multiple linear regression (MLR), k-nearest neighbors (KNNs), and random forest (RF) models) for grain yield prediction. Their cross-validation showed that PP achieved higher PA compared to GP.

Spectral reflectance has also been found to be able to estimate wheat health and environmental stress (Du and Noguchi 2017; Francesconi et al. 2021; Qin et al. 2022), suggesting that the inclusion of these effects into the estimation may improve the PA of PP.

To date, most studies on GP and PP approaches focused on traits related to seed yield, quality, or plant phenology mainly in cereal crops (Lane et al. 2020; Weiß et al. 2022). However, PP has recently been applied to other species such as soybean (Zhu et al. 2021) and rapeseed (Laurençon et al. 2024; Roscher-Ehrig et al. 2024).

Despite its promise, the NIRS-based PP methodology has been mainly applied to annual crop species, and its use in perennial species remains limited. Moreover, PP has only been tested on a narrow range of traits, mainly related to yield or phenology. Recently, Brault et al. (2022) applied PP to grapevine, a perennial crop, using NIRS data from two tissues to predict a broad set of traits, including berry composition, phenology, morphological, and vigor. Furthermore, in order to minimize genotype-by-environment effects, spectra and phenotypes collected across multiple years were used. The achieved results showed that similarity between NIRS and genomic relationship matrices was stable across tissues or years, though variable across populations. In addition, applying a mixed model to NIRS data improved PA, while using data collected from only one tissue had a limited effect.

Nowadays, technological advancements allow to collect spectral data at different growth stages using portable spectrometers or remote sensing platforms. These tools offer rapid, reliable, and high-throughput alternatives to traditional methods of collecting phenotypic data, such as visual scoring, which are often slow and subjective (Atkinson et al. 2018). This has facilitated the expansion of high-throughput phenotyping in plant breeding and the improvement of its significance to plant genetics, as reported by several authors, such as Araus et al. (2018), Atkinson et al. (2018), and Rebetzke et al. (2019).

Recently, phenomics prediction methods have been used to complement genomic prediction approaches to overcome the limitation of linear GP in capturing genotype-by-environment interactions. For instance, Jackson et al. (2023) demonstrated that environmental variation in winter wheat can be effectively captured using a large number of phenomic markers, thereby improving PA of GP. Several other studies also reported an increase in PA when combining genomic and phenomic matrices in a unique prediction model (Cuevas et al. 2019; Galán et al. 2020).

2 Integrating Omics Data for Phenomics Prediction

Advances in genome sequencing and gene expression analysis, particularly through the use of next-generation sequencing (NGS) technologies, have significantly enhanced our understanding of the regulatory mechanisms and networks underlying complex traits (Edwards and Snowdon 2013; Sirangelo et al. 2022). For instance, gene expression analysis and genome-wide transcript profiling expanded our knowledge of the functions and expression patterns of thousands of stress-responsive genes in plants (Langridge and Reynolds 2015; Sirangelo et al. 2023; Sirangelo 2024a). Furthermore, with the rapid evolution of these technologies, multi-omics datasets have become increasingly available, enabling innovative applications. Initially, such approaches were mainly applied to human health purposes. For instance, in human health, Vazquez et al. (2016) reported that integrating multi-omics profiles increased the PA of survival in breast cancer patients compared to the models obtained using only genomic information. Similarly, Amiri Roudbar et al. (2020) enhanced the prediction of the risk of rheumatoid arthritis, integrating methylation profiles into genomic prediction models.

More recently, multi-omics data have been used in agriculture to predict breeding values for agronomically important traits (Fig. 1). The development of omics and multi-omics technologies has opened new frontiers for PP, since the expression of complex traits in plants is the result of interactions of mechanisms happening at multiple molecular levels.

Wheat (*Triticum aestivum* (L.) and *T. turgidum* (L.)) was among the first crops to benefit from the integration of phenotypic, biochemical, and genomics-assisted selection methodologies, particularly for traits associated with drought tolerance (Mwadzingeni et al. 2016). Drought and climate variability are among the principal constraints on the global productivity of wheat. Furthermore, wheat has a complex and large genome that necessitates the use of interdisciplinary approaches for effective breeding. Mwadzingeni et al. (2016) highlighted the latest advances in

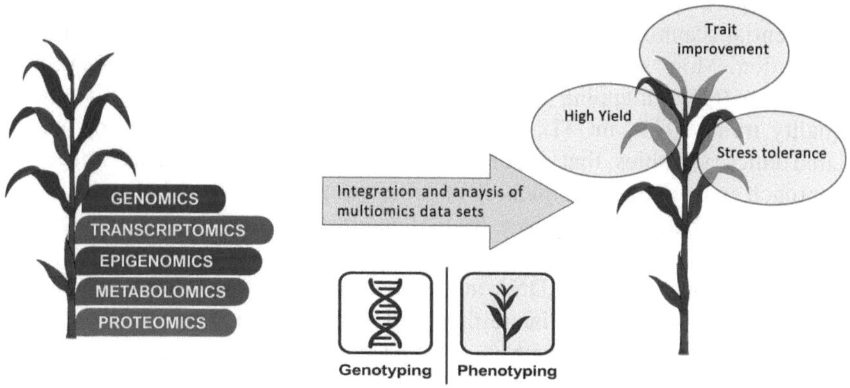

Fig. 1 Schematic representation of the integration of genomics, transcriptomics, epigenomics, metabolomics, and proteomics with high-throughput phenotyping to improve trait selection, yield, and stress tolerance in crops

phenomic and genomic technologies and illustrated their integration with other omics approaches to address these challenges.

Diseases also pose significant threats to wheat production. Particularly, fungal pathogens, such as *Fusarium* spp. causing Fusarium head blight (FHB), demand substantial resources in breeding programs (Sirangelo 2024b). A comprehensive review by Haile et al. (2020) discusses GP studies focusing on the prediction of resistance to wheat pathologies, including those based on the integration of multiple omics datasets to improve GP applicability for disease resistance breeding in wheat.

Resistance to complex stress conditions has also been an important subject of study in barley (*Hordeum vulgare*). Shen et al. (2016) investigated ionomic, metabolomic, and proteomic profiles, identifying several molecular mechanisms associated with the salt stress adaptation in barley shoots. Similarly, Ho et al. (2020) employed transcriptomic, metabolomic, and lipidomic analyses to explore salt tolerance mechanisms in barley. However, neither of these studies involved multi-omics prediction models. Only later, Hansen et al. (2022) demonstrated the predictive potential of whole-genome multi-omics profiles for complex traits in barley. They evaluated multi-omics models, based on genomic, transcriptomic, and DNA methylation data, to predict complex traits such as grain yield, kernel weight, protein content, and nitrogen uptake. Their results showed that multi-omics models explained a greater proportion of phenotypic variance (~0.70–0.90) compared to genomic models alone (~0.55–0.85), highlighting the added value of integrating diverse molecular layers.

Multi-omics approaches have also been applied to predict agronomic traits beyond stress tolerance. For example, genetic and metabolic pathways were explored in order to predict phenotypic traits in potato (*Solanum tuberosum*) (Acharjee et al. 2016). Here, gene expression data, metabolite, and proteomics data, along with selected phenotypic traits, were analyzed with an integrated omics approach for predicting four phenotypic quality traits: tuber flesh color, differential scanning calorimetry (DSC) onset, tuber shape, and enzymatic discoloration. More specifically, using an RF regression model and quantitative trait locus (QTL) mapping approach to identify variables co-segregating with a quality trait (eQTL, mQTL, or pQTL), they constructed networks of interrelated omics variables that accurately predicted these traits. This integrative approach enabled the identification of genes, proteins, and metabolites that were not only predictive for a given trait of interest but also explanatory of trait variation.

In *Arabidopsis thaliana*, DNA methylation data have also shown promise for whole-genome prediction of interesting traits. Hu et al. (2015) reported that epigenetic variation could explain the 65% of the phenotypic variance in plant height. This concept has been considered again by Wang et al. (2024), who investigated whether integrating several omics data, consisting of genomic, transcriptomic, and methylomic data, could improve the prediction of traits such as flowering time in *Arabidopsis thaliana*. Results showed that models based on individual omics datasets (genome, transcriptome, or methylome) performed comparably. Notably, transcriptome- and methylome-based models achieved similar accuracy to genome-based models. Moreover, models built using different types of omics data identified different sets of flowering genes. Importantly, integrating all three omics datasets improved the performance for RF models, whose algorithms are capable of capturing nonlinear interactions among features. Overall, this integrative modeling approach not only enhanced prediction accuracy compared to the models based on a single dataset but also revealed both known and novel gene interactions, thereby expanding our understanding of the regulatory networks underlying flowering time in *Arabidopsis thaliana*.

GP has significantly contributed to accelerating hybrid breeding, which has been shown to effectively increase the productivity of many plant species (Chacko et al. 2023). However, recent advances in multi-omics technologies have brought increasing attention to hybrid prediction strategies. For example, Westhues et al. (2017) evaluated several omics predictors, based on genomic, transcriptomic, and metabolic data, demonstrating that the integration of transcriptomic with genomic data improved the prediction rate of hybrid performance in maize (*Zea mays* L.). Their study showed significant improvements in traits such as yield and protein content compared to traditional GS methods, highlighting the value of combining multiple biological layers to better capture complex genetic effects and enhance hybrid breeding efficiency.

Building on this, Schrag et al. (2018) used genomic, transcriptomic (mRNA and sRNA), and metabolomic data from parental lines to predict important agronomic traits in maize hybrids. Their findings indicated that mRNA data were the most effective predictor for grain yield, while genomic data were superior for grain dry matter content. In contrast, sRNA data showed poor PA for both traits. Notably, combining mRNA and genomic data resulted in higher predictive accuracy across traits than using any single omics dataset, underscoring the complementary nature of multi-omics GP in hybrid selection.

In hybrid rice (*Oryza sativa* L.), Wang et al. (2019) systematically evaluated prediction strategies for traits such as yield, grain weight, number of grains per panicle, and number of tillers per plant. By testing all possible combinations of omics datasets with different prediction methods, they found that combining genomic and metabolomic data in prediction strategies outperformed single-omics approaches and other omics combinations. Among the tested methods, the BLUP model emerged as the most effective. Similarly, Yang et al. (2021) integrated parental phenotypic data into multi-omics prediction models. Using BLUP and a modified fast HAT method (Xu 2017), they evaluated various combinations of predictors. Their results demonstrated that integrating parental phenotypic data significantly improved the predictability of traits such as grain yield, number of tillers per plant, number of grains per panicle, and grain weight, compared to models based on individual omics sources.

Further studies have demonstrated the superiority of multi-omics prediction over single-omics prediction approaches, particularly in multiple environment trials (MET). For instance, Hu et al. (2021) explored the integration of multi-omics data to improve the PA of 17 agronomic and seed nutritional traits in oat, across METs and genetically distant related populations. Their results demonstrated that metabolomics generally contributed more to predictive accuracy than transcriptomics when combined with genomics data. Furthermore, the study revealed that multi-omics models in METs outperformed single-environment omics models. This improvement is attributed to the incorporation of environmentally sensitive biological layers, which better reflect G × E interactions. As a result, multi-omics models provide predictions that are more generalizable across diverse environmental conditions and climate scenarios while maintaining high accuracy.

About innovative prediction models able to manage multi-omics datasets, the potential of integrating phenomics and machine learning (ML) was underlined and discussed in several works (Li et al. 2018; Xu and Jackson 2019; Yan and Wang 2023). Some traditional ML approaches, such as RF, have enabled improvements in PA across various crop traits (Holliday et al. 2012). However, these algorithms need complex data pre-processing before modeling, and several noisy features can significantly reduce their PA (Ren et al. 2024). Despite these challenges, ML excels at identifying patterns in large datasets and analyzing complex interactions among variables, making it well suited for high-dimensional omics data. The development of high-throughput phenotyping techniques, when combined with ML approaches, can provide excellent advantages over other prediction methodologies for large and heterogeneous datasets (Xu and Jackson 2019; Yan and Wang 2023).

Deep learning (DL) model, based on the ML paradigm, has also demonstrated enhanced predictive performance, providing an effective means for exploring the relationship between omics and multi-omics data and phenotypic traits (Ma et al. 2018; Liu et al. 2020; Kang et al. 2022; Wang et al. 2023). For instance, Wang et al. (2023) developed a CNN-based DL model using multi-omics data in plants and compared its performance with classical approaches, including GBLUP. This model achieved higher PA than the other methods, particularly when large-scale breeding datasets were used. Furthermore, the incorporation of self-attention mechanisms within neural networks has enabled models to assign varying importance to different input features when making predictions or decisions (Wang et al. 2025). To combine the advantages of both approaches, architectures incorporating self-attention mechanisms and CNN for phenotypic prediction and functional gene mining of complex traits were designed and developed very recently (Ren et al. 2024). Their effectiveness in predicting pleiotropic genes that influence both flowering time and rosette leaf number was also established. Table 1 summarizes the studies discussed in this section.

Table 1 Phenomic prediction studies based on multi-omics data integration

Omics data	Plant species	Agronomic traits	Brief description and aims	References
All	Wheat	Drought tolerance	To illustrate advances in phenomic and genomic technologies and to introduce their integration with other omics approaches	Mwadzingeni et al. (2016)
All	Wheat	Disease resistance	To discuss the integration of multiple omics data in GP to improve its applicability for disease resistance breeding in wheat	Haile et al. (2020)
Genomic, transcriptomic, and methylomic data	Barley	Grain yield, kernel weight, protein content, and nitrogen uptake	To demonstrate that whole-genome multi-omics profiles contain significant information for the characterization and prediction of barley complex traits	Hansen et al. (2022)
Transcriptomic, metabolomic, and proteomic data	Potato	Tuber flesh color, DSC onset, tuber shape, and enzymatic discoloration	To analyze with an integrated omics approach, the prediction of four phenotypic quality traits	Acharjee et al. (2016)
Genomic and methylomic data	Arabidopsis	Plant height	To describe how DNA methylation data can be useful in the whole-genome prediction of some interesting traits in Arabidopsis	Hu et al. (2015)
Genomic, transcriptomic, and methylomic data	Arabidopsis	Flowering time	To highlight the need to explore different types of omics data when predicting complex traits, as well as to build models based on multi-omics data integration	Wang et al. (2024)

(continued)

Table 1 (continued)

Omics data	Plant species	Agronomic traits	Brief description and aims	References
Genomic, transcriptomic, and metabolomic data	Maize	Heterotic trait, dry matter yield	To demonstrate that the integration of multi-omics data leads to higher success rates in predicting traits in hybrid combinations in maize	Westhues et al. (2017)
Genomic, transcriptomic, and metabolomic data	Maize	Grain yield, grain dry matter content	To show that omics can complement GP and contribute to a more efficient selection of hybrid candidates in maize	Schrag et al. (2018)
Genomic and metabolomic data	Rice	Grain weight, number of grains per panicle, and number of tillers	To demonstrate that prediction approaches based on a combination of genomic and metabolomic data allowed for obtaining better results than single-omics methods	Wang et al. (2019)
Genomic, transcriptomic, and metabolomic data	Rice	Yield-related traits	To integrate parental phenotypic information into various multi-omics prediction models	Yang et al. (2021)
Genomic, transcriptomic, and metabolomic data	Oat	Agronomic and seed nutritional traits	To explore the integration of multi-omics data in order to improve the PA of oat agronomic and seed nutritional traits in METs and distantly related populations	Hu et al. (2021)
All	–	–	To describe a new innovative tool supporting phenomics prediction using multi-omics data	Ren et al. (2024)
All	–	–	To illustrate the DL model advantage in multi-omics data analysis	Kang et al. (2022)
All	–	–	To describe a deep neural network-based method for genomic prediction using multi-omics data in plants	Wang et al. (2023)
All	–	–	To describe the advantages of ML models to manage complex biological data, such as multi-omics data	Xu and Jackson (2019)
All	–	–	To underline DL applications in plant phenomics and ML in genomic selection-assisted breeding	Yan and Wang (2023)

3 Conclusions

This work highlights the growing prevalence of multi-omics-based prediction approaches, which have now been successfully applied across a range of plant species, including rice, corn, and potato. Notably, in maize, these methods have also demonstrated effectiveness in hybrid breeding, illustrating how the integration of multiple biological layers can enhance productivity, disease resistance, and crop quality.

Significant progress has been made in the development of computational tools and methodologies capable of handling the complexity of multi-omics data. In particular, ML and DL techniques have emerged as powerful tools for extracting meaningful patterns and improving prediction accuracy. Their potential has been underlined in numerous studies, especially in the context of predicting complex traits such as grain yield, kernel weight, protein content, and nitrogen uptake. Moreover, multi-omics prediction has increasingly been evaluated in multi-environment trials, expanding its applicability to diverse agricultural contexts. Beyond trait prediction, multi-omics models offer the ability to simulate not only what a plant does, but why it does it, thus revealing which pathways are activated, which metabolites fluctuate, and which genes modulate physiologically relevant responses. In this way, these models provide insights into the internal physiological state of the plant in real time.

Looking ahead, the integration of multi-omics data with predictive systems holds promise for the development of plant "digital twins," virtual representations of plants that simulate growth, development, and responses to environmental conditions. Such systems could revolutionize precision agriculture by enabling real-time decision-making and optimized crop management.

Conflict of Interest Disclosure The authors declare no conflicts of interest.

References

Acharjee A, Kloosterman B, Visser RGF et al (2016) Integration of multi-omics data for prediction of phenotypic traits using random forest. BMC Bioinf 17(5):180

Aguate FM, Trachsel S, Pérez LG, Burgueño J, Crossa J, Balzarini M, Gouache D, Bogard M, de los Campos. G. (2017) Use of hyperspectral image data outperforms vegetation indices in prediction of maize yield. Crop Sci 57(5):2517–2524

Amiri Roudbar M, Mohammadabadi MR, Mehrgardi AA, Abdollahi-Arpanahi R, Momen M, Morota G, Lopes FB, Gianola D, Rosa GJM (2020) Integration of single nucleotide variants and whole-genome DNA methylation profiles for classification of rheumatoid arthritis cases from controls. Heredity 124:658–674

Araus JL, Kefauver SC, Zaman-Allah M, Olsen MS, Cairns JE (2018) Translating high-throughput phenotyping into genetic gain. Trends Plant Sci 23:451–466

Atkinson JA, Jackson RJ, Bentley AR, Ober E, Wells DM (2018) Field phenotyping for the future. Annu Plant Rev Online. https://doi.org/10.1002/9781119312994.apr0651

Bernardo R (2020) Reinventing quantitative genetics for plant breeding: something old, something new, something borrowed, something BLUE. Heredity 125(6):375–385

Bernardo R (2021) Upgrading a maize breeding program via two-cycle genome wide selection: same cost, same or less time, and larger gains. Crop Sci 61(4):2444–2455

Brault C, Lazerges J, Doligez A, Thomas M, Ecarnot M, Roumet P, Bertrand Y, Berger G, Pons T, François P et al (2022) Interest of phenomic prediction as an alternative to genomic prediction in grapevine. Plant Methods 18(1):108

Chacko A, Jayalekshmy VG, Shahiba AM (2023) Hybrid breeding: exploiting heterosis for improved yield and quality. Zenodo. In: Modern approaches in plant breeding enhancing crop genetics by Dr. Shahiba A.M. G. Pranay Pallavi Sonaniya Jeetendra Singh Kushwaha Dr. Sonali Ambadas aware. Elite Publishing House, New Delhi, p 214

Crossa J, Pérez-Rodríguez P, Cuevas J, Montesinos-López O, Jarquín D, de los Campos G, Burgueño J, González-Camacho JM, Pérez-Elizalde S, Beyene Y et al (2017) Genomic selection in plant breeding: methods, models, and perspectives. Trends Plant Sci 22(11):961–975

Crossa J, Fritsche-Neto R, Montesinos-Lopez OA, Costa-Neto G, Dreisigacker S, Montesinos-Lopez A, Bentley AR (2021) The modern plant breeding triangle: optimizing the use of genomics, phenomics, and enviromics data. Front Plant Sci 12:651480

Cuevas J, Montesinos-López O, Juliana P, Guzmán C, Pérez-Rodríguez P, González-Bucio J, Burgueño J, Montesinos-López A, Crossa J (2019) Deep kernel for genomic and near infrared predictions in multi-environment breeding trials. G3 9(9):2913–2924

Du M, Noguchi N (2017) Monitoring of wheat growth status and mapping of wheat yield's within-field spatial variations using color images acquired from UAV-camera system. Remote Sens 9(3):289

Edwards D, Snowdon RJ (2013) Accessing complex crop genomes with next-generation sequencing. Theor Appl Genet 126:1–11

Francesconi S, Harfouche A, Maesano M, Balestra GM (2021) UAV-based thermal, RGB imaging and gene expression analysis allowed detection of fusarium head blight and gave new insights into the physiological responses to the disease in durum wheat. *Frontiers*. Front Plant Sci 12:628575

Galán RJ, Bernal-Vasquez AM, Jebsen C, Piepho HP, Thorwarth P, Steffan P, Gordillo A, Miedaner T (2020) Integration of genotypic, hyperspectral, and phenotypic data to improve biomass yield prediction in hybrid rye. Theor Appl Genet 133(11):3001–3015

Galán RJ, Bernal-Vasquez AM, Jebsen C, Piepho HP, Thorwarth P, Steffan P, Gordillo A, Miedaner T (2021) Early prediction of biomass in hybrid rye based on hyperspectral data surpasses genomic predictability in less-related breeding material. Theor Appl Genet 134(5):1409–1422

Haile JK, N'Diaye A, Sari E, Walkowiak S, Rutkoski JE, Kutcher HR, Pozniak CJ (2020) Potential of genomic selection and integrating "omics" data for disease evaluation in wheat. Crop Breed Genet Genomics 2(4):e200016

Hansen PB, Ruud AK, de Los Campos G, Malinowska M, Nagy I, Svane SF, Thorup-Kristensen K, Jensen JD, Krusell L, Asp T (2022) Integration of DNA methylation and transcriptome data improves complex trait prediction in *Hordeum vulgare*. Plants (Basel) 11(17):2190

Hayes BJ, Panozzo J, Walker CK, Choy AL, Kant S, Wong D, Tibbits J, Daetwyler HD, Rochfort S, Hayden MJ et al (2017) Accelerating wheat breeding for end-use quality with multi-trait genomic predictions incorporating near infrared and nuclear magnetic resonance-derived phenotypes. Theor Appl Genet 130(12):2505–2519

Ho WWH, Hill CB, Doblin MS, Shelden MC, van de Meene A, Rupasinghe T, Bacic A, Roessner U (2020) Integrative multi-omics analyses of barley rootzones under salinity stress reveal two distinctive salt tolerance mechanisms. Plant Commun 1(3):100031

Holliday JA, Wang T, Aitken S (2012) Predicting adaptive phenotypes from multilocus genotypes in Sitka spruce (*Picea sitchensis*) using random forest. G3 (Bethesda) 2:1085–1093

Hu Y, Morota G, Rosa GJ, Gianola D (2015) Prediction of plant height in *Arabidopsis thaliana* using DNA methylation. Data Genet 201:779–793

Hu H, Campbell MT, Yeats TH et al (2021) Multi-omics prediction of oat agronomic and seed nutritional traits across environments and in distantly related populations. Theor Appl Genet 134:4043–4054

Jackson R, Buntjer JB, Bentley AR, Lage J, Byrne E, Burt C, Jack P, Berry S, Flatman E, Poupard B, Smith S, Hayes C, Barber T, Love B et al (2023) Phenomic and genomic prediction of yield on multiple locations in winter wheat. Front Genet 14:1164935

Kang M, Ko E, Mersha TB (2022) A roadmap for multi-omics data integration using deep learning. Brief Bioinform 23:bbab454

Krause MR, González-Pérez L, Crossa J, Pérez-Rodríguez P, Montesinos-López O, Singh RP, Dreisigacker S, Poland J, Rutkoski J, Sorrells M, Gore MA, Mondal S (2019) Hyperspectral reflectance-derived relationship matrices for genomic prediction of grain yield in wheat. G3: Genes Genomes Genet 9(4):1231–1247

Lane HM, Murray SC, Montesinos-López OA, Montesinos-López A, Crossa J, Rooney DK, Barrero-Farfan ID, Fuente GNDL, Morgan CLS (2020) Phenomic selection and prediction of maize grain yield from near-infrared reflectance spectroscopy of kernels. Plant Phenome J 3:e20002

Langridge P, Reynolds MP (2015) Genomic tools to assist breeding for drought tolerance. Curr Opin Biotechnol 32:130–135

Laurençon M, Legrix J, Wagner MH et al (2024) Genomic and phenomic predictions help capture low-effect alleles promoting seed germination in oilseed rape in addition to QTL analyses. Theor Appl Genet 137:156

Li B, Zhang N, Wang Y-G, George AW, Reverter A, Li Y (2018) Genomic prediction of breeding values using a subset of SNPs identified by three machine learning methods. Front Genet 9:237

Liu J, Li J, Wang H, Yan J (2020) Application of deep learning in genomics. Life Sci 63:1860–1878

Ma W, Qiu Z, Song J, Li J, Cheng Q, Zhai J, Ma C (2018) A deep convolutional neural network approach for predicting phenotypes from genotypes. Planta 248:1307–1318

Mwadzingeni L, Shimelis H, Dube E, Laing MD, Tsilo TJ (2016) Breeding wheat for drought tolerance: progress and technologies. J Integr Agric 15(5):935–943

Parmley K, Nagasubramanian K, Sarkar S, Ganapathysubramanian B, Singh AK (2019) Development of optimized phenomic predictors for efficient plant breeding decisions using phenomic-assisted selection in soybean. Plant Phenomics 2019:5809404

Qin W, Wang J, Ma L, Wang F, Hu N, Yang X, Xiao Y, Zhang Y, Sun Z, Wang Z, Yu K (2022) UAV-based multi-temporal thermal imaging to evaluate wheat drought resistance in different deficit irrigation regimes. Remote Sens 14(21):5608

Rebetzke GJ, Jimenez-Berni J, Fischer RA, Deery DM, Smith DJ (2019) Review: High-throughput phenotyping to enhance the use of crop genetic resources. Plant Sci 282:40–48

Ren Y, Wu C, Zhou H, Hu X, Miao Z (2024) Dual-extraction modeling: a multi-modal deep-learning architecture for phenotypic prediction and functional gene mining of complex traits. Plant Commun 5(9):101002

Rincent R, Charpentier JP, Faivre-Rampant P, Paux E, le Gouis J, Bastien C, Segura V (2018) Phenomic selection is a low-cost and high-throughput method based on indirect predictions: proof of concept on wheat and poplar. G3: Genes Genomes Genet 8(12):3961–3972

Robert P, Auzanneau J, Goudemand E, Oury FX, Rolland B, Heumez E, Bouchet S, le Gouis J, Rincent R (2022) Phenomic selection in wheat breeding: identification and optimisation of factors influencing prediction accuracy and comparison to genomic selection. Theor Appl Genet 135(3):895–914

Roscher-Ehrig L, Weber SE, Abbadi A, Malenica M, Abel S, Hemker R, Snowdon RJ, Wittkop B, Stahl A (2024) Phenomic selection for hybrid rapeseed breeding. Plant Phenomics 6:0215

Schrag TA, Westhues M, Schipprack W, Seifert F, Thiemann A, Scholten S, Melchinger AE (2018) Beyond genomic prediction: combining different types of omics data can improve prediction of hybrid performance in maize. Genetics 208(4):1373–1385

Shen Q, Fu L, Dai F, Jiang L, Zhang G, Wu D (2016) Multi-omics analysis reveals molecular mechanisms of shoot adaption to salt stress in Tibetan wild barley. BMC Genomics 17:889

Sirangelo TM (2024a) NLR- and *mlo*-based resistance mechanisms against powdery mildew in *Cannabis sativa*. Plants 13(1):105

Sirangelo TM (2024b) Molecular investigations to improve fusarium head blight resistance in wheat: an update focusing on multi-omics approaches. Plants 13:2179

Sirangelo TM, Ludlow RA, Spadafora ND (2022) Multi-omics approaches to study molecular mechanisms in *Cannabis sativa*. Plants 11:2182

Sirangelo TM, Ludlow RA, Spadafora ND (2023) Molecular mechanisms underlying potential pathogen resistance in *Cannabis sativa*. Plants 12:2764

Vazquez AI, Veturi Y, Behring M, Shrestha S, Kirst M, Resende MFR, de los Campos, G. (2016) Increased proportion of variance explained and prediction accuracy of survival of breast cancer patients with use of whole-genome Multiomic profiles. Genetics 203:1425–1438

Wang S, Wei J, Li R, Qu H, Chater JM, Ma R, Li Y, Xie W, Jia Z (2019) Identification of optimal prediction models using multi-Omic data for selecting hybrid Rice. Heredity 123:395–406

Wang K, Abid MA, Rasheed A, Crossa J, Hearne S, Li H (2023) DNNGP, a deep neural network-based method for genomic prediction using multi-omics data in plants. Mol Plant 16(1):279–293

Wang P, Lehti-Shiu MD, Lotreck S et al (2024) Prediction of plant complex traits via integration of multi-omics data. Nat Commun 15:6856

Wang H, Yan S, Wang W, Chen Y, Hong J, He Q, Diao X, Lin Y et al (2025) Cropformer: an interpretable deep learning framework for crop genomic prediction. Plant Commun 6(3):101223

Weiß TM, Zhu X, Leiser WL, Li D, Liu W, Schipprack W, Melchinger AE, Hahn V, Würschum T (2022) Unraveling the potential of phenomic selection within and among diverse breeding material of maize (*Zea mays* L.). G3: Genes Genomes Genet 12(3):jkab445

Westhues M, Schrag TA, Heuer C, Thaller G, Utz HF, Schipprack W, Thiemann A, Seifert F, Ehret A, Schlereth A et al (2017) Omics-based hybrid prediction in maize. Theor Appl Genet 130:1927–1939

Winn ZJ, Amsberry AL, Haley SD, DeWitt ND, Mason RE (2023) Phenomic versus genomic prediction—a comparison of prediction accuracies for grain yield in hard winter wheat lines. Plant Phenome J 6:e20084

Xu S (2017) Predicted residual error sum of squares of mixed models: an application for genomic prediction. G3: Genes Genomes Genet 7:895–909

Xu C, Jackson SA (2019) Machine learning and complex biological data. Genome Biol 20:76

Yan J, Wang X (2023) Machine learning bridges omics sciences and plant breeding. Trends Plant Sci 28:199–210

Yang X, Zhao Y, Wang X, Ma Y, Li P, Yang Z, Zhang X, Xu C, Xu S (2021) Incorporation of parental phenotypic data into multi-omic models improves prediction of yield-related traits in hybrid rice. Plant Biotechnol J 19:261–272

Zhu X, Leiser WL, Hahn V, Würschum T (2021) Phenomic selection is competitive with genomic selection for breeding of complex traits. Plant Phenome J 4(1):e20027

MIX
Papier aus verantwortungsvollen Quellen
Paper from responsible sources
FSC® C105338

If you have any concerns about our products,
you can contact us on
ProductSafety@springernature.com

In case Publisher is established outside the EU,
the EU authorized representative is:
Springer Nature Customer Service Center GmbH
Europaplatz 3, 69115 Heidelberg, Germany

Printed by Libri Plureos GmbH
in Hamburg, Germany